Springer Complexity

Springer Complexity is an interdisciplinary program publishing the best research and academic-level teaching on both fundamental and applied aspects of complex systems – cutting across all traditional disciplines of the natural and life sciences, engineering, economics, medicine, neuroscience, social and computer science.

Complex Systems are systems that comprise many interacting parts with the ability to generate a new quality of macroscopic collective behavior the manifestations of which are the spontaneous formation of distinctive temporal, spatial or functional structures. Models of such systems can be successfully mapped onto quite diverse "real-life" situations like the climate, the coherent emission of light from lasers, chemical reaction-diffusion systems, biological cellular networks, the dynamics of stock markets and of the internet, earthquake statistics and prediction, freeway traffic, the human brain, or the formation of opinions in social systems, to name just some of the popular applications.

Although their scope and methodologies overlap somewhat, one can distinguish the following main concepts and tools: self-organization, nonlinear dynamics, synergetics, turbulence, dynamical systems, catastrophes, instabilities, stochastic processes, chaos, graphs and networks, cellular automata, adaptive systems, genetic algorithms and computational intelligence.

The three major book publication platforms of the Springer Complexity program are the monograph series "Understanding Complex Systems" focusing on the various applications of complexity, the "Springer Series in Synergetics", which is devoted to the quantitative theoretical and methodological foundations, and the "SpringerBriefs in Complexity" which are concise and topical working reports, case-studies, surveys, essays and lecture notes of relevance to the field. In addition to the books in these two core series, the program also incorporates individual titles ranging from textbooks to major reference works.

Editorial and Programme Advisory Board

Henry Abarbanel, Institute for Nonlinear Science, University of California, San Diego, USA

Dan Braha, New England Complex Systems Institute and University of Massachusetts Dartmouth, USA

Péter Érdi, Center for Complex Systems Studies, Kalamazoo College, USA and Hungarian Academy of Sciences, Budapest, Hungary

Karl Friston, Institute of Cognitive Neuroscience, University College London, London, UK

Hermann Haken, Center of Synergetics, University of Stuttgart, Stuttgart, Germany

Viktor Jirsa, Centre National de la Recherche Scientifique (CNRS), Université de la Méditerranée, Marseille, France

Janusz Kacprzyk, System Research, Polish Academy of Sciences, Warsaw, Poland

Kunihiko Kaneko, Research Center for Complex Systems Biology, The University of Tokyo, Tokyo, Japan

Scott Kelso, Center for Complex Systems and Brain Sciences, Florida Atlantic University, Boca Raton, USA

Markus Kirkilionis, Mathematics Institute and Centre for Complex Systems, University of Warwick, Coventry, UK

Jürgen Kurths, Nonlinear Dynamics Group, University of Potsdam, Potsdam, Germany

Andrzej Nowak, Department of Psychology, Warsaw University, Poland

Linda Reichl, Center for Complex Quantum Systems, University of Texas, Austin, USA

Peter Schuster, Theoretical Chemistry and Structural Biology, University of Vienna, Vienna, Austria

Frank Schweitzer, System Design, ETH Zurich, Zurich, Switzerland

Didier Sornette, Entrepreneurial Risk, ETH Zurich, Zurich, Switzerland

Stefan Thurner, Section for Science of Complex Systems, Medical University of Vienna, Vienna, Austria

Springer Series in Synergetics

Founding Editor: H. Haken

The Springer Series in Synergetics was founded by Herman Haken in 1977. Since then, the series has evolved into a substantial reference library for the quantitative, theoretical and methodological foundations of the science of complex systems.

Through many enduring classic texts, such as Haken's *Synergetics and Information and Self-Organization*, Gardiner's *Handbook of Stochastic Methods*, Risken's *The Fokker Planck-Equation* or Haake's *Quantum Signatures of Chaos*, the series has made, and continues to make, important contributions to shaping the foundations of the field.

The series publishes monographs and graduate-level textbooks of broad and general interest, with a pronounced emphasis on the physico-mathematical approach.

For further volumes:
http://www.springer.com/series/712

Vadim S. Anishchenko • Tatyana E. Vadivasova •
Galina I. Strelkova

Deterministic Nonlinear Systems

A Short Course

Vadim S. Anishchenko
Tatyana E. Vadivasova
Galina I. Strelkova
Department of Physics
Instiute of Nonlinear Dynamics
Saratov State University
Saratov, Russia

ISSN 0172-7389
ISBN 978-3-319-06870-1   ISBN 978-3-319-06871-8 (eBook)
DOI 10.1007/978-3-319-06871-8
Springer Cham Heidelberg New York Dordrecht London

Library of Congress Control Number: 2014942075

© Springer International Publishing Switzerland 2014

This work is subject to copyright. All rights are reserved by the Publisher, whether the whole or part of the material is concerned, specifically the rights of translation, reprinting, reuse of illustrations, recitation, broadcasting, reproduction on microfilms or in any other physical way, and transmission or information storage and retrieval, electronic adaptation, computer software, or by similar or dissimilar methodology now known or hereafter developed. Exempted from this legal reservation are brief excerpts in connection with reviews or scholarly analysis or material supplied specifically for the purpose of being entered and executed on a computer system, for exclusive use by the purchaser of the work. Duplication of this publication or parts thereof is permitted only under the provisions of the Copyright Law of the Publisher's location, in its current version, and permission for use must always be obtained from Springer. Permissions for use may be obtained through RightsLink at the Copyright Clearance Center. Violations are liable to prosecution under the respective Copyright Law.

The use of general descriptive names, registered names, trademarks, service marks, etc. in this publication does not imply, even in the absence of a specific statement, that such names are exempt from the relevant protective laws and regulations and therefore free for general use.

While the advice and information in this book are believed to be true and accurate at the date of publication, neither the authors nor the editors nor the publisher can accept any legal responsibility for any errors or omissions that may be made. The publisher makes no warranty, express or implied, with respect to the material contained herein.

Printed on acid-free paper

Springer is part of Springer Science+Business Media (www.springer.com)

Preface

Dear reader!

We are pleased to present the book *Nonlinear Deterministic Systems: A Short Course*. Nonlinear dynamics has gradually become an independent area of study over the last 30–35 years. However, this scientific field still continues to develop intensively. Nonlinear dynamics is based on input from a range of fundamental scientific disciplines, such as the nonlinear theory of oscillations and waves, the theory of dynamical systems, synergetics, statistical physics, and thermodynamics. When this field of research first began, the main problems to be tackled concerned the dynamical processes and effects realized in nonlinear deterministic systems. Research over the last 10–15 years has shown that complex and interesting phenomena caused by fluctuations can also be considered in the framework of nonlinear dynamics. Nowadays, a new direction has arisen as an integral part of nonlinear dynamics. It can be described as the nonlinear dynamics of stochastic systems. It is clear that stochastic processes cannot be analyzed in detail without a deep understanding of the dynamical properties of the relevant deterministic systems. Therefore, the nonlinear dynamics of deterministic systems is a fundamentally important part of modern nonlinear dynamics as a whole.[1]

The book is based on selected lectures from the course entitled *Introduction to Nonlinear Dynamics*, delivered to students of the Physics Department of Saratov State University by one of the authors. The content of the lectures has been significantly revised and supplemented.

The book contains 15 chapters. Each can be considered as a lecture on the relevant problem and studied almost independently of the others. The first three chapters are devoted to the classical problems of the theory of dynamical systems. The definition and classification of dynamical systems are given in Chap. 1, the fundamentals of the linear theory of stability are discussed in Chap. 2 as applied to

---

[1] Note that, in our book, only dissipative dynamical systems are analyzed and conservative (Hamiltonian) systems are not considered.

differential and discrete-time systems, and the elements of the theory of local and nonlocal bifurcations are described in Chap. 3.

Chapter 4 is devoted to the dynamics of self-sustained oscillatory systems with one degree of freedom. This is exemplified by the van der Pol oscillator with soft and hard excitation. The dynamics of systems with phase space dimension $N \geq 3$ is described qualitatively in Chap. 5. We focus on physical explanation of the onset of dynamical chaos and the description of typical properties of strange attractors. The next two chapters are devoted to the mechanisms of transition to chaotic dynamics through a cascade of period-doubling bifurcations and intermittency (Chap. 6) and through the destruction of quasiperiodic oscillations (Chap. 7). The problem of structural stability is discussed in Chap. 8. The notion of a nonhyperbolic attractor is introduced and chaotic attractors are classified.

Chapter 9 is devoted to recent results on the problem of Poincaré recurrence and describes a global approach to this problem, which is related to the Afraimovich–Pesin dimension of a Poincaré recurrence time sequence. The main theoretical features are confirmed by several simple examples of numerical calculations. Chapter 10 serves as a lecture on the fundamentals of the theory of fractals and its application to nonlinear dynamics. The concept of a fractal is introduced and the fractal nature of sets in the phase space of a dynamical system is discussed.

Chapters 11 and 12 are concerned with several basic models of oscillators with chaotic and quasiperiodic oscillations. The dynamics of the Anishchenko–Astakhov oscillator is analyzed in detail in Chap. 11. The theoretical and experimental results presented in the study of oscillator dynamics illustrate Shilnikov's theorem for systems with a saddle-focus separatrix loop. An autonomous two-frequency oscillator is described in Chap. 12. It is shown that the oscillator can realize quasiperiodic two-frequency oscillations and transition to chaos through two-dimensional torus-doubling bifurcations.

The last three chapters examine synchronization effects in periodic (Chap. 13), two-frequency quasiperiodic (Chap. 14), and chaotic (Chap. 15) self-sustained oscillations.

The content of the book should satisfy the needs of students and researchers in the field of nonlinear dynamics. No special mathematical knowledge is assumed. The usual university courses on higher mathematics and the theory of oscillations would serve as a sufficient basis. The book is aimed mainly at masters and PhD students and young scientists working in this field. At the same time, some of the chapters could be used to give lectures to advanced natural science students in other fields.

The book has a list of recommended references that includes monographs and textbooks. These should be helpful to the reader who wishes to study the relevant chapters of the book in greater depth. We do not give references to the original scientific papers. If necessary, the reader can find the relevant information by using the detailed bibliographies provided in the recommended books.

We would like to express our thanks to the staff of the Radiophysics and Nonlinear Dynamics Chair of Saratov State University, whose joint research results have been used in writing this book. Among them, we note in particular

A.B. Neiman, D.E. Postnov, S.M. Nikolaev, S.V. Astakhov, and Ya.I. Boev. We are also very grateful to V.V. Astakhov for the materials used in preparing Chaps. 4, 11, and 13.

We consider it a duty and a pleasure to express our special gratitude to our teachers and colleagues. This book was only possible as a result of our long-term collaboration and interchange with them. Above all, we would like to thank Yu.L. Klimontovich, L.P. Shilnikov, W. Ebeling, L. Schimansky-Geier, J. Kurths, F. Moss, and S.P. Kuznetsov.

The research results used in writing this book were obtained in the framework of research grants provided by the US Civilian Research and Development Foundation, the Russian Federation Ministry of Education and Science, and the Russian Foundation of Basic Research. V.S. Anishchenko acknowledges support from the Alexander von Humboldt Foundation. We express our deep gratitude to the above organizations and foundations for their support.

Saratov, Russia
October 2013

Vadim S. Anishchenko
Tatyana E. Vadivasova
Galina I. Strelkova

Contents

1	**Dynamical Systems**		1
	1.1 Introduction		1
	1.2 Dynamical Systems and Mathematical Models		1
	1.3 Kinematic Interpretation of a System of Differential Equations		3
	1.4 Definition of a Dynamical System: Classification		4
	1.5 Phase Portraits of Typical Oscillatory Systems		7
		1.5.1 Conservative Oscillator	7
		1.5.2 Damped Linear Oscillator	9
	1.6 Self-Sustained Oscillatory Systems		11
	1.7 Regular and Chaotic Attractors		13
	1.8 Discrete-Time Systems: Return Maps		16
		1.8.1 Stretching Map	20
		1.8.2 Logistic Map	20
		1.8.3 Sine Map	20
		1.8.4 Henon Map	20
		1.8.5 Lozi Map	20
	1.9 Summary		21
	References		21
2	**Stability of Dynamical Systems: Linear Approach**		23
	2.1 Introduction		23
	2.2 Definition of Stability		24
	2.3 Linear Analysis of Stability		25
		2.3.1 Stability of Solutions of a First-Order Differential Equation	25
		2.3.2 Stability of a Dynamical System in $\mathbb{R}^N$	27
	2.4 Stability of Phase Trajectories in Discrete-Time Systems		33
	2.5 Summary		34
	References		35

3 Bifurcations of Dynamical Systems ... 37
- 3.1 Introduction ... 37
- 3.2 Double Equilibrium Bifurcation ... 39
- 3.3 Soft and Hard Bifurcations: Catastrophes ... 40
- 3.4 Triple Equilibrium Bifurcation ... 41
- 3.5 Andronov–Hopf Bifurcation ... 43
- 3.6 Bifurcations of Limit Cycles ... 44
  - 3.6.1 Saddle-Node Bifurcation ... 44
  - 3.6.2 Period-Doubling Bifurcation ... 45
  - 3.6.3 Two-Dimensional Torus Birth (Death) Bifurcation (Neimark–Saker Bifurcation) ... 46
  - 3.6.4 Symmetry-Breaking Bifurcation ... 47
- 3.7 Nonlocal Bifurcations: Homoclinic Trajectories and Structures ... 48
  - 3.7.1 Separatrix Loop of a Saddle Equilibrium Point ... 48
  - 3.7.2 Saddle-Node Separatrix Loop ... 50
  - 3.7.3 Homoclinic Trajectory Appearance of a Saddle Limit Cycle ... 51
- 3.8 Summary ... 52
- References ... 52

4 Dynamical Systems with One Degree of Freedom ... 53
- 4.1 Introduction ... 53
- 4.2 Limit Sets and Attractors in the Phase Plane: The Andronov–Poincaré Limit Cycle ... 54
- 4.3 Structural Stability of Systems in the Phase Plane: Andronov–Pontryagin Systems ... 56
  - 4.3.1 Definition of Robustness of a Dynamical System ... 56
  - 4.3.2 Definition of Structural Stability of a Dynamical System ... 57
  - 4.3.3 Andronov–Pontryagin Theorem ... 57
- 4.4 Oscillators with One Degree of Freedom ... 58
  - 4.4.1 Froude Pendulum ... 58
  - 4.4.2 Fastened Weight on a Moving Belt ... 60
  - 4.4.3 *RC*-Oscillator with Wien Bridge ... 62
  - 4.4.4 Oscillatory Circuit with Active Nonlinear Element ... 63
- 4.5 Analysis of the van der Pol Equation: Onset of Self-Sustained Oscillations ... 65
  - 4.5.1 Amplitude and Phase Equations for the Self-Sustained Oscillator ... 66
- 4.6 Oscillator with Hard Excitation of Self-Sustained Oscillations ... 69
  - 4.6.1 Analysis of the Stability of Equilibrium States ... 69
  - 4.6.2 Truncated Equations for the Amplitude and Phase for the Oscillator with Hard Excitation ... 70

		4.6.3	Bifurcation Diagram of the Oscillator with Hard Excitation	71

	4.7	Summary	73
	References		73

5 Systems with Phase Space Dimension $N \geq 3$: Deterministic Chaos ... 75

- 5.1 Introduction ... 75
- 5.2 Determinism and Chaos for Beginners ... 76
  - 5.2.1 Determinism ... 76
  - 5.2.2 Chaos ... 77
  - 5.2.3 Stability and Instability ... 77
  - 5.2.4 Nonlinearity ... 78
  - 5.2.5 Instability and Nonlinear Restriction ... 78
  - 5.2.6 Deterministic Chaos ... 80
- 5.3 Mixing and Probabilistic Properties of Deterministic Systems ... 81
- 5.4 Is Deterministic Chaos a Mathematical Oddity or a Typical Property of the Material World? ... 83
- 5.5 Strange Chaotic Attractors ... 84
- 5.6 Strange Nonchaotic and Chaotic Nonstrange Attractors ... 85
  - 5.6.1 Chaotic Nonstrange Attractors ... 86
  - 5.6.2 Strange Nonchaotic Attractors ... 88
  - 5.6.3 Geometric Characteristics of SNAs ... 88
  - 5.6.4 LCE Spectrum of SNAs ... 89
  - 5.6.5 Spectrum and Autocorrelation Function ... 89
- 5.7 Summary ... 90
- References ... 91

6 From Order to Chaos: Bifurcation Scenarios (Part I) ... 93
- 6.1 Introduction ... 93
- 6.2 Transition to Chaos via a Cascade of Period-Doubling Bifurcations: Feigenbaum Universality ... 94
- 6.3 Crisis and Intermittency ... 102

7 From Order to Chaos: Bifurcation Scenarios (Part II) ... 107
- 7.1 Route to Chaos via Two-Dimensional Torus Destruction ... 107
  - 7.1.1 Two-Dimensional Torus Breakdown Theorem ... 108
  - 7.1.2 Circle Map: Universal Regularities of Soft Transition from Quasiperiodicity to Chaos ... 111
- 7.2 Route to Chaos via Ergodic Torus Destruction: Chaotic Nonstrange Attractors ... 115
- 7.3 Summary ... 121
- References ... 122

8 Robust and Nonrobust Dynamical Systems: Classification of Attractor Types ... 123
- 8.1 Introduction ... 123
- 8.2 Homoclinic and Heteroclinic Curves ... 124

	8.3	Structurally Stable Systems in $\mathbb{R}^N$, $N \geq 3$: Hyperbolicity.........	126
		8.3.1 Morse–Smale Systems	126
		8.3.2 Hyperbolic Sets	127
		8.3.3 Anosov Systems	128
		8.3.4 Smale Systems with Nontrivial Hyperbolicity: Strange Attractors	129
	8.4	Structurally Unstable Dynamical Systems	130
	8.5	Quasihyperbolic Attractors: Lorenz-Type Attractors	131
		8.5.1 Quasihyperbolic Attractor in the Lozi Map	132
		8.5.2 The Lorenz Attractor	134
	8.6	Nonhyperbolic Attractors and Their Properties....................	136
		8.6.1 Nonhyperbolic Attractor in the Henon Map..............	137
		8.6.2 Nonhyperbolic Attractor in the Oscillator with Inertial Nonlinearity	141
	8.7	Summary...	142
	References..		143
9	**Characteristics of Poincaré Recurrences**...............................		145
	9.1	Introduction...	145
	9.2	Local Approach	146
		9.2.1 Kac's Lemma	146
		9.2.2 Exponential Law for Distribution of First Recurrence Times...	148
		9.2.3 Numerical Examples	148
	9.3	Global Approach: Afraimovich–Pesin Dimension of Recurrence Times	151
	9.4	Afraimovich–Pesin Dimension and Lyapunov Exponents	154
	9.5	Summary...	156
	Reference...		156
10	**Fractals in Nonlinear Dynamics**		157
	10.1	Introduction...	157
	10.2	Definition of a Fractal: Classic Examples of Fractal Sets	158
	10.3	The Nature of Fractality in Dynamical Systems....................	162
	10.4	Fractal Dimensions of Sets	165
		10.4.1 The Hausdorff–Besicovitch Dimension	165
		10.4.2 Capacity D_C	166
		10.4.3 Information Dimension D_I	167
		10.4.4 Correlation Dimension D_{cor}..............................	168
		10.4.5 Generalized Dimension D_q	169
		10.4.6 Lyapunov Dimension D_L	170
	10.5	Relationship Between Different Dimensions	171
	10.6	Summary...	172
	References..		172

11 The Anishchenko–Astakhov Oscillator of Chaotic Self-Sustained Oscillations ... 175
- 11.1 Introduction ... 175
- 11.2 Theodorchik's Oscillator ... 177
- 11.3 Modification of the Oscillator with Inertial Nonlinearity: The Anishchenko–Astakhov Oscillator ... 182
  - 11.3.1 Periodic Regimes of Self-Sustained Oscillations and Their Bifurcations ... 184
  - 11.3.2 Period-Doubling Bifurcations: Feigenbaum Universality ... 192
  - 11.3.3 Chaotic Attractor and Homoclinic Trajectories in the Oscillator ... 194
- 11.4 Summary ... 200
- References ... 201

12 Quasiperiodic Oscillator with Two Independent Frequencies ... 203
- 12.1 Introduction ... 203
- 12.2 Methods for Realizing Two-Frequency Oscillations and Their Properties ... 204
- 12.3 Statement of Oscillator Equations ... 208
- 12.4 Bifurcation Diagram of the Quasiperiodic Oscillator ... 211
- 12.5 Two-Dimensional Torus-Doubling Bifurcation ... 212
- 12.6 Summary ... 215

13 Synchronization of Periodic Self-Sustained Oscillations ... 217
- 13.1 Introduction ... 217
- 13.2 Forced Synchronization of the van der Pol Oscillator: Truncated Equations for the Amplitude and Phase ... 218
  - 13.2.1 Analysis of Synchronization in the Phase Approximation ... 221
  - 13.2.2 Bifurcational Analysis of the System of Truncated Equations ... 225
  - 13.2.3 Bifurcational Analysis of the Nonautonomous van der Pol Oscillator ... 230
- 13.3 Mutual Synchronization: Effect of Oscillation Death in Dissipatively Coupled van der Pol Oscillators ... 236
- 13.4 Summary ... 242
- References ... 243

14 Synchronization of Two-Frequency Self-Sustained Oscillations ... 245
- 14.1 Introduction ... 245
- 14.2 Influence of an External Periodic Force on a Resonant Limit Cycle in a System of Coupled Oscillators ... 245
- 14.3 Basic Bifurcations of Quasiperiodic Regimes When Synchronizing a Resonant Limit Cycle ... 248
  - 14.3.1 Peculiarities in the Synchronization of Resonant Limit Cycles ... 251

		14.3.2	Phase Synchronization of a System of Coupled van der Pol Oscillators by an External Harmonic Signal	254

        14.3.2 Phase Synchronization of a System
of Coupled van der Pol Oscillators
by an External Harmonic Signal 254
        14.3.3 Bifurcations of Equilibrium States........................ 257
        14.3.4 Bifurcations of Invariant Curves 260
        14.3.5 Synchronization of Two-Frequency
Oscillations in a Self-Sustained Quasiperiodic
Oscillator... 263
  14.4 Summary... 270

15 Synchronization of Chaotic Oscillations 273
  15.1 Introduction... 273
  15.2 Phase–Frequency Synchronization of Chaotic
Self-Sustained Oscillations 274
  15.3 Experimental Investigation of Forced Synchronization
of an Oscillator with Spiral Chaos............................... 283
  15.4 Complete Synchronization of Interacting Chaotic Systems 285
  15.5 Quantitative Characteristics of the Degree
of Synchronization of Chaotic Self-Sustained Oscillations 290
  15.6 Summary... 294
  References... 294

Chapter 1
Dynamical Systems

1.1 Introduction

The temporal and spatial behavior of a system can be predicted if initial conditions are known. This task is one of the most important problems in the natural sciences. It amounts to finding a law that enables us to define the future state of the system at a time $t > t_0$ when given some information on the system at the initial time t_0. Depending on the complexity of the system, this law can be deterministic or probabilistic, and it can describe either the temporal or the spatio-temporal evolution of the system.

The problem of predicting the evolution of systems in the natural sciences is undoubtedly a mathematical task, and mathematical logic requires us to specify the subject and the problem as precisely as possible. We must therefore formulate a definition of the system under investigation and indicate its properties. Our analysis here concerns so-called *dynamical systems* and the mathematical interpretation of this term.

1.2 Dynamical Systems and Mathematical Models

A *dynamical system* (DS) is understood to be an object or process whose state is uniquely defined by a set of quantities or functions at a given time and which evolves in time according to a certain law. The latter allows us to predict the future state of the dynamical system from its initial state and is called the *evolution law*. Dynamical systems can be instantiated by mechanical, physical, chemical, and biological objects, as well as computational processes and processes of information transformation performed according to specific algorithms. The evolution law (or evolution operator) that describes dynamical systems can be specified in different ways, such as differential equations, discrete maps, graph theory, the theory of topological Markov chains, etc. The specific form of *mathematical model* used to study a DS depends on which method of description is chosen.

The mathematical model of a dynamical system is given by introducing dynamical variables (*coordinates*) which uniquely define the system *state* and specifying the *evolution law*.

Depending on the degree of approximation, the same system can be associated with different mathematical models. Real systems are explored by considering their corresponding mathematical models. They are improved and developed through the comparative analysis of experimental and theoretical results. In this connection a dynamical system can be understood as equivalent to its mathematical model. When studying the same dynamical system (for example, pendulum motion), different mathematical models can be obtained depending on which factors or conditions are taken into account.

This situation can be exemplified by considering a nonlinear conservative oscillator:

$$\ddot{x} + \sin x = 0 , \qquad \ddot{x} = \frac{d^2 x}{dt^2} . \tag{1.1}$$

It is known that the function $\sin x$ is analytical and can be expanded in a Taylor series as follows:

$$\sin x = x - \frac{x^3}{3!} + \frac{x^5}{5!} - \cdots = \sum_{n=0}^{\infty} \frac{x^{4n+1}}{(4n+1)!} - \sum_{n=1}^{\infty} \frac{x^{4n-1}}{(4n-1)!} . \tag{1.2}$$

For small $x \ll 1$, we have $\sin x \simeq x$. As x increases, the second, third, and further terms of the series must be taken into account to approximate $\sin x$ within a given accuracy. Therefore, when $x \ll 1$, we get the simplest model of a mathematical pendulum:

$$\ddot{x} + x = 0 . \tag{1.3}$$

The following approximation yields the model of a nonlinear pendulum:

$$\ddot{x} + x - \frac{x^3}{6} = 0 , \tag{1.4}$$

and so forth. Figure 1.1 depicts approximation results for the function $\sin x$, taking a finite number of terms $n = 0, 1, \ldots, 43$ in the series (1.2). For each particular value of n, we get a new dynamical system, which describes, to a given approximation, the oscillating process of a physical pendulum.

1.3 Kinematic Interpretation of a System of Differential Equations

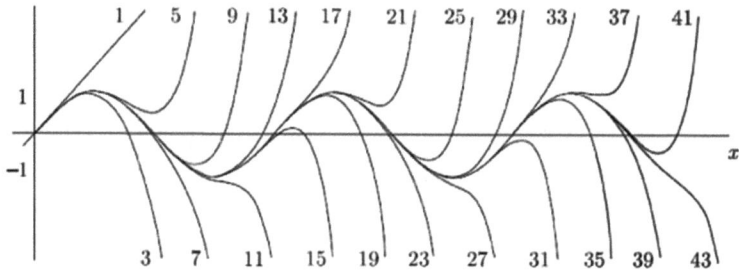

Fig. 1.1 Approximation of the function $\sin x$ by a finite number of terms $n = 1, \ldots, 43$ in the series (1.2)

1.3 Kinematic Interpretation of a System of Differential Equations

Here we consider dynamical systems which are modeled by a finite number of ordinary differential equations. Such systems can be analyzed using the terminology and methods of mechanics. In our case a dynamical system is defined if there is such an object whose state is described by quantities $x_1, x_2, \ldots, x_N$ at time $t = t_0$. These quantities x_i can take arbitrary values, but two different sets of x_i and x_i' correspond to two different states. The temporal evolution of the DS is described by a system of ordinary differential equations of the form

$$\frac{dx_i}{dt} = \dot{x}_i = f_i(x_1, x_2, \ldots, x_N), \quad i = 1, 2, \ldots, N. \quad (1.5)$$

If the quantities $x_1, x_2, \ldots, x_N$ are considered to be coordinates of a point $\mathbf{x}$ in an N-dimensional space, the system state can be represented geometrically by this point. It is called a *representative* or *phase point*, and the state space is called the *phase space* of the DS. The motion of a phase point along a certain curve called a *phase trajectory* corresponds to the time evolution of a state of the system. Equations (1.5) define a velocity vector field in the system phase space. This field associates each point $\mathbf{x}$ with a velocity vector $\mathbf{F}(\mathbf{x})$ whose components are given by the right-hand sides of (1.5):

$$\left[f_1(x_1, x_2, \ldots, x_N), f_2(x_1, x_2, \ldots, x_N), \ldots, f_N(x_1, x_2, \ldots, x_N) \right].$$

The dynamical system (1.5) can be written in the vector form

$$\dot{\mathbf{x}} = \mathbf{F}(\mathbf{x}), \quad (1.6)$$

where $\mathbf{F}(\mathbf{x})$ is a vector function of dimension N.

We need to specify the relationship between the number of degrees of freedom and the phase space dimension of a DS. The *number of degrees of freedom* is understood as the minimal number of independent coordinates needed to unambiguously define a system state. Coordinates were originally considered to be spatial variables characterizing the relative positions of bodies and objects. However, to get an unambiguous solution to the equations of motion, it is also necessary, besides the coordinates, to specify the corresponding initial values of momenta or velocities. In this case a system with n degrees of freedom is characterized by a phase space of dimension $N = 2n$.

1.4 Definition of a Dynamical System: Classification

A dynamical system is formally defined if the following three elements are given:

1. A set of states X that forms the full metric space (phase space);
2. A set of times Θ;
3. An *evolution operator* $T_{t_0}^{\tau}$ representing some map $T_{t_0}^{\tau} : X \to X$ such that each state $\mathbf{x}_0 \in X$ at the initial time $t_0 \in \Theta$ is unambiguously associated with the state $\mathbf{x}_t \in X$ at another time $t = t_0 + \tau \in \Theta$.

Accordingly, we can write

$$\mathbf{x}_t = T_{t_0}^{\tau} \mathbf{x}_0 , \qquad t = t_0 + \tau . \tag{1.7}$$

The evolution operator is continuous in X and possesses the following properties:

$$T_{t_0}^{0} \mathbf{x}_0 = \mathbf{x}_0 , \tag{1.8}$$

$$T_{t_0}^{\tau+s} \mathbf{x}_0 = T_{t_0+s}^{\tau} \circ T_{t_0}^{s} \mathbf{x}_0 = T_{t_0+\tau}^{s} \circ T_{t_0}^{\tau} \mathbf{x}_0 , \tag{1.9}$$

where the symbol $\circ$ denotes composition of operators.

Dynamical systems can be classified by taking into consideration the nature of the sets X, Θ and the properties of the evolution operator. If $\Theta = \mathbb{R}^1$, i.e., time is represented by a continuous set of values, the evolution operator is continuous in τ and the corresponding dynamical system is called a *continuous-time system*, or a *flow* in the case of fluid current. If the set Θ is countable, the DS is called a *discrete-time system* or a *cascade*.

The set of states X, as well as the set of times, can assume various forms. It can be a finite or countable set, as is typical for a class of dynamical systems called *cellular automata*. The set X can be an arithmetical space with a finite dimension N (real $\mathbb{R}^N$ or complex $\mathbb{C}^N$). This definition is typical for the phase space of a DS given by ordinary differential equations. Systems with a finite-dimensional phase space (and hence with a finite number of degrees of freedom) are called *lumped* or

1.4 Definition of a Dynamical System: Classification

point systems. Finally, X can be a function space. In this case a DS is described by partial differential equations, integral equations, integro-differential equations, or time-delay ordinary differential equations. One often has to deal with systems whose state is described by functions of spatial coordinates. Such systems are called *distributed* systems or *media*. If spatial coordinates are defined on a continuous set of values, the number of degrees of freedom of a system is infinite, and an infinite number of quantities is then required to define its states. In the theory of electric oscillations a system can be treated as lumped when the wavelength of oscillations significantly exceeds geometrical measurements of the system itself. If the dimensions of a device are commensurable with the wavelength of generated oscillations, the system must be treated as distributed. Hereafter we consider flow systems with state space $X = \mathbb{R}^N$.

The evolution operator can possess characteristic properties which enable one to distinguish different classes of dynamical systems. For example, there is a class of *linear dynamical systems* for which the evolution operator is linear and satisfies the superposition rule

$$T_{t_0}^\tau(\mathbf{x} + \mathbf{y}) = T_{t_0}^\tau \mathbf{x} + T_{t_0}^\tau \mathbf{y} \:. \tag{1.10}$$

If the operator is nonlinear, i.e., if it does not obey a rule like (1.10), the corresponding dynamical system is said to be *nonlinear*.

If the evolution operator $T_{t_0}^\tau$ is specified for all values of the time shift τ, both for $\tau \geq 0$ and $\tau < 0$, then it is reversible, i.e., there is an operator that is inverse to it, $T_{t_0+\tau}^{-\tau}$, which can be used to find the system state at the preceding moment t_0 from the system state at $t = t_0 + \tau$. Such a dynamical system is also said to be *reversible in time*. If the evolution operator is determined only for $\tau \geq 0$, it is irreversible and the preceding state of the system cannot be unambiguously defined. In this case the system is said to be *irreversible in time*.

If the evolution operator $T_{t_0}^\tau$ does not depend on the initial time t_0 and is defined only by the initial state $\mathbf{x}_0$ and the interval τ, the corresponding dynamical system is said to be *autonomous*, otherwise the system is *nonautonomous*. The initial time can be omitted in the notation of the evolution operator of an autonomous system, i.e., $T_{t_0}^\tau = T^\tau$, and the property (1.9) reads

$$T^{\tau+s}\mathbf{x}_0 = T^\tau \circ T^s \mathbf{x}_0 = T^s \circ T^\tau \mathbf{x}_0 \:.$$

From the physical point of view, the autonomy of a system means that the system is not subjected to external forces and its parameters are temporally constant. In what follows we consider autonomous systems or systems that can be reduced to autonomous ones by adding some additional state variables. It is important in this case that typical trajectories of the DS remain bounded. For example, a harmonically driven system can be easily reduced to an autonomous one by introducing the phase of an external force, which is defined in a bounded interval, for example, within the range $[-\pi, +\pi]$.

In terms of energy, dynamical systems are divided into conservative and nonconservative. *Conservative systems* are characterized by an invariant store of energy. In mechanics they are called *Hamiltonian systems*. A so-called *Hamiltonian* $H(\mathbf{p}, \mathbf{q})$ is defined for conservative systems with n degrees of freedom, where q_i are generalized coordinates, p_i are generalized momenta of the system, $i = 1, 2, \ldots, n$. The Hamiltonian fully characterizes the dynamical nature of a system and, from a physical viewpoint, it represents its total energy in most cases. The temporal evolution of conservative systems is described by the Hamiltonian equations

$$\dot{q}_i = \frac{\partial H(\mathbf{p}, \mathbf{q})}{\partial p_i}, \qquad \dot{p}_i = -\frac{\partial H(\mathbf{p}, \mathbf{q})}{\partial q_i}. \tag{1.11}$$

These imply

$$\sum_{i=1}^{n} \left(\frac{\partial \dot{q}_i}{\partial q_i} + \frac{\partial \dot{p}_i}{\partial p_i} \right) = 0. \tag{1.12}$$

This means that the divergence of the vector field is equal to zero. The motion of representative points in phase space is treated in this case as a stationary flow of an incompressible liquid which obeys a continuity equation. Hence it follows that, in conservative systems, a phase volume element does not change in time.

Dynamical systems in which the amount of energy changes in time are said to be *nonconservative*. Systems whose energy decreases in time due to friction or dissipation are *dissipative*. Systems in which the energy increases in time are systems with negative friction or negative dissipation. Such systems can be treated like dissipative ones if time is counted in the opposite direction. A peculiarity of dissipative systems is that a phase volume element is always time-dependent. The phase volume decreases in systems with energy absorption and increases in those with negative friction.

An important group of dynamical systems are those which can generate oscillations. Regarding mathematical models, oscillatory systems can be divided into certain classes. The following oscillatory systems are distinguished: *lumped* and *distributed*, *conservative* and *dissipative*, *autonomous* and *nonautonomous*.

The majority of real oscillatory systems in physics, radiophysics, biology, chemistry and other fields of investigation are nonconservative. Among them, a special class includes so-called *self-sustained oscillatory systems*, which are typically nonconservative and nonlinear. A system is said to be *self-sustained oscillatory* if it transforms the energy of an external source into the energy of nondecaying oscillations. The basic characteristics of such oscillations (amplitude, frequency, waveform, etc.) are determined by system parameters and are independent, within certain limits, of initial conditions.

1.5 Phase Portraits of Typical Oscillatory Systems

The method for analyzing oscillations of a DS by means of their graphical representation was introduced into the theory of oscillations by L.I. Mandelstam and A.A. Andronov. Since then, this method has become a standard tool for studying various oscillatory phenomena. Here we discuss several simple but typical examples of dynamical process representation in the form of a representative point trajectory in phase space.

1.5.1 Conservative Oscillator

Consider a linear oscillator without losses whose equations can be formulated using the oscillatory *LC*-circuit shown in Fig. 1.2a, assuming the amplitude of the oscillations to be sufficiently small. Let the charge q on the capacitor be a variable. Then according to Kirchhoff's equations, we have

$$\ddot{q} + (LC)^{-1} q = 0 . \tag{1.13}$$

Multiplying (1.13) by $L\dot{q}$, we get

$$\frac{d}{dt} \left(\frac{L\dot{q}^2}{2} + \frac{q^2}{2C} \right) = 0 , \tag{1.14}$$

i.e., at any time the following equalities hold:

$$E = E_L + E_C = \text{const.} , \qquad E_L = L\dot{q}^2/2 , \qquad E_C = q^2/2C . \tag{1.15}$$

They account for the temporal consistency of the total energy of the oscillator (the sum of magnetic and electric energies, E_L and E_C, respectively). The equations of the conservative oscillator can be written more conveniently by replacing the time by $\tau = t/\sqrt{LC}$:

$$\ddot{x} + x = 0 , \qquad \dot{x}^2 + x^2 = a^2 , \qquad a = \text{const.} \tag{1.16}$$

Phase coordinates $x_1 = x$ and $x_2 = \dot{x}$ then satisfy

$$\dot{x}_1 = x_2 , \qquad \dot{x}_2 = -x_1 , \qquad x_1^2 + x_2^2 = a^2 . \tag{1.17}$$

The phase portrait of the system looks like a circle with radius a, centered at the coordinate origin. A point in phase space at which a phase velocity vector vanishes is called a *singular point* or an *equilibrium point*. It corresponds to the equilibrium

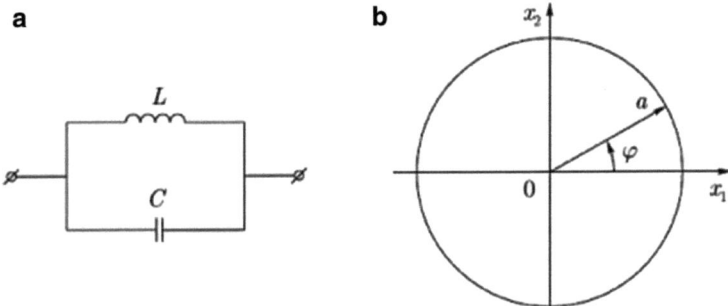

Fig. 1.2 Conservative oscillator: (**a**) the oscillatory circuit modelled by (1.16) and (**b**) the phase portrait of oscillations for a given energy level

state of the system when dynamical variables are temporally constant. In this case the coordinate origin is a singular point called a *center*.

In this example the integral of motion of the second-order conservative system indicates conservation of energy (1.15). This fact enables us to describe the system by means of a first-order equation. Indeed, introducing a new variable φ by the relations

$$x_1 = a \cos \varphi, \qquad x_2 = -a \sin \varphi, \qquad (1.18)$$

we derive the equations

$$\dot{\varphi} = 1, \qquad \dot{a} = 0, \qquad (1.19)$$

which constitute the law of motion of a phase point. Only one variable φ evolves in time, and the conservative oscillator with a given energy has a one-dimensional phase space. Harmonic oscillations of the oscillator are associated with the uniform motion of a representative point along a circle with radius a, as shown in Fig. 1.2b.

If a conservative system is nonlinear, its phase portrait is more complicated. This can be illustrated by the following equation:

$$\ddot{x} + \sin x = 0. \qquad (1.20)$$

In phase variables $x_1 = x$, $x_2 = \dot{x}$, (1.20) can be written in the form

$$\dot{x}_1 = x_2, \qquad \dot{x}_2 = -\sin x_1. \qquad (1.21)$$

Equilibria of the nonlinear pendulum on the phase plane are arranged along the axis x_1 ($x_2 = 0$) at points $x_1 = 0, \pm\pi, \pm 2\pi, \ldots$. The corresponding phase portrait of the system is shown in Fig. 1.3. We see that the singular points $x_1 = 0, \pm 2\pi, \pm 4\pi, \ldots$

1.5 Phase Portraits of Typical Oscillatory Systems

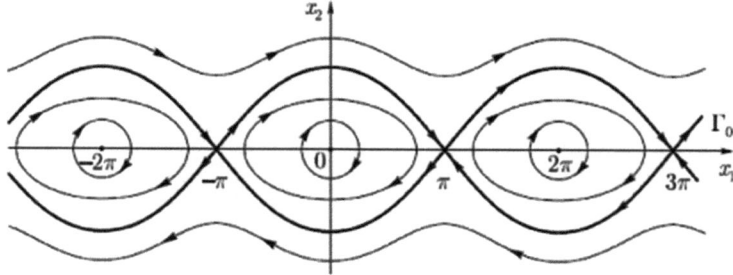

Fig. 1.3 Phase portrait of the oscillator described by (1.20)

are center-type equilibria, while $x_1 = \pm\pi, \pm 3\pi, \ldots$ correspond to unstable singular points called *saddles*.

Near the centers the phase portrait corresponds to that of a linear oscillator: trajectories are closed circle-like curves reflecting small-amplitude oscillations that are close to harmonic. Singular integral curves Γ_0 passing through unstable points are called *separatrices of a saddle*. They divide the phase space into regions with different behavior. As the energy of the pendulum increases, its harmonic oscillations near the center-type points evolve to nonlinear periodic oscillations near the separatrices. As the energy grows further, the oscillatory motion becomes rotational (the motion outside the separatrices). The situation when the energy of the pendulum corresponds to motion along a separatrix is said to be nonrobust. Even infinitesimal energy deviations to one side or the other can lead to qualitatively different types of motion, i.e., oscillatory or rotational.

As can be seen from Fig. 1.3, the state of the pendulum is defined by the angle x_1 of its deviation from equilibrium and by the velocity x_2. But for x_1 values that differ by an integer number of 2π, the system dynamics is identical. Oscillatory motions inside a separatrix loop are bounded. But if the energy of the system exceeds a critical value and the motion is rotational, the trajectories become unbounded along the axis x_1. In this case a *cylindrical* phase space is introduced by sewing together the values $x_1 = -\pi$ and $x_1 = \pi$. All the trajectories on the cylinder will be bounded.

1.5.2 Damped Linear Oscillator

Energy dissipation due to the presence of losses has a significant effect on the nature of the motion. The simplest regularities are manifested in systems with full energy dissipation when friction forces act over all degrees of freedom and there is no input energy. Consider the processes in a linear dissipative oscillator when a friction force is proportional to the rate of coordinate change. An example of such a system is an oscillatory circuit containing a resistance R. The circuit is then described by

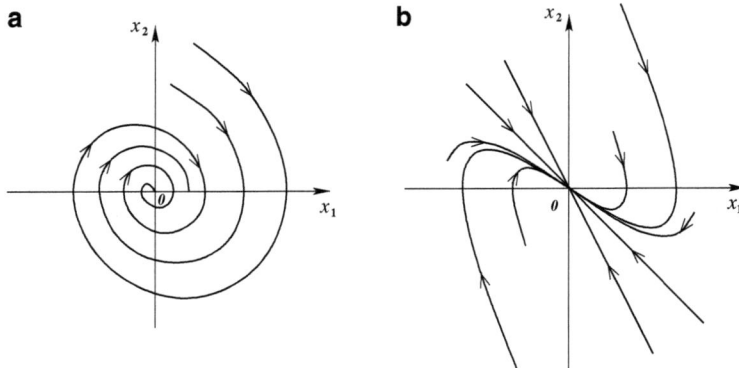

Fig. 1.4 Phase portrait of the dissipative oscillator (1.23) with (**a**) $\delta < 1$ and (**b**) $\delta > 1$

$$L\ddot{q} + R\dot{q} + q/C = 0 \, . \tag{1.22}$$

This can be reduced to its dimensionless form by the change of variables

$$\ddot{x} + 2\delta\dot{x} + x = 0 \, , \qquad 2\delta = R\sqrt{L/C} \, , \qquad \tau = t/\sqrt{LC} \, . \tag{1.23}$$

For $\delta = 0$ we have the conservative linear oscillator considered above. When low friction is introduced into the system, its phase portrait changes qualitatively. For $0 < \delta < 1$, (1.23) has solution

$$x = A \exp(-\delta\tau)\cos(\omega\tau + \psi) \, , \qquad \omega = (1 - \delta^2)^{1/2} \, , \tag{1.24}$$

where A and ψ are arbitrary constants specified by initial conditions. Phase trajectories starting from any initial point in the phase plane look like twisted spirals along which phase points asymptotically approach the coordinate origin, thus describing the damped oscillatory process. The origin is a singular point of the system which is a *stable focus* for $\delta < 1$ (Fig. 1.4a). If the friction coefficient $\delta > 1$, the process is aperiodic:

$$x = A_1 \exp(\lambda_1\tau) + A_2 \exp(\lambda_2\tau) \, , \qquad \lambda_{1,2} = \frac{1}{2}\left[-\delta \pm (\delta^2 - 1)^{1/2}\right] . \tag{1.25}$$

The phase trajectories constitute a family of typical curves along which, as in the previous case, the representative points tend to the coordinate origin (Fig. 1.4b). The singular point in these conditions is a *stable node*.

Thus, for all values of the physical parameters of the system and when $\delta > 0$, the dissipative pendulum is characterized by a single globally stable equilibrium at the phase coordinate origin. Independently of the choice of initial conditions, either damped oscillations or aperiodic motion can be observed. For $t \to \infty$ each (!) phase point tends to the origin in a stable focus or node.

The property described above is common for dynamical systems with full energy dissipation. The equilibria of stable focus- or node-type are here *globally attracting* in the sense that phase trajectories from any point in phase space approach them asymptotically. Stationary undamped oscillations appear to be impossible in linear dissipative systems. The reason is quite clear from a physical viewpoint: there are no conditions that would sustain oscillations. The energy needed to overcome friction forces is not being supplied.

1.6 Self-Sustained Oscillatory Systems

The possibility of a periodic asymptotically stable motion existing in an autonomous system can only be realized in nonlinear dissipative systems. This motion is represented by an isolated closed curve in phase space that attracts trajectories from some vicinity regardless of initial conditions. This type of dynamical system is so important in the study of oscillatory processes that A.A. Andronov suggested the special term *self-sustained oscillatory systems*. A mathematical image of periodic self-sustained oscillations is the *Andronov–Poincaré limit cycle*, an isolated closed trajectory in phase space corresponding to stable periodic motion.

A dynamical system with the Poincaré limit cycle is exemplified by the classical nonlinear van der Pol oscillator described by

$$\ddot{x} - (\varepsilon - x^2)\dot{x} + x = 0 . \tag{1.26}$$

The parameter ε characterizing the energy pumped into the system from an external source is a significant parameter of the oscillator and is called the *excitation parameter*. Comparing (1.26) and (1.23), it follows that the van der Pol oscillator describes a more complicated oscillatory circuit in which the character and the value of dissipation depend on the variable x. The oscillator equation (1.26) can be written in phase coordinates as

$$\dot{x}_1 = x_2 , \qquad \dot{x}_2 = (\varepsilon - x_1^2)x_2 - x_1 , \tag{1.27}$$

with

$$\varepsilon - x_1^2 \neq 0 . \tag{1.28}$$

Equations (1.27) cannot be solved analytically and are therefore analyzed numerically. In the important case ($\varepsilon > 0$), the system (1.27) has a unique stable solution in the form of a *limit cycle* Γ shown in Fig. 1.5a (see also Chap. 4).

The equilibrium at the origin, for which the nonlinearity can be neglected in the vicinity of zero, is an unstable focus. Trajectories from the vicinity of the equilibrium tend asymptotically to the limit cycle. Analysis shows that the limit

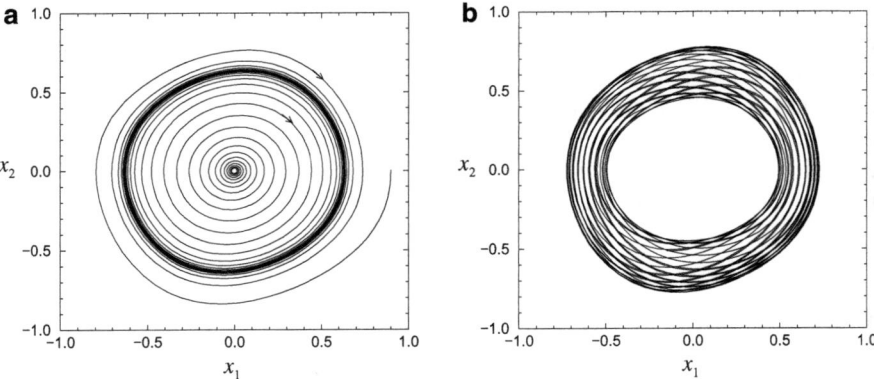

Fig. 1.5 Limit sets of the van der Pol oscillator (1.26): (**a**) limit cycle Γ for $\varepsilon = 0.1$ and (**b**) projection of a two-dimensional torus on the plane $(x_1 = x, x_2 = \dot{x})$. Equation (1.29) is integrated numerically for $\varepsilon = 0.1$, $B = 0.1$, $p = 1.35$, and $\varphi_0 = 0$

cycle is a stable isolated curve attracting trajectories from any point in the phase plane.

Thus, dynamical systems in which energy dissipation depends nonlinearly on an oscillating variable constitute a fundamentally new type of attracting set of phase trajectories: a limit cycle. Calculations show that, during the oscillation period on the limit cycle, the contributions of outgoing and incoming energy are strongly compensated.

Finally, consider another case of a typical structure in the phase space of a dynamical system which can emerge, for instance, in a periodically driven system with a stable limit cycle. We add a harmonic source with a relatively small amplitude B and frequency p to (1.26). p is assumed to be irrationally related to the frequency of periodic oscillations of the autonomous oscillator (1.26). Then we have

$$\ddot{x} - (\varepsilon - x^2)\dot{x} + x = B\sin(p\tau + \varphi_0) \,. \tag{1.29}$$

The periodic modulation of the limit cycle of the autonomous system leads to a situation where the phase trajectory rotates with a given frequency p around the limit cycle and lies on a two-dimensional surface which happens to be the *surface of a torus*. As for the limit cycle case, this surface will be a limit set that attracts all trajectories from the vicinity of the torus (from both inside and outside!). It is easy to imagine that the minimal dimension of the phase space which a *two-dimensional torus* can be embedded is three. Figure 1.5b shows the projection of a phase trajectory on a two-dimensional torus obtained numerically for the system (1.29).

1.7 Regular and Chaotic Attractors

All trajectories in the phase space of a dissipative system can be divided into trajectories corresponding to transient processes (relaxation of the system to certain stationary regimes) and trajectories belonging to invariant limit sets. The second type of trajectories correspond to stationary (steady-state) regimes of system operation. We have already mentioned a limit cycle and a limit toroidal set. Now we shall formulate a general definition of limit sets for dynamical systems.

The point $p \in \mathbb{R}^N$ is called an *ω-limit point* of a trajectory $\mathbf{x}(t)$, $t \geq t_0$, if there is a sequence $t_k \to \infty$ as $k \to \infty$ such that the sequence of states $\mathbf{x}(t_k)$ converges to the point p. Similarly, the point $q \in \mathbb{R}^N$ is called an *α-limit point* of a trajectory $\mathbf{x}(t)$, $t \leq t_0$, if there is a sequence $t_k \to -\infty$ as $k \to \infty$ such that the sequence of states $\mathbf{x}(t_k)$ converges to the point q. The set of all ω-limit points of a trajectory $\mathbf{x}(t)$ is called an *ω-limit set of the given trajectory* and denoted by $\omega(\mathbf{x}(t))$. The set of all α-limit points of a trajectory $\mathbf{x}(t)$ is an *α-limit set of the given trajectory* and denoted by $\alpha(\mathbf{x}(t))$. Limit sets of any phase trajectory themselves consist of phase trajectories. Furthermore, ω- and α-limit sets are invariant with respect to the evolution operator. This means that the evolution operator maps any point of a limit set onto a point of the same set.

Having considered where different trajectories of a DS tend to forwardly and backwardly in time, all invariant limit sets in phase space can be distinguished. Equilibria, periodic motions, and singular trajectories of separatrix-contour-type being double-asymptotic to saddle equilibria represent typical limit sets of trajectories in the phase plane. These limit sets completely exhaust all possible situations in the phase plane. They correspond to three different solutions. Separatrix contours and loops are singular curves which are not *structurally stable (robust)* in dissipative systems. They exist only for certain parameter values and disappear when the evolution operator is slightly perturbed. In contrast, when certain conditions are fulfilled, equilibria and limit cycles are structurally stable limit sets and can exist in a certain region of parameter space.

Consider structurally stable limit sets in $\mathbb{R}^N$. If a set Q is a limit set in the phase space of a DS, this implies one of the three possibilities:

1. Q is an ω-limit set for all trajectories from a phase space region U which do not belong to Q. In this case, Q is an *attracting limit set* or an *attractor* of the dynamical system.
2. Q is an α-limit set for all trajectories from U not belonging to Q. Then Q is called a *repelling limit set* or a *repeller*.
3. There are trajectories in U, not belonging to Q, for which Q is an ω-limit set, as well as trajectories for which Q is an α-limit set. In this case Q is a *saddle* (for example, a saddle equilibrium, a saddle limit cycle, a saddle torus).

Phase trajectories for which the saddle Q is an ω-limit set and phase trajectories for which Q is an α-limit set belong to smooth invariant manifolds with dimension less than N. They are called the *stable and unstable manifolds* of the saddle limit set Q, respectively. For a saddle equilibrium in the plane, stable and unstable manifolds consist of a pair of separatrices incoming to the saddle and outgoing from it, respectively (see Fig. 1.3).

When time is reversed ($t \to -t$), attractors of a DS become repellers and repellers are transformed into attractors. Saddles remain saddles, but invariant manifolds reverse their directions: stable manifolds become unstable and unstable manifolds become stable.

Although repellers and saddles occupy an important place in the structure of a system's phase portrait, attractors are of most interest as they correspond to experimentally observed stable modes of system operation. That is why we turn our attention to the definition of the attractor. The above definition of an attractor (and a repeller) is neither completely strict nor unique. Generally speaking, there is no commonly accepted definition of an attractor. Most often, especially in the mathematical literature, an attractor is defined as a *maximal attractor of an absorbing region*. We shall now give this definition.

Let a DS be given by the evolution operator $T^\tau : \mathbb{R}^N \to \mathbb{R}^N$ which at some fixed time τ specifies the map $F : B \to B$, where B is the *absorbing region* in $\mathbb{R}^N$. A *maximal attractor* A_{max} in the absorbing region B is the set

$$A_{max} = \bigcap_{n=0}^{\infty} F^n B \ .$$

A certain invariant set A is called an *attractor* of the DS if there exists an absorbing region for which A is a maximal attractor. The *basin of attraction* of attractor A is a set U from which all trajectories tend to A as $t \to \infty$.

The maximal attractor does not always coincide with an ω-limit set of trajectories in U. For example, consider a two-dimensional torus in $\mathbb{R}^N$ with a resonant structure consisting of stable and saddle cycles. In this case the torus is formed by the unstable manifold of the saddle cycle. The entire toroidal surface is a maximal attractor for some absorbing region, but the ω-limit set is the stable resonant cycle alone. The latter is the attractor from a 'physical viewpoint', since it corresponds to an experimentally observed mode of steady-state oscillations.

Attractors of systems whose state is given by one real scalar variable (i.e., the state set is $\mathbb{R}^1$) can only be equilibrium points. For systems in a phase plane, attractors can be represented by both equilibrium points and limit cycles. What would result if the dimension of a system were increased, for example, to $N = 3$, i.e., if one passed from the plane to the three-dimensional phase space? Not so long ago, before the beginning of the 1960s, the increase in the phase space dimension of dissipative systems was related to the possibility of appearance (in addition to the above) of quasiperiodic attractors which correspond to motions on k-dimensional tori ($k = 2, 3, \ldots$).

An important outcome of studies in recent years has been the discovery of fundamentally new types of motion in dynamical systems. Such motions in a phase space of dimension $N \geq 3$ correspond to intricately structured attracting sets. The

1.7 Regular and Chaotic Attractors

trajectories of their phase points belong to none of the above types of attractor. Phase trajectories are presented here in the form of an infinite nowhere-crossing curve. As $t \to \infty$, a trajectory does not leave a closed region and is not attracted to the known types of attractor. The existence of such trajectories is connected to the possibility of chaotic behavior in deterministic dynamical systems with phase space dimension $N \geq 3$. Similar properties of dynamical systems were first discovered by E. Lorenz in 1963 when he carried out numerical studies of the dynamics of a three-dimensional model of thermal convection. Eight years later, in a theoretical paper by D. Ruelle and F. Takens, the attracting region in the phase space of a DS characterized by a regime of steady-state nonperiodic oscillations was called a *strange attractor*. This term was immediately adopted by researchers and taken to refer to the mathematical image of the irregular oscillations of deterministic dynamical systems.

Attractors in the form of equilibrium points, limit cycles, or k-dimensional tori are called *simple* or *regular*, thus emphasizing that the motions on them meet the established ideas on Lyapunov-stable deterministic behavior of a dynamical system. A *strange* attractor is associated with an irregular oscillatory regime, in the sense of lack of periodicity, that is similar in many respects to our concepts of stationary random processes.

However, the term 'random' has a definite meaning. A random motion is unpredictable or is predictable with a certain probability. In other words, trajectories of a random motion cannot be repeatedly and uniquely reproduced numerically or experimentally. An example is the classical motion of a Brownian particle. In the case of a strange attractor, there is strict predictability in terms of determination of the evolution law. The solution of equations (as for regular attractors) satisfies a uniqueness theorem and is unambiguously reproduced for fixed initial conditions. Therefore, complex 'noise-like' self-sustained oscillations whose mathematical image is a strange attractor can be specified using the terms *dynamical stochasticity*, *deterministic chaos*, and such like. It is important to distinguish them from stochastic processes in the classical sense, whose description requires us to consider fluctuations in initial dynamical equations or which are directly governed by the equations for the probability density of the statistical theory.

A system with a chaotic attractor can be exemplified by the equations of an oscillator with inertial nonlinearity, viz., the Anishchenko–Astakhov oscillator (see Chap. 11). This system is a generalization of the van der Pol equations to three-dimensional space, described by the equations

$$\dot{x} = mx + y - xz, \qquad \dot{y} = -x, \qquad \dot{z} = -gz + gI(x)x^2, \qquad (1.30)$$

$$I(x) = \begin{cases} 1, & x > 0, \\ 0, & x \leq 0. \end{cases}$$

Results for the numerical solution of (1.30) with parameter values $m = 1.5, g = 0.2$ are shown in Fig. 1.6, which illustrates a chaotic attractor.

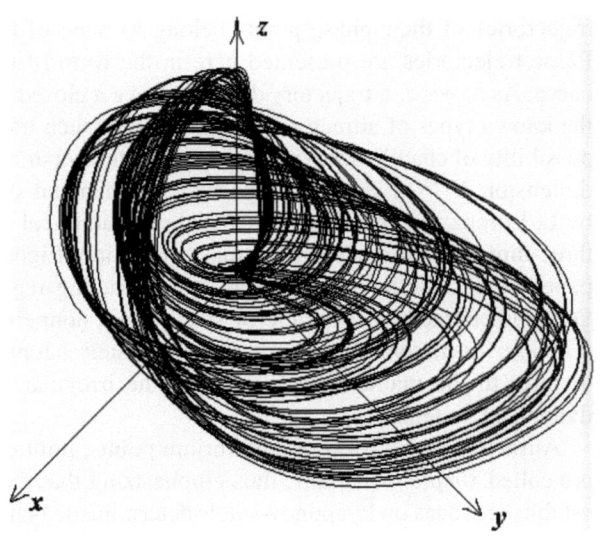

Fig. 1.6 Chaotic attractor in the Anishchenko–Astakhov oscillator model (1.30)

1.8 Discrete-Time Systems: Return Maps

Discrete-time systems (cascades) play an important role in nonlinear dynamics. They are also called *point maps* or *return maps*.[1] A return map can be written in the form

$$\mathbf{x}_{n+1} = \mathbf{P}(\mathbf{x}_n) \,, \tag{1.31}$$

where $\mathbf{x} \in \mathbb{R}^N$ is the state vector, n is a discrete time variable, and $\mathbf{P}(\mathbf{x})$ is a vector function called the *return function*. The return function explicitly sets the evolution operator in one time step (one iteration): $T_n^1 \mathbf{x} = \mathbf{P}(\mathbf{x}_n)$. If the state vector $\mathbf{x}$ is given at some time moment n, the system state can be found at any other time moment $m = n + k, k > 0$:

$$\mathbf{x}_{n+k} = T_n^k \mathbf{x} = \mathbf{P}^{(k)}(\mathbf{x}_n) = \mathbf{P}\big(\mathbf{P}(\ldots \mathbf{P}(\mathbf{x}_n)\ldots)\big) \,.$$

Thus, (1.31) defines a dynamical system. For return maps, we do not require the reversibility of the evolution operator in time. Otherwise, we have to exclude from consideration the irreversible one-dimensional maps that are widely used in nonlinear dynamics as basic models. As well as flows, return maps can be dissipative (contracting) and conservative (volume-preserving). Attractors and other limit sets can be distinguished in the phase space of dissipative maps. A phase trajectory of (1.31) consists of a sequence of points $\mathbf{x}_1, \mathbf{x}_2, \ldots, \mathbf{x}_i, \ldots$. A phase trajectory

---

[1] The commonly used term 'discrete map' is not entirely appropriate since a map is usually given on a continuous set of states and only time moments form a discrete set.

1.8 Discrete-Time Systems: Return Maps

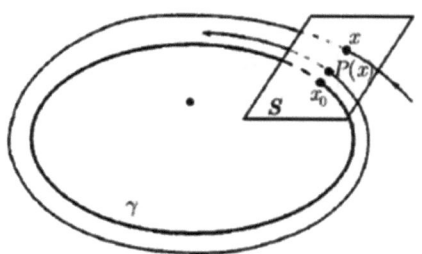

Fig. 1.7 Construction of the Poincaré intersecting surface S

belonging to a limit set of system (1.31) may include only one point $\mathbf{x}_0$ called a *fixed point*. It satisfies

$$\mathbf{P}(\mathbf{x}_0) = \mathbf{x}_0 \ .$$

If the trajectory is closed and consists of k points $\mathbf{x}_i$, $i = 1, 2, \ldots, k$, such that for any i

$$\mathbf{P}^{(k)}(\mathbf{x}_i) = \mathbf{x}_i \ ,$$

the set of points $\mathbf{x}_i$ is a *cycle of period k* or a *period-k cycle*, or simply a *k-cycle*. The points of the k-cycle are also called *fixed points of multiplicity k*. Quasiperiodic and chaotic trajectories of the return map are represented by nonclosed sequences of points which never return strictly to their previous positions.

Reversible return maps can be directly related to flow systems given by ordinary differential equations. In order to pass from a flow system to a discrete-time system, one must introduce an intersecting plane S (a hypersurface in the multi-dimensional case) such that all trajectories intersect it strictly transversally. If we consider the points appearing when the trajectories intersect the surface S in one direction, the flow generates a return map in S, also known as the *Poincaré map* (Fig. 1.7). The Poincaré map is necessarily reversible and is uniquely (but not one-to-one) related to the initial flow. The serial number of the intersection of a given trajectory with the intersecting surface is the discrete time n.

If a dynamical system is subjected to a periodic external force, a return map can be constructed by using a so-called *stroboscopic section*. One considers the points of a phase trajectory that appear in the period of the external force T: $\mathbf{x}(t_0 + nT)$, $n = 0, 1, 2, \ldots$. The stroboscopic section is in fact a special case of the Poincaré section. Indeed, if the external force is periodic, a nonautonomous system can be represented as an autonomous one by introducing the phase of the external force $\Psi \in [-\pi, +\pi]$. In this case the stroboscopic section is reduced to the section of trajectories by the plane $\Psi = $ const.

The Poincaré map reduces the dimension of all limit sets in the system phase space by one and thus makes the phase portrait more descriptive. Limit cycles of a flow system correspond to fixed points or cycles of the Poincaré map,

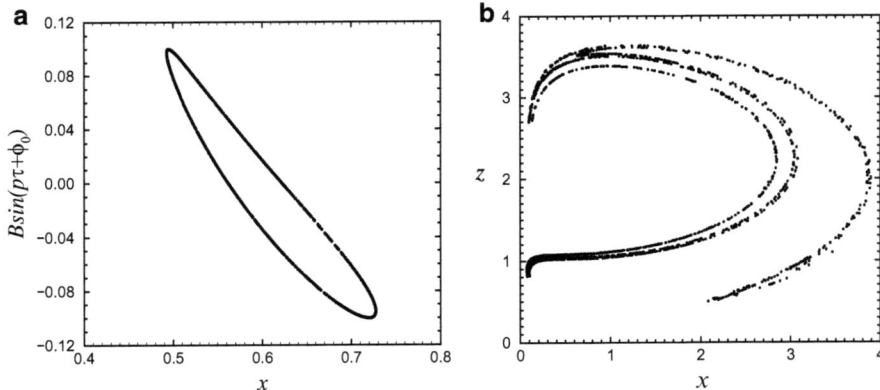

Fig. 1.8 Poincaré sections of complex limit sets: (**a**) the invariant closed curve in the section of a two-dimensional torus in (1.29) for the parameters $a = 1$, $\varepsilon = 0.1$, $b = 0.3$, $B = 0.1$, $p = 1.35$, and $\varphi_0 = 0$, and (**b**) the set with a complex geometric structure in the section of the chaotic attractor in (1.30) for $m = 1.5$, $g = 0.2$. The plane $y = 0$ is used as the intersecting surface S in both cases

consisting of k points (Fig. 1.7). The number k is determined by the number of times a trajectory crosses the surface S in a given direction. Quasiperiodic and chaotic trajectories of the flow system correspond to quasiperiodic and chaotic nonclosed trajectories of the map. Quasiperiodic trajectories generated by ergodic trajectories on a two-dimensional torus fill an invariant closed curve which is the image of the two-dimensional torus under the map (Fig. 1.8a). Chaotic trajectories of the Poincaré map belong to sets with a complex geometric structure. If the dimension of a chaotic attractor is not high, its geometrical structure in S is more illustrative than the 'ball of thread' observed in the phase space of a flow system (Fig. 1.8b).

If a DS has phase space dimension $N = 3$, the Poincaré map is two-dimensional. In this case it can be reduced to a map of the plane, which makes the system dynamics particularly evident. If the phase space of a flow system is significantly contracted in certain directions, one may observe a situation where the points in the section S lie virtually on a curve. In fact, this is not a one-dimensional curve, but rather a fractal set with a complex transverse structure which is not visible due to the large contraction. Within a given accuracy, the system state can be specified by one coordinate $x = s$ which is the distance from a given point on the curve to the fixed point chosen as the origin (Fig. 1.9a). Thus, in some cases with large contraction, the behavior of a DS can be described by the one-dimensional point map of an interval into itself:

$$x_{n+1} = P(x_n) \,. \tag{1.32}$$

1.8 Discrete-Time Systems: Return Maps

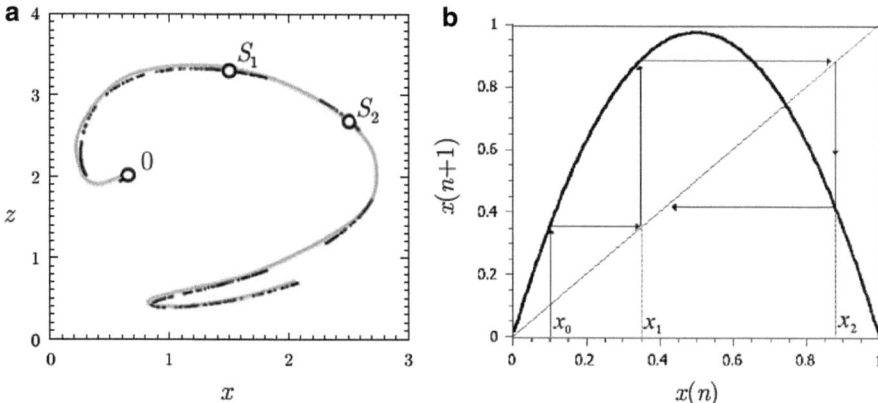

Fig. 1.9 One-dimensional return map in the section of the chaotic attractor in (1.30) for $m = 1.45$, $g = 0.35$ (**a**), and construction of the iterative Lamerey diagram for a model one-dimensional map (**b**)

Such a map is usually irreversible but its forward behavior in time can be used to analyze qualitatively some important features of the behavior of the original flow system. It is much easier to study a one-dimensional map, both numerically and theoretically.

A one-dimensional map can also be explored by drawing an iterative *Lamerey diagram*. This is constructed using the graph of the return function $P(x)$ and the diagonal of the first coordinate angle (see Fig. 1.9b). A vertical line is drawn from the initial point x_0 to the intersection with the return function graph. The intersection point is projected horizontally onto the diagonal, and this point is then projected vertically onto the x-axis to produce the point x_1, which is the image of point x_0. Successive steps with this procedure lead to the construction of the Lamerey staircase, which is a broken line consisting of vertical and horizontal segments (these segments are marked in Fig. 1.9b by arrows indicating the direction of motion on the diagram). The Lamerey staircase can 'lead' to a fixed point of the map or it can be closed to form a cycle. In the case of quasiperiodic or chaotic behavior of the system (1.32), the staircase consists of an infinite number of steps. The Lamerey diagram thus gives a very visual representation of the system dynamics.

Unfortunately, getting two-dimensional and one-dimensional return maps explicitly (in the form of systems (1.31) or (1.32)) for a specific flow DS is in most cases a complex problem or even impossible. But there are many model one-dimensional and two-dimensional point maps which, although they are not directly related to specific flow systems, are nevertheless widely used in nonlinear dynamics to study and describe various fundamental phenomena. We list the most popular model maps.

1.8.1 Stretching Map

$$x_{n+1} = \alpha x_n, \quad \text{mod } 1.$$

This map is given on the interval [0, 1] and mod 1 means that the fractional part is taken. $\alpha > 0$ is the parameter of the map.

1.8.2 Logistic Map

$$x_{n+1} = \alpha x_n (1 - x_n).$$

This map is defined on the interval [0, 1] and $\alpha \in [0, 4]$ is the parameter. The logistic map can also be rewritten in the following forms:

$$x_{n+1} = a - x_n^2 \quad \text{or} \quad x_{n+1} = 1 - \varepsilon x_n^2.$$

1.8.3 Sine Map

$$x_{n+1} = \Omega + x_n + K \sin(x_n), \quad \text{mod } 1.$$

This map is defined on the interval [0, 1], while $\Omega \in [0, 1]$ and $K \geq 0$ are the parameters.

1.8.4 Henon Map

$$x_{n+1} = 1 - a x_n^2 + y_n,$$
$$y_{n+1} = b y_n.$$

a and b are the parameters of the map.

1.8.5 Lozi Map

$$x_{n+1} = 1 - a|x_n| + y_n,$$
$$y_{n+1} = b y_n.$$

The Henon and Lozi maps of the plane are reversible.
The study of these and similar maps provides a way to reveal the universal regularities of chaos, examine the structure and bifurcations of chaotic attractors, explore the specifics of the interaction of chaotic systems, and model the behavior of oscillatory ensembles and nonlinear distributed media with complex dynamics. Many results obtained for simple model return maps which seem to be unrelated

to real systems have turned out to be fundamental and have been confirmed experimentally.

1.9 Summary

In this chapter we have given a general definition of a dynamical system and provided a detailed description of dynamical systems specified by ordinary differential equations. We have established that the behavior of such dynamical systems falls into four types of stationary regime: an equilibrium state, periodic motion, quasiperiodic motion, and chaotic motion. These types of solution correspond to attractors in the form of a stable equilibrium point, a limit cycle, a quasiperiodic attractor (a k-dimensional torus), and a chaotic or strange attractor. The important thing is that the simplest types of quasiperiodic and chaotic attractor can be realized in dynamical systems with phase space dimension not less than three. We have also examined discrete-time systems (return maps), shown their relationship with dynamical systems given by ordinary differential equations, and analyzed the possible types of phase trajectories of return maps.

There are many books devoted to the mathematical and physical aspects of the theory of dynamical systems [1–43]. Answers on almost any question can be found in these books. However, we recommend the beginners in the field of nonlinear dynamics to start their detailed and deep study with books [1, 20, 21, 28, 39].

References

1. Afraimovich, V.S., Arnold, V.I., Ilyashenko, Yu.S., Shilnikov, L.P.: Dynamical Systems V. Encyclopedia of Mathematical Sciences. Springer, Heidelberg (1989)
2. Afraimovich, V.S., Shilnikov, L.P.: Strange attractors and quasiattractors. In: Barenblatt, G.I., Iooss, G., Joseph, D.D. (eds.) Nonlinear Dynamics and Turbulence, p. 1. Pitman, Boston (1983)
3. Andronov, A.A., Vitt, E.A., Khaikin, S.E.: Theory of Oscillations. Pergamon, Oxford (1966)
4. Anishchenko, V.S.: Dynamical Chaos – Models and Experiments. World Scientific, Singapore (1995)
5. Anishchenko, V.S., Astakhov, V.V., Neiman, A.B., Vadivasova, T.E., Schimansky-Geier, L.: Nonlinear Dynamics of Chaotic and Stochastic Systems. Springer, Berlin (2002)
6. Arnold, V.I.: Mathematical Methods of Classical Mechanics. Springer, Heidelberg (1974)
7. Arnold, V.I.: Geometrical Methods in the Theory of Ordinary Differential Equations. Springer, New York (1983)
8. Berge, P., Pomeau, I., Vidal, C.G.: Order Within Chaos. Wiley, New York (1984)
9. Binney, J.: The Theory of Critical Phenomena: An Introduction to the Renormalization Group. Oxford University Press, Oxford (1992)/World Scientific, Singapore (2010)
10. Bowen, R.: Equilibrium States and the Ergodic Theory of Anosov Diffeomorphisms. Lecture Notes in Mathamatics, vol. 470. Springer, Berlin (1975)
11. Bohr, T., Jensen, M.H., Paladin, G., Vulpiani, A.: Dynamical System Approach to Turbulence. Cambridge University Press, Cambridge (1998)

12. Chen, G., Dong, X.: From Chaos to Order: Perspectives, Methodologies, and Applications. World Scientific, Singapore (1998)
13. Collins, J.: Renormalization: An Introduction to Renormalization. Cambridge University Press, Cambridge (1984/1986)
14. Crownover, R.M.: Introduction to Fractals and Chaos. Jones and Bartlett, London (1995)
15. Devaney, R.L.: An Introduction to Chaotic Dynamical Systems. Westview, Boulder (1989/2003)
16. Drazin, P.G.: Nonlinear Systems. Cambridge University Press, Cambridge (1992)
17. Eckmann, J.-P., Collet, P.: Iterated Maps as Dynamical Systems. Birkhauser, Basel (1980)
18. Glendinning, P.: Stability, Instability, and Chaos: An Introduction to the Theory of Nonlinear Differential Equations. Cambridge University Press, Cambridge (1994)
19. Grebogi, C., Yorke, J.A. (eds.): The Impact of Chaos on Science and Society. United Nations University Press, Tokyo (1997)
20. Guckenheimer, J., Holmes, P.: Nonlinear Oscillations, Dynamical Systems, and Bifurcations of Vector Fields. Springer, New York (1983)
21. Haken, H.: Synergetics: Introduction and Advanced Topics. Springer, Heidelberg (2004)
22. Hilborn, R.C.: Chaos and Nonlinear Dynamics. An Introduction for Scientists and Engineers. Oxford University Press, Oxford (2002/2004)
23. Jackson, E.A.: Perspectives of Nonlinear Dynamics, vols. 1 and 2. Cambridge University Press, Cambridge (1989/1990)
24. Katok, A., Hasselblatt, B.: Introduction to the Modern Theory of Dynamical Systems. Cambridge University Press, Cambridge (1995)
25. Kapitaniak, T.: Chaos for Engineers: Theory, Applications, and Control. Springer, New York (1998)
26. Kapitaniak, T.: Chaotic Oscillators: Theory and Applications. World Scientific, Singapore (1992)
27. Lakshmanan, M., Murali, K.: Chaos in Nonlinear Oscillators: Controlling and Synchronization. World Scientific, Singapore (1996)
28. Lichtenberg, A., Lieberman, M.A.: Regular and Stochastic Motion. Springer, New York (1983)
29. Marek, M., Schreiber, I.: Chaotic Behaviour of Deterministic Dissipative Systems. Cambridge University Press, Cambridge (1991/1995)
30. Marsden, L.E., McCraken, V.: The Hopf Bifurcation and Its Applications. Springer, New York (1976)
31. Moon, F.C.M.: Chaotic and Fractal Dynamics: An Introduction for Applied Scientists and Engineers. Wiley, New York (1992)
32. Moon, F.C.M.: Chaotic Vibration: An Introduction for Applied Scientists and Engineers. Wiley, New York (2004)
33. Mosekilde, E., Maistrenko, U., Postnov, D.: Chaotic Synchronization: Applications to Living Systems. World Scientific, Singapore (2002)
34. Nicolis, G.: Introduction to Nonlinear Science. Cambridge University Press, Cambridge (1995)
35. Ogorzalek, M.J.: Chaos and Complexity in Nonlinear Electronic Circuits. World Scientific, Singapore (1997)
36. Ott, E.: Chaos in Dynamical Systems. Cambridge University Press, Cambridge (1993/2002)
37. Pesin, Ya.B.: Dimension Theory in Dynamical Systems: Contemporary Views and Applications. University of Chicago Press, Chicago (1997)
38. Schroeder, M.: Fractals, Chaos, Power Laws. Freeman, New York (1991)
39. Schuster, H.G.: Deterministic Chaos. Physik-Verlag, Weinheim (1988)
40. Seydel, R.: Practical Bifurcation and Stability Analysis: From Equilibrium to Chaos. Springer, New York (1994/2009)
41. Shilnikov, L.P., Shilnikov, A.L., Turaev, D.V., Chua, L.O.: Methods of Qualitative Theory in Nonlinear Dynamics. World Scientific, Singapore (2001)
42. Thompson, J.M.T., Stewart, H.B.: Nonlinear Dynamics and Chaos. Wiley, New York (1986)
43. Zaslavsky, G.M.: Chaos in Dynamical Systems. Harwood, New York (1985)

Chapter 2
Stability of Dynamical Systems: Linear Approach

2.1 Introduction

Our understanding of the stability of a particular operating mode of a dynamical system is formed intuitively as we build up our experience and understanding of everyday life and nature. The first steps of a small child give him or her very real representations of the stability of walking, although these representations may not yet enter consciousness. Looking at the famous painting entitled *Young Acrobat on a Ball* by P. Picasso, we have a distinct feeling that the girl's equilibrium is not quite stable. As adults, we can already discuss the stability of a ship on a stormy sea, the stability of economic trends in relation to the actions of managers and politicians, the stability of our nervous system with regard to stressful perturbation, etc. In each case, we talk about different properties that are specific to the considered systems. However, if we think about it carefully, we can find something in common, inherent in any system. The common feature is that, when we talk about stability, we understand the way the dynamical system reacts to a small perturbation of its state. If arbitrarily small changes in the system state begin to grow in time, the system is unstable. Otherwise, small perturbations decay with time and the system is stable.

It is extremely important from a practical point of view to be able to analyse the stability of the operating modes of dynamical systems. Stability of such systems as a car, an aircraft, or an ocean liner to perturbations is certainly a vital factor in the truest sense of the word, since such perturbations are always going to be present in one form or another.

These arguments are qualitative and can be made precise only if we manage to translate them into the formal language of mathematics. The fundamentals of the rigorous mathematical theory of stability were laid down in the works of the prominent Russian mathematician A.M. Lyapunov 100 years ago, while the development of the qualitative theory and bifurcation theory of dynamical systems is associated with Russian scientists A.A. Andronov, V.I. Arnold, and their pupils.

In this chapter we shall define the stability of a dynamical system and, with the help of some simple and clear examples, attempt to illustrate its content and also some methods for solving stability problems.

2.2 Definition of Stability

There are in fact many different definitions of stability, among which the following are the most frequently encountered: stability according to Poisson, stability according to Lyapunov, and asymptotic stability. Let a DS be described by the system of ordinary differential equations (1.5) or by (1.6) in the vector form. We are interested in the stability of a trajectory $\mathbf{x}^0(t)$.

Stability according to Poisson means that, after a while, the phase trajectory returns to an arbitrarily small neighbourhood of the initial point $\mathbf{x}_0^0 = \mathbf{x}^0(t_0)$. Moreover, if the system is reversible, return occurs both forward and backward in time. The time interval after which the trajectory returns to a neighborhood of the point $\mathbf{x}_0^0$ with given radius ε is called the *Poincaré recurrence time*. Recurrence times may correspond to the period or quasiperiod of a regular motion and represent a random sequence in the regime of dynamical chaos (Fig. 2.1).

Stability according to Poisson is an important but weak form of stability. We can say nothing about the behavior of neighboring trajectories, initially close to $\mathbf{x}^0(t)$. In practical problems we are often interested in another property of stability, associated with a small perturbation of a given trajectory. Depending on the temporal dynamics of the perturbation, we distinguish stability according to Lyapunov and asymptotic stability.

The trajectory $\mathbf{x}^0(t)$ is said to be *stable according to Lyapunov* if, for any arbitrarily small $\varepsilon > 0$, there is $\delta(\varepsilon) > 0$ such that, for any trajectory $\mathbf{x}(t)$ for which $\|\mathbf{x}(t_0) - \mathbf{x}^0(t_0)\| < \delta$, the inequality $\|\mathbf{x}(t) - \mathbf{x}^0(t)\| < \varepsilon$ is satisfied for all $t > t_0$. The symbol $\|\ldots\|$ denotes the vector norm in $\mathbb{R}^N$. Thus, a small initial perturbation does not grow in time for a phase trajectory that is stable according to Lyapunov. If the small perturbation δ vanishes as time goes by, i.e., $\|\mathbf{x}(t) - \mathbf{x}^0(t)\| \to 0$ as $t \to \infty$, the trajectory possesses a stronger stability property, namely, *asymptotic stability*. Any asymptotically stable phase trajectory is stable according to Lyapunov. The opposite is not generally true.

The stability properties of phase trajectories belonging to limit sets, e.g., attractors, are of special importance for understanding the system dynamics. In many cases, a change in the kind of stability of one or another limit set can change the operating mode of the system.

Fig. 2.1 Poisson-stable non-closed trajectory

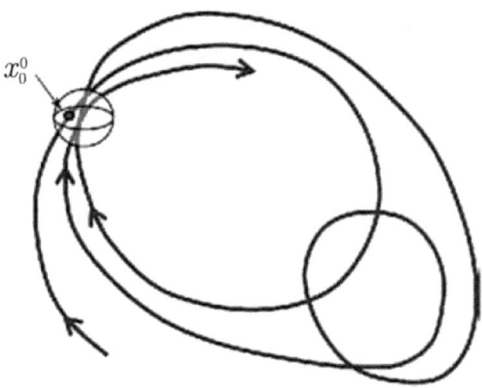

2.3 Linear Analysis of Stability

2.3.1 Stability of Solutions of a First-Order Differential Equation

Any dynamical system (physical, chemical, mechanical, etc.) is associated in our minds with an evolution in time. Anticipating objections, we note that an equilibrium state, i.e., a stationary state, in which the rate of the process under study is equal to zero, can also be treated as a limiting case of the temporal evolution of the system. Consider a simple model of a DS described by the single first-order ordinary differential equation

$$\frac{dx(t)}{dt} = \dot{x} = F(x), \qquad (2.1)$$

where $x(t)$ is the state variable and F is a function characterizing the evolution law. The state space of such a system is a set of real numbers $\mathbb{R}^1$. If the initial condition $x(t_0)$ is given, there is a unique solution of (2.1) that defines the state $x(t)$ at any time t.

Because the problem of Lyapunov stability and asymptotic stability involves analysis of the way the system reacts to a small perturbation, it can be studied in a linear approximation. Let us explain this. Suppose $x^0(t)$ is a particular solution of (2.1) whose stability we would like to investigate. We introduce a variable $y(t)$ which specifies a small deviation from $x^0(t)$, i.e.,

$$y(t) = x(t) - x^0(t). \qquad (2.2)$$

Here $x(t)$ is a perturbed solution.

Our task is to study the time evolution of the small perturbation $y(t)$ which obeys (2.1). We expand the function F in a series in the neighborhood of $x^0(t)$:

$$F(x_0 + y) = \left.\frac{dF}{dx}\right|_{x=x^0(t)} y(t) + \frac{1}{2}\left.\frac{d^2F}{dx^2}\right|_{x=x^0(t)} y^2(t) + \ldots \quad (2.3)$$

The derivatives of F must be calculated at points corresponding to the particular solution. We now rewrite (2.1) for the perturbation $y(t)$ using (2.3), whence

$$\dot{y}(t) = \left.\frac{dF}{dx}\right|_{x=x^0(t)} y(t) + \Phi(y), \quad (2.4)$$

where

$$\Phi(y) = \frac{1}{2}\left.\frac{d^2F}{dx^2}\right|_{x=x^0(t)} y^2(t) + \ldots \quad (2.5)$$

The terms in $\Phi(y)$ include all terms going as y^n ($n \geq 2$), i.e., they account for all the nonlinear components. By definition, the variable $y(t)$ is a small deviation from the particular solution. Therefore, the nonlinear terms in (2.4) can be neglected in a first approximation. The evolution of the small perturbation can thus be described by the linear equation

$$\dot{y} = A(t)y, \quad \text{where } A(t) = \left.\frac{dF}{dx}\right|_{x=x^0(t)}. \quad (2.6)$$

Consider now the following example. Let a dynamical system be described by

$$\dot{x} = a - bx^2, \quad a > 0, \, b > 0. \quad (2.7)$$

We find stationary states x^0 of the system and analyze their stability. Since there are no temporal changes in a stationary state, $dx/dt|_{x^0} = 0$ and we obtain

$$x^0_{1,2} = \pm\sqrt{\frac{a}{b}}. \quad (2.8)$$

We now apply Eq. (2.6) for the perturbation to the first stationary state x^0_1, yielding

$$\dot{y} = -(2bx^0_1)y = (-2\sqrt{ab})y = sy, \quad s = \left.\frac{dF}{dx}\right|_{x^0_1} = -2\sqrt{ab}. \quad (2.9)$$

Equation (2.9) has solution $y = \exp(st)$. The perturbation y decays exponentially in time because s is negative. This means that the state x^0_1 is stable. Since the second

2.3 Linear Analysis of Stability

state x_2^0 differs from the first only by its sign, the solution of (2.9) increases in time. Hence, the stationary state x_2^0 is unstable.

A sufficiently simple idea for predicting stability in the linear approximation has proved to be very fruitful. Equation (2.6) for the perturbation can also be generalized to N state variables.

2.3.2 Stability of a Dynamical System in $\mathbb{R}^N$

Consider a DS given by a vector differential equation of the form

$$\dot{\mathbf{x}} = \mathbf{F}(\mathbf{x}) , \qquad (2.10)$$

where $\mathbf{x} \in \mathbb{R}^N$. We analyze the stability of a particular solution $\mathbf{x}^0(t)$. Whereas the one-dimensional equation (2.1) describes the evolution exclusively in the neighborhood of equilibria, solutions of (2.10) can include equilibrium points and periodic, quasiperiodic, and chaotic orbits.

We introduce a perturbation vector $\mathbf{y} = \mathbf{x}(t) - \mathbf{x}^0(t)$, assuming that its length $\|\mathbf{y}\|$ is small. For $\mathbf{y}$ we may write

$$\dot{\mathbf{y}} = \mathbf{F}(\mathbf{x}^0 + \mathbf{y}) - \mathbf{F}(\mathbf{x}^0) . \qquad (2.11)$$

Expanding $\mathbf{F}(\mathbf{x}^0 + \mathbf{y})$ in a series in the vicinity of $\mathbf{x}^0$ and taking into account the fact that the perturbation is small in norm, we arrive at the following equations, linearized with respect to $\mathbf{y}$:

$$\dot{\mathbf{y}} = \hat{A}(t)\mathbf{y} , \qquad (2.12)$$

where $\hat{A}(t)$ is a matrix with elements

$$a_{jk}(t) = \left. \frac{\partial f_j}{\partial x_k} \right|_{\mathbf{x}(t) = \mathbf{x}^0(t)} , \quad j, k = 1, 2, \ldots, N , \qquad (2.13)$$

called the *linearization matrix* of the system in the vicinity of the solution $\mathbf{x}^0(t)$, and f_j are the components of the vector function $\mathbf{F}$. As the elements of the matrix $\hat{A}$ depend on a point on the studied trajectory, they generally vary in time. The matrix is characterized by eigenvalues $s_i(t)$ which are also time dependent. The eigenvalues are roots of the characteristic equation

$$\det\left[\hat{A}(t) - s\hat{E}\right] = 0 , \qquad (2.14)$$

where $\hat{E}$ is the unit matrix. The N eigenvalues (counting multiple eigenvalues) are associated with N linearly independent eigenvectors $\mathbf{e}_i(t)$ which change direction as one moves along the trajectory $\mathbf{x}^0(t)$. These eigenvectors satisfy

$$\hat{A}(t)\mathbf{e}_i(t) = s_i(t)\mathbf{e}_i(t), \quad i = 1, 2, \ldots, N. \tag{2.15}$$

The increase or decrease of the perturbation $\mathbf{y}(t)$ is determined by the sign of the real part of $s_i(t)$. As one moves along the trajectory $\mathbf{x}^0(t)$, it may be that the perturbation grows at some points of the given trajectory and decreases at others. The problem is to specify those characteristics of the perturbation behavior that would define it as a whole along the given trajectory, and the relevant tool for this is the *Lyapunov characteristic exponent*.

According to Lyapunov's theorem, if the matrix $\hat{A}(t)$ is bounded, then for each nontrivial solution $\mathbf{y}(t)$ of the system (2.12), there is a finite *Lyapunov characteristic exponent*, i.e., the real number defined by

$$\lambda[\mathbf{y}(t)] = \overline{\lim_{t\to\infty}} \frac{1}{t} \ln \|\mathbf{y}(t)\|, \tag{2.16}$$

where the bar indicates the upper limit and $\|\ \|$ denotes the vector norm. Linearly independent solutions are characterized, in general, by different Lyapunov exponents. For N linearly independent solutions $\mathbf{y}^i(t)$, $i = 1, 2, \ldots, N$ making up a fundamental matrix of solutions $\hat{Y}(t)$ of the system (2.12), there are N Lyapunov characteristic exponents:

$$\lambda_i = \overline{\lim_{t\to\infty}} \frac{1}{t} \ln \|\mathbf{y}^i(t)\|, \quad i = 1, \ldots, N. \tag{2.17}$$

Arranged in decreasing order, the real numbers $\lambda_1 \geq \lambda_2 \geq \ldots \geq \lambda_N$ form the *Lyapunov characteristic exponent spectrum (LCE spectrum)*. λ_1 is called the maximal Lyapunov exponent. For certain sufficiently general conditions, the LCE spectrum does not depend on the choice of the fundamental matrix of solutions and completely defines the local stability properties of trajectory $\mathbf{x}^0(t)$. Each exponent in the LCE spectrum determines the rate of exponential contraction or stretching of a perturbation component in the direction of a relevant eigenvector of the fundamental matrix $\hat{Y}(t)$, on average, along the trajectory.

If $\hat{A}(t)$ is a bounded real matrix, the Lyapunov inequality is satisfied:

$$\sum_{i=1}^{N} \lambda_i \geq \overline{\lim_{t\to\infty}} \frac{1}{t - t_0} \int_{t_0}^{t} \text{Tr}\,\hat{A}(t')dt', \tag{2.18}$$

where $\text{Tr}\,\hat{A}(t)$ is the trace of the matrix $\hat{A}(t)$. Equality holds in (2.18) for *systems that are said to be tame according to Lyapunov*. According to the Ostrogradsky–Liouville formula, the trace of the linearization matrix determines the evolution of a small volume element of the phase space along the trajectory $\mathbf{x}^0(t)$:

$$V(t) = V(t_0) \exp\left[\int_{t_0}^{t} \text{Tr}\,\hat{A}(t')dt'\right]. \tag{2.19}$$

2.3 Linear Analysis of Stability

It is obvious that $\operatorname{Tr} \hat{A}(t) = \operatorname{div} \mathbf{F}(\mathbf{x}(t))$, where $\mathbf{F}(\mathbf{x}(t))$ is a phase velocity field. Accordingly, the mean divergence along the trajectory $\mathbf{x}^0(t)$ satisfies the inequality

$$\langle \operatorname{div} \mathbf{F}(\mathbf{x}(t)) \rangle \leq \sum_{i=1}^{N} \lambda_i . \tag{2.20}$$

If the sum of Lyapunov exponents is negative, the divergence of $\mathbf{F}$ is on average negative and the phase volume vanishes with time. This indicates the presence of dissipation in the system.

If the trajectory $\mathbf{x}^0(t)$ is stable according to Lyapunov, then an arbitrary initial perturbation $\mathbf{y}(t_0)$ does not grow, on average, along the trajectory. A necessary and sufficient condition for this is that the LCE spectrum should not contain positive exponents.

If an arbitrary bounded trajectory $\mathbf{x}^0(t)$ belongs to a limit set of the autonomous system (2.10) which is not an equilibrium or a saddle separatrix, then at least one of the Lyapunov exponents is always equal to zero. Indeed, the small perturbation remains on average unchanged along the direction tangent to the trajectory.

A phase volume element must be contracted for phase trajectories located near the attractor. In this case the dissipative dynamical system has a negative average divergence $\mathbf{F}(\mathbf{x}(t))$ and the sum of the Lyapunov exponents satisfies the inequality

$$\sum_{i=1}^{N} \lambda_i < 0 . \tag{2.21}$$

Stability of Equilibrium States in $\mathbb{R}^N$

If the particular solution $\mathbf{x}^0(t)$ of a system (2.10) is an equilibrium point, i.e., $\mathbf{F}(\mathbf{x}^0) = 0$, the linearization matrix $\hat{A}$ is considered at only one point of phase space, so it is a matrix with constant elements a_{ij}. The eigenvectors and eigenvalues of the matrix $\hat{A}$ are constant in time and the Lyapunov exponents coincide with the real parts of the eigenvalues, i.e., $\lambda_i = \operatorname{Re} s_i$. The signature of the LCE spectrum indicates whether the equilibrium is stable or not. To analyze the behavior of phase trajectories in a local neighborhood of an equilibrium, one also needs to know the imaginary parts of the linearization matrix eigenvalues. In a phase plane, $N = 2$, the equilibrium is characterized by the two eigenvalues of the matrix $\hat{A}$, namely, s_1 and s_2. The following cases can be realized in the phase plane:

1. s_1 and s_2 are real negative numbers, in which case the equilibrium is a stable node.
2. s_1 and s_2 are real positive numbers, in which case the equilibrium is an unstable node.
3. s_1 and s_2 are real numbers but with different signs, in which case the equilibrium is a saddle.

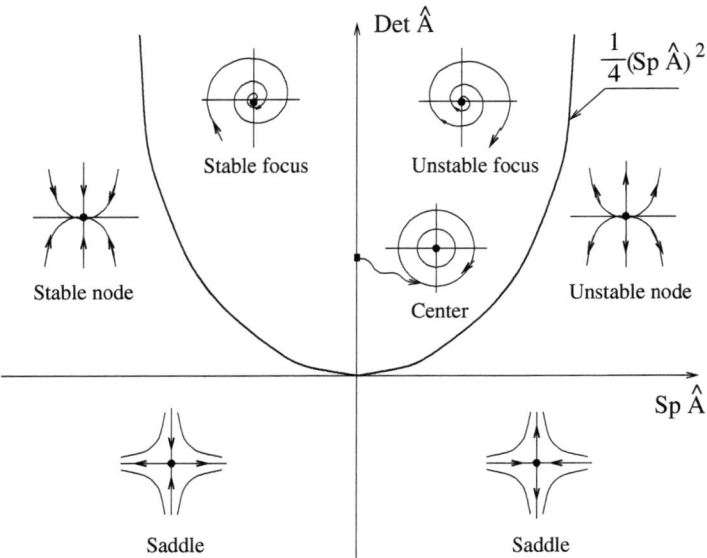

Fig. 2.2 Equilibria in the plane. Phase portraits are shown in transformed coordinates

4. s_1 and s_2 are complex conjugates with $\operatorname{Re} s_{1,2} < 0$ and the equilibrium is a stable focus.
5. s_1 and s_2 are complex conjugates with $\operatorname{Re} s_{1,2} > 0$ and the equilibrium is an unstable focus.
6. s_1 and s_2 are pure imaginary, so can be written $s_{1,2} = \pm i\omega$, in which case the equilibrium is a center.

Figure 2.2 shows the equilibria realized in the plane for different values of the determinant and trace of the matrix $\hat{A}$, i.e., $\det \hat{A} = s_1 s_2$ and $\operatorname{Tr} \hat{A} = s_1 + s_2$.

Besides the aforementioned states, other kinds of equilibria are possible in a phase space with dimension $N \geq 3$, e.g., an equilibrium state called a saddle-focus which is unstable according to Lyapunov. Figure 2.3 shows two possible types of saddle-focus in a three-dimensional phase space. These are distinguished by the dimensions of their stable and unstable manifolds.

To identify which type of limit set the equilibrium corresponds to, it is enough to know the Lyapunov exponents. The equilibrium is considered to be an attractor if it is asymptotically stable in all directions and its LCE spectrum consists only of negative exponents (stable node and stable focus). If the equilibrium is unstable in all directions, it is a repeller (unstable node and unstable focus). If the LCE spectrum includes both positive and negative exponents, the equilibrium is of saddle type (simple saddle or a saddle-focus). In addition, the exponents $\lambda_i \geq 0$ ($\lambda_i \leq 0$) determine the dimension of the unstable (stable) manifold.

2.3 Linear Analysis of Stability

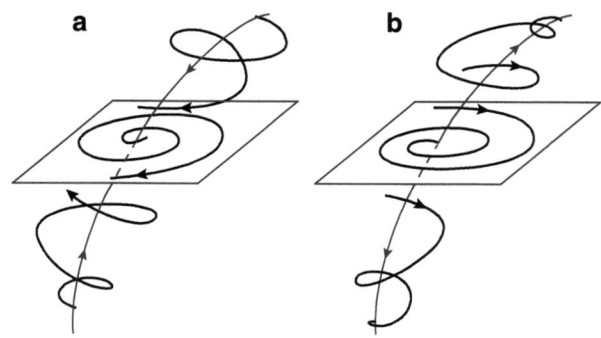

Fig. 2.3 Saddle-foci in three-dimensional phase space: (**a**) s_1 is real and negative, while $s_{2,3}$ are complex conjugate with Re $s_{2,3} > 0$. (**b**) s_1 is real and positive, while $s_{2,3}$ are complex conjugates with Re $s_{2,3} < 0$

Stability of Periodic Solutions

Any periodic solution $\mathbf{x}^0(t)$ of the system (2.10) satisfies the condition

$$\mathbf{x}^0(t) \equiv \mathbf{x}^0(t+T), \qquad (2.22)$$

where T is the period of the solution. The linearization matrix $\hat{A}(t)$ calculated at points of the trajectory corresponding to the periodic solution $\mathbf{x}^0(t)$ is also periodic:

$$\hat{A}(t) = \hat{A}(t+T). \qquad (2.23)$$

In this case, Eq. (2.12) for perturbations is linear with periodic coefficients. The stability of a periodic solution can be estimated once it is known how a small perturbation $\mathbf{y}(t_0)$ evolves over the period T. Its evolution can be represented by

$$\mathbf{y}(t_0 + T) = \hat{M}_T \mathbf{y}(t_0), \qquad (2.24)$$

where $\hat{M}_T$ is the *monodromy matrix*. It is independent of time. The eigenvalues of the monodromy matrix, i.e., the roots of the characteristic equation

$$\det\left[\hat{M}_T - \mu \hat{E}\right] = 0, \qquad (2.25)$$

are called *multipliers* of the periodic solution $\mathbf{x}^0(t)$ and define its stability. Indeed, the monodromy operator (2.24) acts as follows. The initial perturbation of a periodic solution, considered via its projections onto the eigenvectors of the matrix $\hat{M}_T$, is multiplied by an appropriate multiplier μ_i over the period T. Thus, a necessary and sufficient requirement for the periodic solution $\mathbf{x}^0(t)$ to be stable according to Lyapunov is that its multipliers should satisfy $|\mu_i| \leq 1$, $i = 1, 2, \ldots, N$. At least one of the multipliers is equal to $+1$. Since they are the eigenvalues of the monodromy matrix, the multipliers obey the relations

$$\sum_{i=1}^{N} \mu_i = \operatorname{Tr} \hat{M}_T, \qquad \prod_{i=1}^{N} \mu_i = \det \hat{M}_T. \qquad (2.26)$$

They are related to the Lyapunov exponents of the periodic solution by

$$\lambda_i = \frac{1}{T} \ln |\mu_i|. \qquad (2.27)$$

One of the LCE spectrum exponents of a limit cycle is always zero and corresponds to a unit multiplier. The limit cycle is an attractor if all the other exponents are negative. If the LCE spectrum includes exponents of different sign, the limit cycle is a saddle. The dimension of its unstable manifold is equal to the number of non-negative exponents in the LCE spectrum, and the dimension of its stable manifold is equal to the number of exponents for which $\lambda_i \leq 0$. If, besides the zero exponent, all the other exponents satisfy $\lambda_i > 0$, then the limit cycle is absolutely unstable (a repeller).

Stability of Quasiperiodic and Chaotic Solutions

Let a particular solution $\mathbf{x}^0(t)$ of the system (2.10) correspond to quasiperiodic oscillations with k independent frequencies ω_j, $j = 1, 2, \ldots, k$. Then

$$\begin{aligned}\mathbf{x}^0(t) &= \mathbf{x}^0\big(\varphi_1(t), \varphi_2(t), \ldots, \varphi_k(t)\big) \\ &= \mathbf{x}^0\big(\varphi_1(t) + 2\pi m, \varphi_2(t) + 2\pi m, \ldots, \varphi_k(t) + 2\pi m\big), \end{aligned} \qquad (2.28)$$

where m is an arbitrary integer and $\varphi_j(t) = \omega_j t$, $j = 1, 2, \ldots, k$. The stability of the quasiperiodic solution is characterized by the LCE spectrum. The linearization matrix $\hat{A}(t)$ is quasiperiodic, and the Lyapunov exponents are strictly defined only in the limit as $t \to \infty$. In the case of *ergodic quasiperiodic oscillations*, the periodicity of the solution with respect to each of the arguments φ_j results in the LCE spectrum containing k zero exponents. If all other exponents are negative, the toroidal k-dimensional hypersurface (which we shall refer to as the k-dimensional torus for simplicity) on which the relevant quasiperiodic trajectory lies is an attractor. When all other exponents are positive, the k-dimensional torus is a repeller. The torus is said to be a saddle[1] if the LCE spectrum of quasiperiodic trajectories on the torus has, besides zero exponents, both positive and negative ones.

A chaotic trajectory which belongs to a chaotic attractor is always unstable in at least one direction. The LCE spectrum of a chaotic solution always has at least one positive Lyapunov exponent. There is no contradiction between instability of

---

[1] This situation should be distinguished from the case of chaos on a k-dimensional torus, which is observed for $k \geq 3$.

phase trajectories and the attracting nature of the limit set to which they belong. Phase trajectories starting from close initial points in the basin of attraction tend to the attractor but they are separated on it. Hence, chaotic trajectories are unstable according to Lyapunov, but stable according to Poisson.

2.4 Stability of Phase Trajectories in Discrete-Time Systems

Let a discrete time system be described by the return map

$$\mathbf{x}_{n+1} = \mathbf{P}(\mathbf{x}_n) \,, \tag{2.29}$$

where $\mathbf{x} \in \mathbb{R}^N$ is the state vector, n is a discrete time variable, and $\mathbf{P}(\mathbf{x})$ is a vector function with components P_j, $j = 1, 2, \ldots, N$. Let us analyze the stability of an arbitrary solution $\mathbf{x}_n^0$. Introducing a small perturbation $\mathbf{y}_n = \mathbf{x}_n - \mathbf{x}_n^0$ and linearizing the map in the vicinity of the solution $\mathbf{x}_n^0$, we deduce the linear equation for the perturbation:

$$\mathbf{y}_{n+1} = \hat{M}(n)\mathbf{y}_n \,, \tag{2.30}$$

where $\hat{M}(n)$ is the linearization matrix with elements

$$m_{jk} = \left. \frac{\partial P_j(\mathbf{x})}{\partial x_k} \right|_{\mathbf{x} \in \mathbf{x}_n^0}. \tag{2.31}$$

It follows from (2.30) that the initial perturbation evolves according to the law

$$\mathbf{y}_{n+1} = \hat{M}(n)\hat{M}(n-1)\ldots\hat{M}(1)\mathbf{y}_1 \,. \tag{2.32}$$

By analogy with differential systems, we consider the Lyapunov exponents of the solution $\mathbf{x}_n^0$:

$$\lambda_i = \overline{\lim_{n \to \infty}} \frac{1}{n} \ln \|\mathbf{y}_n^i\| \,, \tag{2.33}$$

where $\mathbf{y}_n^i$, $i = 1, \ldots, N$ are linearly independent solutions of the system (2.30).

The stability of fixed points and cycles of the map is characterized by multipliers. The sequence of states $\mathbf{x}_1^0, \mathbf{x}_2^0, \ldots, \mathbf{x}_l^0$ is called a *period-l cycle* of the map, or simply an *l-cycle*, if the following condition is satisfied:

$$\mathbf{x}_1^0 = \mathbf{P}^l(\mathbf{x}_1^0) \,. \tag{2.34}$$

If $l = 1$, i.e.,

$$\mathbf{x}^0 = \mathbf{P}(\mathbf{x}^0) \,, \tag{2.35}$$

the state $\mathbf{x}^0$ is a *fixed point* or *period-1 cycle*. The linearization matrix $\hat{M}$ along the periodic solution $\mathbf{x}_n^0$ is periodic, i.e., $\hat{M}(n+l) = \hat{M}(n)$. The perturbation component $\mathbf{y}_1^i$ transforms as follows over the period l:

$$\mathbf{y}^i l + 1 = \hat{M}(l)\hat{M}(l-1)\ldots\hat{M}(1)\mathbf{y}_1^i = \hat{M}_l \mathbf{y}_1^i . \tag{2.36}$$

The matrix $\hat{M}_l$ does not depend on the initial point and is an analogue of the monodromy matrix in a differential system. The eigenvalues μ_i^l of the matrix $\hat{M}_l$ are called *multipliers of the l-cycle* of the map. They characterize how projections of the perturbation vector onto the eigenvectors of the linearization matrix $\hat{M}_l$ change over the period l. The multipliers μ_i^l are related to the Lyapunov exponents by

$$\lambda_i = \frac{1}{l} \ln |\mu_i^l| . \tag{2.37}$$

The l-cycle of the map is asymptotically stable if its multipliers satisfy $|\mu_i^l| < 1$, $i = 1, 2, \ldots, N$. Thus, the LCE spectrum involves only negative numbers.

If the map has the phase space dimension $(N-1)$ and is the Poincaré map of some N-dimensional continuous-time system, then it has the following property: the eigenvalues μ_i^l, $i = 1, 2, \ldots, (N-1)$, of the matrix $\hat{M}_l$ for the l-cycle, supplemented by the unit multiplier $\mu_N^l = 1$, are strictly equal to the eigenvalues of the monodromy matrix of the corresponding limit cycle in this continuous-time system. On this basis, the stability of periodic oscillations in differential systems can be described quantitatively by the multipliers of the relevant cycle in the Poincaré map.

2.5 Summary

In this chapter we have given a brief and simplified description of the basic ideas and methods of the theory of stability. The main focus has been on linear analysis of the stability of trajectories. The theory of stability is essential for nonlinear dynamics. By studying the stability of trajectories, one can determine the character of the system's limit sets and obtain a qualitative phase portrait. In addition, when the system parameters are varied, the resulting change in the stability of trajectories belonging to a particular limit set can be used to diagnose bifurcations. The most typical bifurcations of dynamical systems will be discussed in the next chapter. We note that, even though the linear analysis of stability is very important, it is not sufficient to provide a complete picture of the system behavior or describe the possible bifurcations in the system. The stability of dynamical systems is described in more detail in [1–17].

References

1. Andronov, A.A., Vitt, E.A., Khaikin, S.E.: Theory of Oscillations. Pergamon, Oxford (1966)
2. Anishchenko, V.S.: Dynamical Chaos – Models and Experiments. World Scientific, Singapore (1995)
3. Anishchenko, V.S., Astakhov, V.V., Neiman, A.B., Vadivasova, T.E., Schimansky-Geier, L.: Nonlinear Dynamics of Chaotic and Stochastic Systems. Springer, Berlin (2002)
4. Barreira, L., Pesin, Y.: Nonuniform Hyperbolicity: Dynamics of Systems With Nonzero Lyapunov Exponents. Encyclopedia of Mathematics and Its Applications. Cambridge University Press, Cambridge (2007)
5. Drazin, P.G.: Nonlinear Systems. Cambridge University Press, Cambridge (1992)
6. Glendinning, P.: Stability, Instability, and Chaos: An Introduction to the Theory of Nonlinear Differential Equations. Cambridge University Press, Cambridge (1994)
7. Guckenheimer, J., Holmes, P.: Nonlinear Oscillations, Dynamical Systems, and Bifurcations of Vector Fields. Springer, New York (1983)
8. Hilborn, R.C.: Chaos and Nonlinear Dynamics. An Introduction for Scientists and Engineers. Oxford University Press, Oxford (2002/2004)
9. Marek, M., Schreiber, I.: Chaotic Behaviour of Deterministic Dissipative Systems. Cambridge University Press, Cambridge (1991/1995)
10. Marsden, L.E., McCraken, V.: The Hopf Bifurcation and Its Applications. Springer, New York (1976)
11. Moon, F.C.M.: Chaotic and Fractal Dynamics: An Introduction for Applied Scientists and Engineers. Wiley, New York (1992)
12. Moon, F.C.M.: Chaotic Vibration: An Introduction for Applied Scientists and Engineers. Wiley, New York (2004)
13. Nicolis, G.: Introduction to Nonlinear Science. Cambridge University Press, Cambridge (1995)
14. Ogorzalek, M.J.: Chaos and Complexity in Nonlinear Electronic Circuits. World Scientific, Singapore (1997)
15. Schuster, H.G.: Deterministic Chaos. Physik-Verlag, Weinheim (1988)
16. Seydel, R.: Practical Bifurcation and Stability Analysis: From Equilibrium to Chaos. Springer, New York (1994/2009)
17. Thompson, J.M.T., Stewart, H.B.: Nonlinear Dynamics and Chaos. Wiley, New York (1986)

Chapter 3
Bifurcations of Dynamical Systems

3.1 Introduction

In the natural sciences, it turns out that the formulation of mathematical models leads to temporal evolution laws for state variables that depend on parameters. The values of these parameters are defined by system elements that do not change over time. When described mathematically, a wide class of physical problems lead to differential equations or maps which depend on one or several parameters. Fixing parameter values determines the type of solutions for given initial conditions, while variation of these values may result in both quantitative and qualitative changes in the nature of the solutions.

Consider as an example the equation for the oscillations of a dissipative pendulum or an *RLC*-circuit:

$$\ddot{x} + \alpha \dot{x} + \omega_0^2 x = 0 . \tag{3.1}$$

Equation (3.1) has two parameters: α is the damping parameter, which characterizes friction, while ω_0 defines the frequency of oscillations. We fix $\omega_0 = 1$. If there are no energy losses, the solution of (3.1) for a damping parameter $\alpha = 0$ represents undamped harmonic oscillations. With low friction $0 < \alpha < 1$, the system oscillates with an amplitude that decreases exponentially in time. Finally, for sufficiently high friction ($\alpha > 1$), the pendulum exhibits aperiodic motion decaying in time. Even in this simple example, two special parameter values are distinguished, $\alpha = 0$ and $\alpha = 1$. Any deviations from them can qualitatively change the system properties.

By controlling dynamical system parameters like the coefficient α in (3.1), we can observe a qualitative change in the phase portrait. This phenomenon is called a *bifurcation* of the dynamical system, and the parameter values at which the bifurcation takes place are called *bifurcation values*. By a qualitative change of phase portrait we mean profound structural changes, consisting in the appearance and disappearance of limit sets as well as changes in the stability of trajectories belonging to those limit sets. For example, if there is a stable equilibrium point in the system phase space and it becomes unstable when one or several control parameters are varied, then the behavior of all phase trajectories from the attraction region of the equilibrium changes fundamentally. They will now be attracted to another attractor, i.e., a bifurcation occurs in the system.

Very often the observed state of a system above and below the bifurcation point changes significantly when a system parameter changes. However, there are also bifurcations which do not appear to change the system state. These may include bifurcations of limit sets that are not attractors. The result of such bifurcations is not immediately noticeable in an experimental context since the stationary state existing before the bifurcation continues to exist and remains stable. However, there are changes in the phase space which are somehow essential for a DS. An example is a saddle-repeller bifurcation of limit cycles in the synchronization region of self-sustained oscillations. This leads to the disappearance of a pair of cycles, viz., saddle and repeller cycles. The stable cycle remains unchanged and the same periodic oscillations are observed in experiments. However, before the bifurcation, the stable limit cycle is located on the surface of a torus, while after the saddle-repeller bifurcation, the torus does not even exist. The disappearance of the torus significantly alters the system phase portrait. Many features of the system behavior change, but this is not immediately visible. However, it can be observed when the control parameters are varied, when the system is subjected to external signals, or when the effects of noise are analyzed.

We distinguish *local* and *nonlocal bifurcations* of a DS. *Local bifurcations* are associated with the local neighborhood of a trajectory on a limit set. They reflect the changed stability of individual trajectories as well as of the entire limit set, and can attest to the disappearance of the limit set if it merges with another one. All the above-listed phenomena can be identified in the context provided by the linear analysis of stability. For example, when one of the Lyapunov exponents of a trajectory on a limit set changes sign, this attests to a local bifurcation of the limit set. *Nonlocal bifurcations* are related to the behavior of manifolds of saddle limit sets. These bifurcations include in particular the formation of separatrix loops, homoclinic and heteroclinic curves, and tangency between an attractor and separatrix curves or surfaces. Such effects cannot be detected using the linear approach alone. In this situation, the nonlinear properties of the system must be taken into consideration.

In mathematics and physics there is a concept of robustness or structural stability. The essence of this concept is that a small change in a parameter of a robust system leads to minor changes in the operating mode of the system. One can say that the phase portrait is not qualitatively modified. From this point of view, for robust systems, crossing the bifurcation point means replacing one structurally stable regime by another. The system is not robust at the bifurcation point: a small change in the system parameter in either direction leads to dramatic changes in the system state.

The analysis of DS bifurcations when varying system parameters enables one to construct a *bifurcation diagram* of the system. This is a set of points, lines, and surfaces in parameter space which corresponds to certain bifurcations of the

3.2 Double Equilibrium Bifurcation

limit sets of the system. If several limit sets are realized in the same parameter range, the bifurcation diagram appears to be multi-sheeted. The co-existence of a large (even infinite) number of limit sets is typical for systems with complex dynamics. In a multi-dimensional parameter space, bifurcations are characterized by a certain number of conditions imposed on system parameters. The number of such conditions defines the *codimension of bifurcation*. For example, codimension 1 means that there is only one bifurcation condition.

3.2 Double Equilibrium Bifurcation

Let us return to our example of the stability of equilibrium points in the system (2.7). We agreed that parameters a and b in (2.7) are positive. Stability is determined by the sign of the derivative of the right-hand side of (2.7) at equilibrium, i.e., by the sign of s in (2.9). This derivative is always different from zero for positive values of a and b. But what would happen if we decreased the value of a? As can be seen from (2.9), for $a = 0$ (regardless of $b > 0$), the value of s goes to zero, so the perturbation y neither grows nor decays! Besides, if $a = 0$, two equilibrium points in the system seem to merge into one ($x = 0$)! Further, if $a < 0$, there are no equilibria at all! Indeed, in this case $x_{1,2}^0 = \pm j\sqrt{|a|/b}$, i.e., they are purely imaginary.

We now present the results of the mathematical analysis of the above bifurcation, which is known as the *double equilibrium bifurcation* or the *saddle-node bifurcation of equilibria*. Consider again (2.7). Let $x^0(a)$ be a robust equilibrium state, i.e., $s(a) \neq 0$. This means that, when the parameter a is slightly varied, the equilibrium $x^0(a)$ continues to exist as a stable or an unstable point.

At a certain parameter value $a = a^*$, the eigenvalue $s(a^*)$ at the equilibrium point goes to zero:

$$s(a) = \left.\frac{dF}{dx}\right|_{x^0} = 0, \quad a = a^*. \tag{3.2}$$

The double equilibrium bifurcation is realized if the second derivative is nonzero:

$$\left.\frac{d^2F}{dx^2}\right|_{x^0} \neq 0. \tag{3.3}$$

To satisfy conditions (3.2) and (3.3) in the general case, the right-hand side of the original equation must include at least a quadratic nonlinear term, as in our example (2.7).

If conditions (3.2) and (3.3) are met, then x^0 is a double root of the original Eq. (2.7). The parameter value a^* at which the condition (3.2) is satisfied is called the bifurcation point. Above it, $a > a^*$ and we have two equilibria: one is stable and the other is unstable. They merge into one equilibrium at the bifurcation point

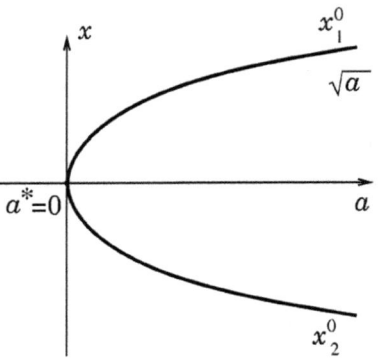

Fig. 3.1 The so-called double equilibrium bifurcation

$a = a^*$, and for $a < a^*$ there are no equilibria in the system! In the case of the system (2.7), we have $a^* = 0$. These results are illustrated in Fig. 3.1. In phase space $\mathbb{R}^N$, the unstable point is more often a saddle, and the above bifurcation is called the saddle-node bifurcation of equilibrium points.

3.3 Soft and Hard Bifurcations: Catastrophes

Despite its long history, the existence and development of the classical theory of stability and bifurcations went largely unnoticed by the general public until there came a moment (as is often the case) when it suddenly attracted worldwide attention. The reason for this was the popular presentation of French mathematician René Thom's work on what became known as catastrophe theory. At the beginning of the 1970s when this theory became fashionable, non-specialists felt that its message was clear, and thanks to its many and varied claims, it came to resemble certain pseudo-scientific theories of the past. But what is the essence of this development? To experts in the field, Thom's catastrophe theory was of course more objectively assessed, and while several related results deserve the deepest respect, there is no 'philosophical' discovery here. Let us explain why.

The essence of the theory lies in the fact that, although we are dealing with the same bifurcations, we choose one type in particular, the so-called *hard bifurcations* or *crises*. This can be explained by considering two simple examples. In the first case (Fig. 3.2a and b), the initial stationary state loses its stability through bifurcation and two new stationary states are born. The new stationary states (Fig. 3.2c) are located close to the initial state that has lost its stability (marked with an asterisk). Bifurcations of this type are said to be *soft*, bearing in mind that a newly born regime of system operation arises from the regime which has lost its stability, and coexists with it.

Another example of bifurcation is presented qualitatively in Fig. 3.3. For $\alpha < \alpha^*$ (Fig. 3.3a), the ball is in its stable stationary state. In addition, there is one more unstable state, marked with an asterisk. At the bifurcation point $\alpha = \alpha^*$, the stable

3.4 Triple Equilibrium Bifurcation

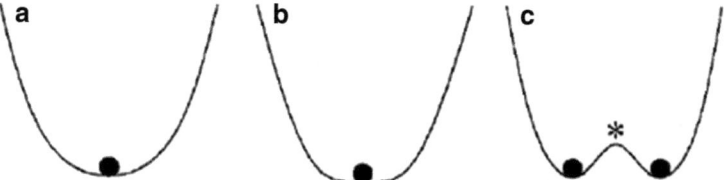

Fig. 3.2 Example of a soft bifurcation. The stationary state (**a**) losses its stability (**b**), and two new stationary states appear near it (**c**)

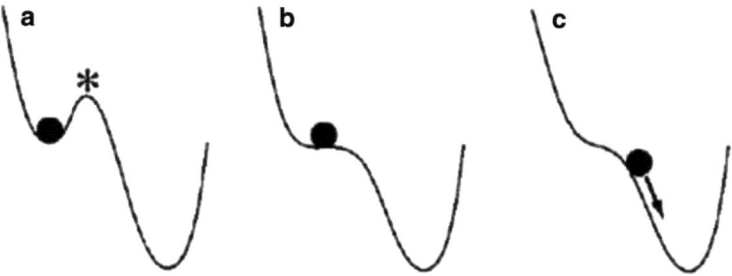

Fig. 3.3 Hard loss of stability by a stationary state, leading to catastrophe. Qualitative illustration of the so-called double equilibrium bifurcation (see Fig. 3.1)

and unstable states merge into one state (Fig. 3.3b). Furthermore, they disappear and the system chooses a new regime, e.g., as shown in Fig. 3.3c, which differs essentially from the previous one and is not located near the initial state. This type of bifurcation is said to be *hard*, and it was the hard bifurcations that were the subject of analysis in catastrophe theory. The above example of a double equilibrium bifurcation in system (2.7) is a typical example of a hard bifurcation, illustrated qualitatively in Fig. 3.3.

3.4 Triple Equilibrium Bifurcation

The so-called double equilibrium bifurcation considered above refers to codimension 1 bifurcations. We now analyze the triple equilibrium bifurcation, which has codimension two and is thus controlled by two parameters. This bifurcation consists in the merging of three equilibria, two nodes Q_1, Q_2, and a saddle Q_0 located between them. As a result, there remains a stable node at point Q_0. This is illustrated in Fig. 3.4.

The model system for this bifurcation can be written in the form

$$\dot{x} = \alpha_1 + \alpha_2 x - x^3 . \tag{3.4}$$

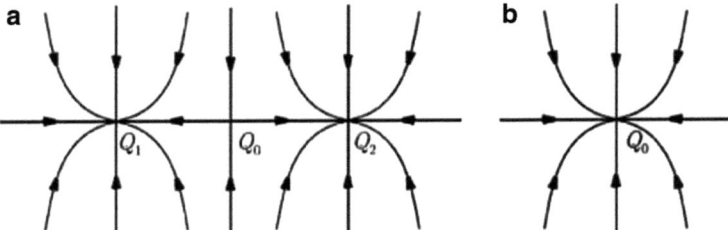

Fig. 3.4 Illustration of a so-called triple equilibrium bifurcation. (**a**) Two stable nodes and a saddle before the bifurcation. (**b**) One stable node after the bifurcation

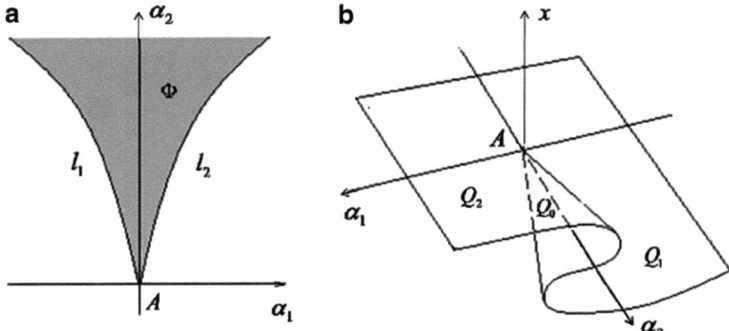

Fig. 3.5 Illustration of the so-called triple equilibrium bifurcation. (**a**) Bifurcation diagram. (**b**) Phase-parameter diagram

Analysis of the equilibria shows that, as long as $\alpha_2 < 0$ and for any α_1, the system has a unique equilibrium state Q_0 with eigenvalue $s_{Q_0} < 0$, i.e., the equilibrium is asymptotically stable. For $\alpha_2 > 0$, there exists a range of values for the parameter α_1 (the shaded region in the bifurcation diagram of Fig. 3.5a) where the system has three equilibria, Q_0, Q_1, and Q_2. One of them, Q_0, is unstable with $s_{Q_0} > 0$, while the other two, Q_1 and Q_2, are stable with $s_{Q_{1,2}} \leq 0$. The bistability region in the bifurcation diagram of Fig. 3.5a is bounded by curves l_1 and l_2, which correspond to saddle-node bifurcations of nodes $Q_{1,2}$ with saddle Q_0. The curves l_1 and l_2 converge to the point A ($\alpha_1 = \alpha_2 = 0$), called the *assembly point*, or the *cusp*. At this point, two bifurcation conditions are satisfied simultaneously: $s_{Q_1}(\alpha_1, \alpha_2) = 0$ and $s_{Q_2}(\alpha_1, \alpha_2) = 0$. This bifurcation is thus called the triple equilibrium and has codimension two. The structure shown in Fig. 3.5b appears in the phase-parameter space of the system (3.4) and is called the *assembly*. In the assembly region, the upper and lower leaves correspond to the stable equilibria, while the central one corresponds to the unstable equilibrium. For the special case $\alpha_1 = 0$, the triple equilibrium bifurcation turns into the pitchfork bifurcation for equilibria.

3.5 Andronov–Hopf Bifurcation

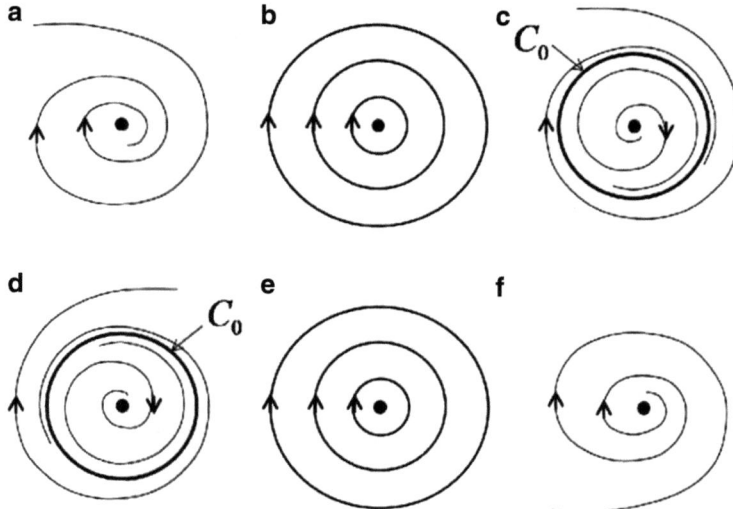

Fig. 3.6 (**a–c**) Supercritical and (**d–f**) subcritical Andronov–Hopf bifurcations

3.5 Andronov–Hopf Bifurcation

In a DS with dimension $N \geq 2$, there is a situation where a pair of complex-conjugate eigenvalues of an equilibrium of stable focus type intersect the imaginary axis. This implies that the bifurcation condition $\operatorname{Re} s_{1,2} = 0$ is satisfied. Moreover, $\operatorname{Im} s_{1,2} \neq 0$. This case corresponds to *Andronov–Hopf bifurcation* or the *limit cycle birth (death) bifurcation*. This bifurcation was first explored by A.A. Andronov for the case $N = 2$ and then generalized by E. Hopf to systems with arbitrary dimension N. There are two different kinds of Andronov–Hopf bifurcation, namely, the *supercritical bifurcation*, and the *subcritical*, or *hard bifurcation*. The supercritical bifurcation is soft, and the subcritical bifurcation is an attractor crisis. The Andronov–Hopf bifurcation is defined by only one bifurcation condition and has thus codimension one.

The *supercritical Andronov–Hopf bifurcation* is illustrated in Fig. 3.6a–c and can be described as follows. For $\alpha < \alpha^*$, there exists a stable focus F which, at the bifurcation point $\alpha = \alpha^*$, turns into a center and has a pair of pure imaginary eigenvalues $s_{1,2} = \pm i\omega_0$. For $\alpha > \alpha^*$, the focus F becomes unstable ($\operatorname{Re} s_{1,2} > 0$) and a stable limit cycle C_0 is born in the near vicinity.

The *subcritical Andronov–Hopf bifurcation* occurs when, for $\alpha = \alpha^*$, the unstable (in the general case for $N > 2$, saddle) limit cycle C_0 'sticks' onto the focus F, being stable for $\alpha < \alpha^*$. As a result, the cycle no longer exists and the focus becomes unstable (Fig. 3.6d–f).

The model system for Andronov–Hopf bifurcation is described by the Stuart–Landau equation:

$$\dot{a} = (\alpha + i\omega_0)a + L_1 a |a|^2, \qquad \omega_0 \neq 0, \quad L_1 \neq 0, \qquad (3.5)$$

where a is the instantaneous complex amplitude and L_1 is called the *first Lyapunov quantity* of equilibrium. If $L_1 < 0$, the bifurcation is supercritical and if $L_1 > 0$, it is subcritical.[1] For the real instantaneous amplitude and phase of oscillations, (3.5) implies

$$\dot{\rho} = \alpha\rho + L_1\rho^3, \qquad \dot{\Phi} = \omega_0, \qquad (3.6)$$

where $\rho = |a|$ and $\Phi = \arg(a)$. From the equation for stationary amplitudes, viz., $\alpha\rho + L_1\rho^3 = 0$, we obtain amplitude values for the equilibrium ($\rho_F = 0$) and for the limit cycle ($\rho_0 = \sqrt{-\alpha/L_1}$). The limit cycle exists if $\alpha/L_1 < 0$. The quantity ω_0 gives its period $T = 2\pi/\omega_0$. The eigenvalues for the solutions $\rho = \rho_F$ and $\rho = \rho_0$, viz., $s_{F,0} = \alpha + 3L_1\rho_{F,0}^2$, are found by examining the linearized equation for amplitude perturbation. If $L_1 < 0$, the cycle is found to exist and be stable for $\alpha > 0$, whereas the focus is stable for $\alpha < 0$ and unstable for $\alpha > 0$. When $L_1 > 0$, both the unstable cycle and the stable focus exist for $\alpha < 0$, while only the unstable focus exists for $\alpha > 0$.

3.6 Bifurcations of Limit Cycles

Consider local codimension-one bifurcations of a nondegenerate limit cycle which has only one multiplier equal to 1. We eliminate the unit multiplier from consideration and arrange the remaining multipliers in decreasing order of absolute value. In this case, limit-cycle bifurcations are related to one or two (complex conjugate) first multipliers, $\mu_{1,2}$. Since codimension-one bifurcation assumes only one bifurcation condition corresponding to the equality $|\mu_1| = 1$, three different outcomes are possible: $\mu_1(\alpha^*) = +1$, $\mu_1(\alpha^*) = -1$, $\mu_{1,2}(\alpha^*) = \exp(\pm i\varphi)$, where α^* is the bifurcation parameter. To analyze limit-cycle bifurcations, it is more convenient to use a Poincaré surface-of-section technique. Fixed points of the corresponding Poincaré map are characterized by the same multipliers and the transition to the Poincaré section makes the analysis more instructive.

3.6.1 Saddle-Node Bifurcation

The multiplier μ_1 of a stable cycle becomes $+1$ when the parameter α attains its bifurcation value $\alpha = \alpha^*$. Figure 3.7 illustrates this bifurcation for a three-

---

[1] Higher powers of a must also be taken into consideration when investigating the nature of bifurcation in the particular (degenerate) case $L_1 = 0$.

3.6 Bifurcations of Limit Cycles

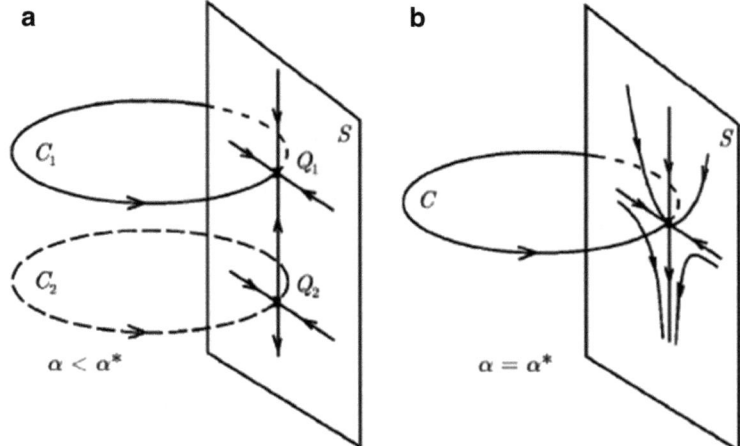

Fig. 3.7 Saddle-node bifurcation of limit cycles: (**a**) before the bifurcation and (**b**) after the bifurcation

dimensional phase space $N = 3$. For $\alpha < \alpha^*$ there are two cycles, a stable cycle C_1 and a saddle cycle C_2 (Fig. 3.7a). These are associated with stable and unstable fixed points Q_1 and Q_2, respectively, in the Poincaré section. The condition $\mu_1 = 1$ determines the bifurcation, which is similar to the saddle-node bifurcation of equilibrium considered above. At the bifurcation point $\alpha = \alpha^*$, cycles C_1 and C_2 merge to form a nonrobust closed trajectory C of saddle-node type (see Fig. 3.7b). This curve disappears for $\alpha > \alpha^*$. When the parameter α is varied in reverse order, a pair of cycles C_1 and C_2 is born from the phase trajectory concentration.

3.6.2 Period-Doubling Bifurcation

The multiplier $\mu_1(\alpha^*)$ becomes -1 at the critical point $\alpha = \alpha^*$ and $d\mu/d\alpha|_{\alpha^*} \neq 0$. The latter expression is the bifurcation condition for a *period-doubling bifurcation*. This bifurcation can be supercritical (internal) or subcritical (crisis). Supercritical bifurcation is described as follows. For $\alpha < \alpha^*$, suppose we have a stable limit cycle C_0 with period T_0. When $\alpha > \alpha^*$, the cycle C_0 becomes a saddle, and a stable limit cycle C with period T close to double the period T_0 ($T \approx 2T_0$) is born in the neighborhood of the former cycle C_0. Phase trajectories C_0 and C and their Poincaré sections near the bifurcation point are plotted in Fig. 3.8a. Figure 3.8b shows how the waveform of one of the dynamical variables changes at the moment of period doubling.

When a subcritical period-doubling bifurcation occurs, the stable cycle C_0 and the saddle cycle C with doubled period, which exist for $\alpha < \alpha^*$, merge at the

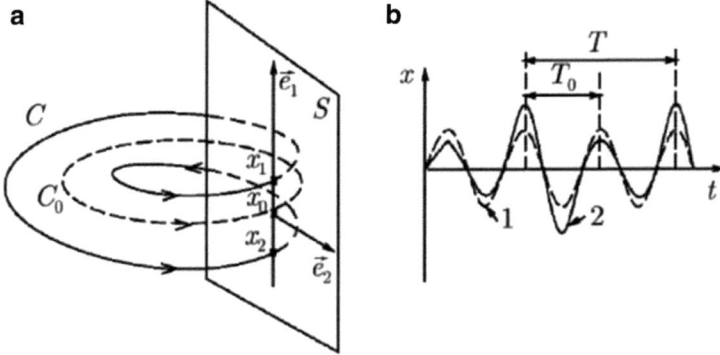

Fig. 3.8 Supercritical period-doubling bifurcation. (**a**) Cycles C_0 and C and their Poincaré sections. (**b**) Waveforms before (*curve 1*) and after (*curve 2*) the bifurcation

bifurcation point, whereupon only the cycle C_0 which has become a saddle cycle remains in the phase space.

3.6.3 Two-Dimensional Torus Birth (Death) Bifurcation (Neimark–Saker Bifurcation)

This bifurcation is realized when a pair of complex-conjugate multipliers of a limit cycle go out to the unit circle. At the bifurcation point $\alpha = \alpha^*$, we have $\mu_{1,2}(\alpha^*) = \exp(\pm i\varphi)$, where $\varphi \in [0, 2\pi]$, and $\varphi(\alpha^*) \neq 0, \pi, 2\pi/3$ (so-called *strong resonances* are excluded). This bifurcation can also be supercritical (internal) and subcritical (crisis). Different situations are realized depending on which kind of bifurcation occurs. In the case of supercritical bifurcation, a stable two-dimensional (2D) torus T^2 is born from a stable limit cycle C_0, the latter becoming unstable thereafter (in general, it is of saddle-focus type). Subcritical bifurcation takes place when an unstable (saddle) torus T^2 'sticks' onto the stable cycle C_0 at the moment when it loses stability. Torus birth from a limit cycle is pictured in Fig. 3.9a. A small perturbation **y** of the cycle C_0 near the bifurcation point $\alpha = \alpha^*$ is rotated through an angle φ in one revolution of the trajectory C_0. At the same time, its magnitude remains unchanged since $|\mu_{1,2}(\alpha^*)| = 1$. Thus, a representative point in the Poincaré section moves along a closed curve L called the *invariant curve*. The quantity $\theta(\alpha) = \varphi/2\pi$ is called the *winding number* on the torus T^2 (or on the corresponding invariant curve). If the winding number $\theta(\alpha^*)$ takes an irrational value, no trajectory C on the torus closes and the resulting torus is *ergodic* (Fig. 3.9b). If $\theta(\alpha^*) = p/q$, where p and q are any positive integers, a resonance phenomenon of order p/q is said to take place on the torus. An example of resonance on a torus is shown in Fig. 3.9c.

3.6 Bifurcations of Limit Cycles

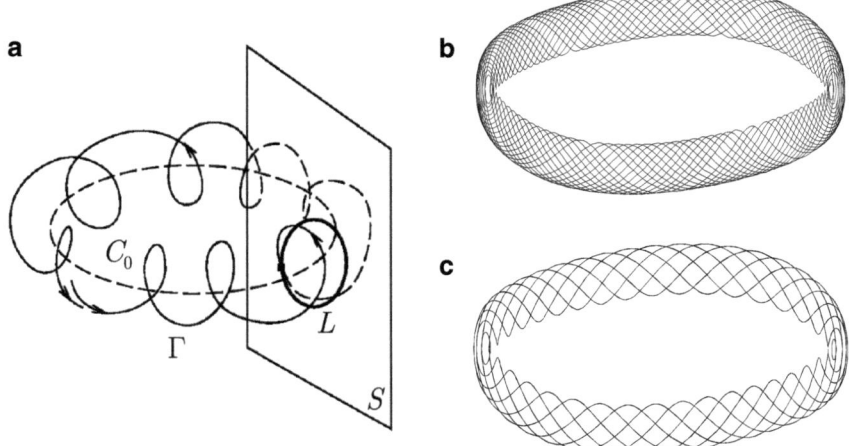

Fig. 3.9 Torus birth bifurcation from the limit cycle C_0. (**a**) Trajectory Γ on the torus in the neighborhood of the unstable cycle C_0. (**b**) Ergodic torus. (**c**) Resonance on a torus

3.6.4 Symmetry-Breaking Bifurcation

The bifurcations of limit cycles described by the conditions $\mu_1(\alpha^*) = \pm 1$ and $\mu_{1,2}(\alpha^*) = \exp(\pm i\varphi)$ may lead to a situation where a limit cycle loses its symmetry. Such bifurcations are typical, for example, for systems consisting of two or more identical partial subsystems. The symmetry property of a limit set is related to the existence of some invariant manifold U in the system phase space. For instance, consider the two identical coupled subsystems

$$\dot{\mathbf{x}} = \mathbf{F}(\mathbf{x}, \alpha) + \gamma \mathbf{g}(\mathbf{y}, \mathbf{x}) ,$$
$$\dot{\mathbf{y}} = \mathbf{F}(\mathbf{y}, \alpha) + \gamma \mathbf{g}(\mathbf{x}, \mathbf{y}) , \quad (3.7)$$

where $\mathbf{x}, \mathbf{y} \in \mathbb{R}^N$ are the state vectors of the subsystems and $\boldsymbol{\alpha}$ is the parameter vector. The function $\mathbf{g}$ is responsible for the coupling between the subsystems and $\mathbf{g}(\mathbf{x}, \mathbf{x}) = 0$. In this case the subspace $\mathbf{x} = \mathbf{y}$ is referred to as an invariant symmetric manifold. Suppose a stable limit cycle is located in U, i.e., it is symmetric. If one of the limit cycle multipliers which corresponds to the eigenvector not lying in U attains the bifurcation value, then a symmetry-breaking bifurcation occurs. It results in the appearance of a nonsymmetric attractor (or attractors) which does not lie in U. The bifurcation is said to result in the loss of attractor symmetry. The symmetry-breaking bifurcations determined by the conditions $\mu_1(\alpha^*) = -1$ and $\mu_{1,2}(\alpha^*) = \exp(\pm i\varphi)$ are very similar to the analogous bifurcations in systems without symmetry. Bifurcation defined by the condition $\mu_1(\alpha^*) = +1$ represents a special case. As a result of this bifurcation, the symmetric cycle $C_0 \in U$ still exists

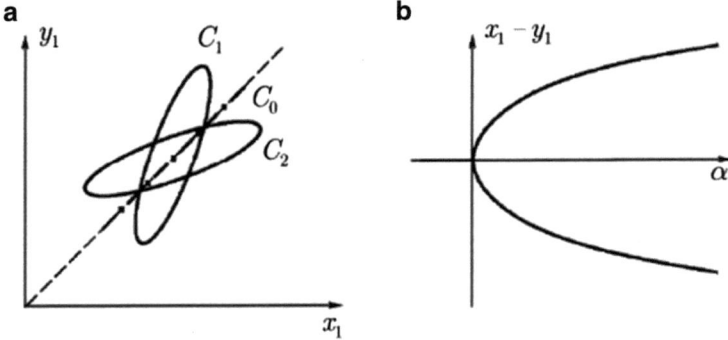

Fig. 3.10 Pitchfork bifurcation in a system with symmetry: (**a**) projection of limit cycles after bifurcation and (**b**) qualitative illustration of the phase-parameter diagram

but becomes a saddle cycle. It gives birth to two stable cycles with the same period. They do not lie in U but are mutually symmetric. Such a bifurcation is referred to as a *pitchfork bifurcation* of a limit cycle. Figure 3.10a shows phase portraits of cycles after a pitchfork bifurcation. A phase-parameter diagram of this bifurcation is sketched in Fig. 3.10b, where the ordinate is the difference between appropriate coordinates $(x_1 - y_1)$ in some section of the cycles.

We have considered the local bifurcations of equilibria and limit cycles. More complicated sets (tori, chaotic attractors) also undergo different bifurcations. However, their study is more often based on experimental results, and a full theory of bifurcations of quasiperiodic and chaotic attractors remains to be developed.

3.7 Nonlocal Bifurcations: Homoclinic Trajectories and Structures

Nonlocal bifurcations are associated with the form of stable and unstable manifolds of saddle limit sets in phase space. They do not cause the saddle limit sets themselves to change topologically, but may significantly affect the system dynamics. Here we consider the basic nonlocal bifurcations.

3.7.1 Separatrix Loop of a Saddle Equilibrium Point

This bifurcation can be realized in its simplest form even in a phase plane. Suppose we have a saddle equilibrium Q whose stable W_Q^s and unstable W_Q^u manifolds come close together with increasing parameter α and touch one another at $\alpha = \alpha^*$. At the tangency point, a bifurcation takes place, leading to the formation of a singular

3.7 Nonlocal Bifurcations: Homoclinic Trajectories and Structures

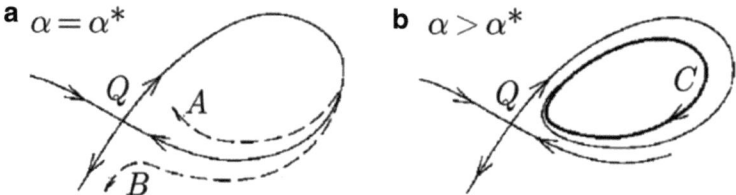

Fig. 3.11 Bifurcation of separatrix loop formation (**a**) at the moment of bifurcation and (**b**) after bifurcation

double-asymptotic phase trajectory Γ_0, called the *separatrix loop of the saddle point* (Fig. 3.11a). The separatrix loop in a dissipative system is a nonrobust structure and collapses at $\alpha \neq \alpha^*$. What will happen thereafter depends on which way the separatrices proceed after splitting, i.e., loop breaking, and also on the *saddle quantity* σ_Q of equilibrium at the bifurcation point. The saddle quantity is specified as $\sigma_Q(\alpha) = s_1(\alpha) + s_2(\alpha)$, where $s_{1,2}$ are the eigenvalues of the linearization matrix at the point Q. If $\sigma_Q(\alpha^*) < 0$, then when the loop is broken in the direction A as shown in Fig. 3.11a, the only stable cycle C arises from the loop (Fig. 3.11b). No cycle is born when the loop is destroyed in the direction B. If $\sigma_Q(\alpha^*) > 0$, then as Γ_0 is destroyed, an unstable limit cycle can arise from it.

Considered in reverse order, the bifurcation of separatrix loop formation may be treated as a crisis of the limit cycle C when it touches the saddle Q. At the moment of tangency, the loop Γ_0 is created. As one approaches the bifurcation point, the cycle period tends to infinity and the cycle multipliers approach zero.

A more complicated but similar type of nonlocal bifurcation is possible in a phase space with dimension $N \geq 3$. We confine ourselves to $N = 3$. Let Q be a saddle-focus with one-dimensional unstable and two-dimensional stable manifolds and with the so-called *first saddle quantity* $\sigma_1(\alpha) = \operatorname{Re} s_{1,2}(\alpha) + s_3(\alpha)$, where $s_{1,2} = \operatorname{Re} s_{1,2} \pm \mathrm{i} \operatorname{Im} s_{1,2}$, and s_3 are the eigenvalues of the linearization matrix at the point Q. Suppose a *saddle-focus separatrix loop* Γ_0 forms at $\alpha = \alpha^*$ and $\sigma_1(\alpha^*) \neq 0$ (see Fig. 3.12). Under these assumptions, we may apply the following theorem due to L.P. Shilnikov:

- $\sigma_1(\alpha^*) < 0$ (the loop is *not dangerous*). If loop destruction corresponds to case A as shown in Fig. 3.12, a stable cycle C arises from it. Nothing happens when the loop is broken in the direction B.
- $\sigma_1(\alpha^*) > 0$ (the loop is *dangerous*). A complicated structure of phase trajectories emerges in the vicinity of the loop Γ_0 when it exists and is then destroyed in any direction. This structure consists of a countable set of periodic attractors, repellers, and saddles, as well as a subset of chaotic trajectories called a *nontrivial hyperbolic subset*. Such a structure is related to the presence of a set of Smale's horseshoe-type maps[2] in the vicinity of the loop in the Poincaré section.

---

[2] Smale's horseshoe is described in detail in Chap. 10.

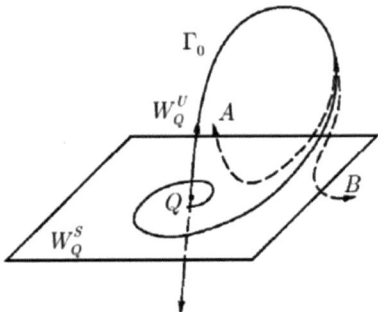

Fig. 3.12 Saddle-focus separatrix loop and possible ways to destroy it

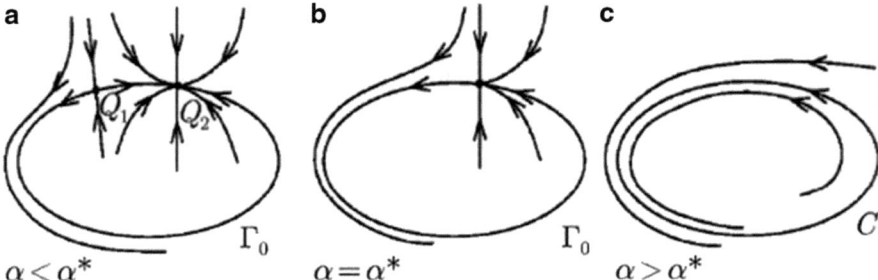

Fig. 3.13 Bifurcation of saddle-node separatrix loop formation: (**a**) before the bifurcation, (**b**) at the moment of bifurcation, and (**c**) after the bifurcation

3.7.2 Saddle-Node Separatrix Loop

This bifurcation is also possible for $N = 2$. Assume that, for $\alpha < \alpha^*$, there are two equilibria, namely, a saddle Q_1 and a stable node Q_2. In addition, a separatrix loop is formed by the unstable separatrices of the saddle approaching the stable node, as shown in Fig. 3.13a. At the bifurcation point $\alpha = \alpha^*$, a saddle-node bifurcation of equilibria occurs, and a nonrobust equilibrium of saddle-node type arises. Here, the saddle-node has a double-asymptotic homoclinic trajectory Γ_0, i.e., the separatrix loop (Fig. 3.13b). When $\alpha > \alpha^*$, the saddle-node disappears and a limit cycle C is generated from the loop (Fig. 3.13c).

When considered in reverse order, this is a bifurcation of cycle C extinction, leading to the creation of a saddle-node on it. The cycle period grows infinitely as $\alpha \to \alpha^*$, and the cycle multipliers tend to zero.

The bifurcation described above preserves the boundaries of an absorbing area (basin of attraction) and is thus an internal bifurcation.

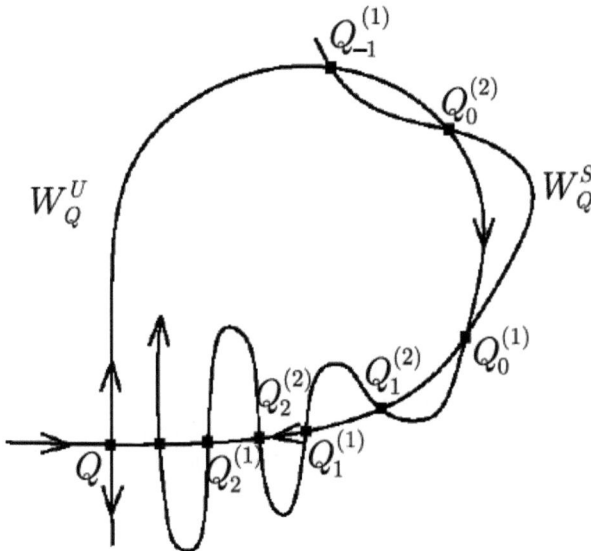

Fig. 3.14 Homoclinic intersection of saddle cycle trajectories (graphical representation in the Poincaré section)

3.7.3 Homoclinic Trajectory Appearance of a Saddle Limit Cycle

This bifurcation is realized when $N \geq 3$. In this case there may exist saddle limit cycles with two-dimensional stable and two-dimensional unstable manifolds, W^s and W^u, respectively. Such a cycle corresponds to a saddle fixed point in a secant plane. When $N = 3$, this fixed point has one-dimensional stable and unstable manifolds. With increasing values of the parameter α, the manifolds of the cycle approach each other and touch at $\alpha = \alpha^*$. Being of codimension one, this bifurcation results in the emergence of a nonrobust double-asymptotic curve Γ_0 called the *Poincaré homoclinic curve*. When $\alpha > \alpha^*$, the manifolds W^s and W^u intersect and two robust homoclinic curves Γ_1^0 and Γ_2^0 are created. In the secant plane each homoclinic curve corresponds to an infinite double-asymptotic sequence of intersection points of separatrices Q_n, $n = 0, \pm 1, \pm 2, \ldots$ (Fig. 3.14). The points Q_n become denser as one approaches the saddle, but they tend to it only in the limit as $n \to \pm \infty$.

It has been shown that a complicated set of trajectories is generated in the vicinity of a homoclinic curve of a saddle cycle. This set is called a *homoclinic structure*. It is similar to the structure arising in the neighborhood of a dangerous saddle-focus separatrix loop and is also connected with the formation of local horseshoe-type maps in the vicinity of the loop. Stable, unstable, and saddle periodic orbits are dense everywhere in the vicinity of the homoclinic trajectory. Besides, the homoclinic structure includes a subset of chaotic trajectories which, under appropriate conditions, may become attracting.

A similar structure is characteristic of the neighborhood of a *heteroclinic trajectory* which emerges when the unstable manifold of one saddle cycle touches and then intersects the stable manifold of another saddle cycle.

3.8 Summary

When the parameters of a dynamical system are changed, this may lead to loss of stability of some operating modes and bifurcation transitions to new modes. Such 'phase transitions' can be accomplished smoothly and softly or can occur abruptly, in the form of catastrophes. Today, a wide range of problems related to bifurcation transitions can be studied in various dynamical systems through rigorous mathematical analysis of stability and bifurcations. One such problem is the transition of a system from regular behavior to chaotic dynamics, something discussed further in Chaps. 6 and 7. More information on bifurcations of dynamical systems can be found in [1–15].

References

1. Anishchenko, V.S.: Dynamical Chaos – Models and Experiments. World Scientific, Singapore (1995)
2. Anishchenko, V.S., Astakhov, V.V., Neiman, A.B., Vadivasova, T.E., Schimansky-Geier, L.: Nonlinear Dynamics of Chaotic and Stochastic Systems. Springer, Berlin (2002)
3. Drazin, P.G.: Nonlinear Systems. Cambridge University Press, Cambridge (1992)
4. Glendinning, P.: Stability, Instability, and Chaos: An Introduction to the Theory of Nonlinear Differential Equations. Cambridge University Press, Cambridge (1994)
5. Guckenheimer, J., Holmes, P.: Nonlinear Oscillations, Dynamical Systems, and Bifurcations of Vector Fields. Springer, New York (1983)
6. Hilborn, R.C.: Chaos and Nonlinear Dynamics. An Introduction for Scientists and Engineers. Oxford University Press, Oxford (2002/2004)
7. Marek, M., Schreiber, I.: Chaotic Behaviour of Deterministic Dissipative Systems. Cambridge University Press, Cambridge (1991/1995)
8. Marsden, L.E., McCraken, V.: The Hopf Bifurcation and Its Applications. Springer, New York (1976)
9. Moon, F.C.M.: Chaotic and Fractal Dynamics: An Introduction for Applied Scientists and Engineers. Wiley, New York (1992)
10. Moon, F.C.M.: Chaotic Vibration: An Introduction for Applied Scientists and Engineers. Wiley, New York (2004)
11. Nicolis, G.: Introduction to Nonlinear Science. Cambridge University Press, Cambridge (1995)
12. Ogorzalek, M.J.: Chaos and Complexity in Nonlinear Electronic Circuits. World Scientific, Singapore (1997)
13. Schuster, H.G.: Deterministic Chaos. Physik-Verlag, Weinheim (1988)
14. Seydel, R.: Practical Bifurcation and Stability Analysis: From Equilibrium to Chaos. Springer, New York (1994/2009)
15. Thompson, J.M.T., Stewart, H.B.: Nonlinear Dynamics and Chaos. Wiley, New York (1986)

Chapter 4
Dynamical Systems with One Degree of Freedom

4.1 Introduction

Consider a class of autonomous continuous-time dynamical systems whose state at any time can be unambiguously given by a variable x and its derivative $y = dx/dt$. The phase space of such a system is the phase plane (x, y). Thus, the phase space dimension is $N = 2$ and the number of degrees of freedom is $N/2 = 1$. In most cases the mathematical model of a DS with one degree of freedom can be reduced to the generalized equation of an oscillator

$$\ddot{x} + p(x, \dot{x})\dot{x} + q(x) = 0 . \tag{4.1}$$

Here $p(x, \dot{x})$ and $q(x)$ are given functions or can be represented by a system of two first-order ordinary differential equations

$$\dot{x} = f(x, y) , \qquad \dot{y} = g(x, y) , \tag{4.2}$$

where functions $f(x, y)$ and $g(x, y)$ determine the components of the phase velocity at each point of the phase plane. The DS with one degree of freedom can have only one oscillatory mode, i.e., there is only one independent frequency of oscillations.

As in the general case, a DS in the phase plane can be linear or nonlinear, conservative or nonconservative. Examples of conservative linear and nonlinear oscillators with one degree of freedom were presented in Chap. 1, along with the dynamics of the linear oscillator with constant losses. In (4.1), the coefficient $p(x, \dot{x})$ is responsible for energy dissipation. If $p = \text{const.} > 0$, we have an oscillator with constant dissipation. Sooner or later, oscillations arising in this oscillator are damped out and the system goes into a stable equilibrium state. However, $p(x, \dot{x})$ can be defined in such a way that the energy spent on dissipation in some region of the phase plane is compensated by pumping energy into another part of the plane. Thus, self-sustained oscillations can arise in the system. The classic example of a

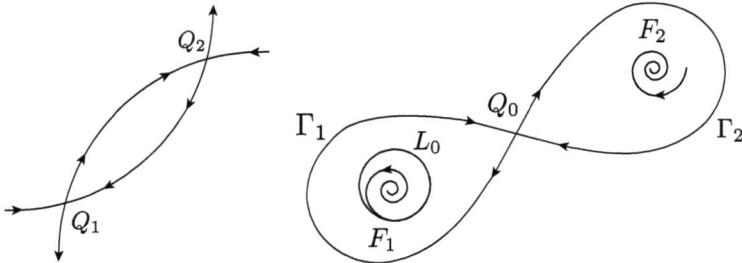

Fig. 4.1 Examples of separatrix loops in the phase plane

self-sustained oscillatory system, the van der Pol oscillator, was also examined in Chap. 1. In the present chapter, given the fundamental importance of this model, we explore the van der Pol oscillator in more detail using an approximate analytical averaging method.

4.2 Limit Sets and Attractors in the Phase Plane: The Andronov–Poincaré Limit Cycle

Only a very limited collection of limit sets can be realized in the phase plane. Poincaré and Bendixon established that a bounded limit set in the plane can belong to one of the following topological types: (i) an equilibrium, (ii) a limit cycle, (iii) a separatrix loop or a separatrix contour consisting of equilibria and trajectories connecting them and tending to equilibria as $t \to \pm\infty$ (Fig. 4.1).

Since a perturbation vector in the plane can be resolved in only two orthogonal directions, the stability of trajectories is characterized by just two Lyapunov exponents, two eigenvalues of the linearization matrix, or two multipliers. The stability of equilibrium points in the plane has already been discussed in Chap. 2, so we omit the details here. We simply note that an equilibrium point can be an attractor (a stable node and a stable focus), a repeller (an unstable node and an unstable focus), or a saddle. A center-type equilibrium point is not typical for nonconservative systems and may exist only for a special choice of parameters. The stability of a limit cycle in the phase plane is determined by just one multiplier, because the second one is always equal to unity (there is no contraction or stretching along the cycle). Therefore, a limit cycle in systems with one degree of freedom can be either an attractor or a repeller. If both multipliers are equal to unity, the system is at the point of tangent bifurcation of stable and unstable cycles. Saddle cycles cannot be realized in the phase plane.

For many years, before computer simulation methods became commonplace, the theory of nonlinear oscillations focused on the phase plane where motions are relatively simple in nature and can to some extent be explored analytically. Despite the limited types of behavior, the study of the phase plane enables us

4.2 Limit Sets and Attractors in the Phase Plane: The Andronov–Poincaré...

to formulate many of the basic concepts of nonlinear dynamics, such as a limit set, an attractor, separatrices, local and nonlocal bifurcations, structural stability of dynamical systems, and so on. Self-sustained oscillatory systems with one degree of freedom have played an especially important role. Self-sustained oscillations can be realized if the phase space dimension N of the system is at least two. Self-sustained oscillatory systems with half a degree of freedom are degenerate and are deduced from a certain idealization of a DS when one-dimensional sets of 'fast' and 'slow' motions can be distinguished in a two-dimensional phase space.

The terms 'self-sustained oscillations' and 'self-sustained oscillatory systems' were first introduced by A.A. Andronov almost a 100 years ago. Andronov emphasized that self-sustained oscillatory systems should be set in their own special class of numerous and practically important systems. The common feature of these systems, according to Andronov, is that they can engage in *self-sustained oscillations*, i.e., oscillations whose amplitude can remain constant for a long time, but does not generally depend on the initial conditions, and is determined not by the initial conditions but rather by the properties of the system itself. Andronov notes that the independence of the oscillation parameters from the initial conditions is a characteristic feature of self-sustained oscillations. However, this property is not absolute and may not extend to all initial states but only to some finite region of phase space. That is, there are several stationary processes with different oscillatory characteristics, depending on which region the initial state is chosen in. Nowadays, this phenomenon is called multistability. According to Andronov, another typical feature of self-sustained oscillations is the pumping of energy from a constant source, which is performed at certain times and controlled by the system itself, i.e., a periodic process is created by a nonperiodic source of energy.[1]

A.A. Andronov developed the theory of self-sustained oscillatory systems with one degree of freedom. Self-sustained oscillations of systems with one degree of freedom are, according to Andronov, structurally stable periodic oscillations whose image in the phase plane is an asymptotically stable isolated closed trajectory, the Andronov–Poincaré attracting limit cycle Γ_0. By an isolated closed trajectory we understand a trajectory whose neighborhood contains no other closed trajectories, apart from itself. Asymptotic stability means that a small perturbation of a trajectory on a cycle decreases exponentially and asymptotically approaches zero at $t \to \infty$. In other words, an asymptotically stable limit cycle attracts trajectories from some neighborhood, i.e., it is an attractor. Whether initial conditions are chosen inside the cycle Γ_0 or outside it, phase trajectories tend to Γ_0 as $t \to \infty$ and stay on it. It is absolutely clear that a limit cycle (a regime of self-sustained periodic oscillations) can be realized only in nonlinear dissipative systems. The mean divergence of the vector field for trajectories inside the cycle must be positive, and outside it negative. This is possible only when both dissipation and pumping of energy are present in the system. Furthermore, the relationship between dissipation and pumping depends on

---

[1] Andronov et al. [2].

the instantaneous state. In the case of periodic self-sustained oscillations, the energy spent is compensated strictly over the period of oscillation.

4.3 Structural Stability of Systems in the Phase Plane: Andronov–Pontryagin Systems

The notions of *robustness* and *structural stability* of dynamical systems play an important role in nonlinear dynamics. In the previous chapters these terms have already been used, although definitions of these properties have not been given yet.[2] These notions were first formulated by Andronov and Pontryagin for systems with one degree of freedom and then generalized to the case of a dynamical system with an arbitrary finite phase space dimension. The definitions given below are valid for any N.

Before defining the properties of robustness and structural stability of a DS we need to introduce the notion of topological equivalence. Two dynamical systems are said to be *topologically equivalent* if there is a homeomorphism of the phase space which maps trajectories of one system into trajectories of the other and preserves the topological structure of the phase space partition, i.e., all the limit sets, their topology, and stability properties are preserved.

Consider a dynamical system in the form

$$\dot{\mathbf{x}} = \mathbf{F}(\mathbf{x}), \qquad \mathbf{x} \in \mathbb{R}^N,$$

where the trajectories $\mathbf{x}(t)$ are smooth functions defined in a closed bounded domain $V \subset \mathbb{R}^N$. On a set of possible dynamical systems of the given dimension N, we can introduce the norm[3] $\|\mathbf{F}(\mathbf{x})\|$ and the associated metric that characterizes the difference between two systems.

4.3.1 Definition of Robustness of a Dynamical System

The system $\mathbf{F}(\mathbf{x})$ is said to be *robust* in the region V if, for any arbitrarily small $\varepsilon > 0$, there is $\delta(\varepsilon) > 0$ such that, whenever

$$\|\mathbf{F}(\mathbf{x}) - \tilde{\mathbf{F}}(\mathbf{x})\| < \delta,$$

---

[2] These terms were somewhat different in the original definition, although now they are often used synonymously.

[3] We will not consider the methods for introducing the norm, in order to avoid excessive mathematics.

for the initial and perturbed systems $\mathbf{F}(\mathbf{x})$ and $\tilde{\mathbf{F}}(\mathbf{x})$, respectively, i.e., the perturbation is small in norm, the two systems are topologically equivalent: $\tilde{\mathbf{F}}(\mathbf{x}) = \mathbf{H}\mathbf{F}(\mathbf{x})$, and the homeomorphism $\mathbf{H}$ connecting them is close to the identity transformation $\mathbf{I}$ so that $\|\mathbf{H} - \mathbf{I}\| < \varepsilon$.

4.3.2 Definition of Structural Stability of a Dynamical System

The system $\mathbf{F}(\mathbf{x})$ is *structurally stable* in V if there is $\delta(\varepsilon) > 0$ such that, whenever

$$\|\mathbf{F}(\mathbf{x}) - \tilde{\mathbf{F}}(\mathbf{x})\| < \delta ,$$

it follows that the systems $\mathbf{F}(\mathbf{x})$ and $\tilde{\mathbf{F}}(\mathbf{x})$ are topologically equivalent.

4.3.3 Andronov–Pontryagin Theorem

For a DS with one degree of freedom, we have the *Andronov–Pontryagin theorem*: the dynamical system given by (4.2) with smooth right-hand sides is robust if and only if the following conditions are fulfilled:

- All the equilibria of the DS (4.2) are simple, i.e., none of the roots of the characteristic equation of the linearized system lie on the imaginary axis.
- All the periodic trajectories of the DS are simple, i.e., only one of the multipliers (the eigenvalues of the monodromy matrix) lies on the unit circle.
- There are no trajectories doubly asymptotic to a saddle or going from a saddle to a saddle (i.e., there are no separatrix loops).

As already mentioned in Chap. 1, robust limit sets in the phase plane can include equilibrium points with no purely imaginary eigenvalues and periodic orbits having only one unit magnitude multiplier (equal to $+1$). Structurally unstable sets in the phase plane include loops and contours formed by doubly asymptotic trajectories of saddle equilibrium points. The definitions of robustness and structural stability and the Andronov–Pontryagin theorem can also be formulated for cascades (discrete-time maps).

Systems with one degree of freedom which satisfy the theorem conditions are called *Andronov–Pontryagin systems*. They form a subset of a certain class of structurally stable systems in $\mathbb{R}^N$ called Morse–Smale systems. These will be described in the following chapters.

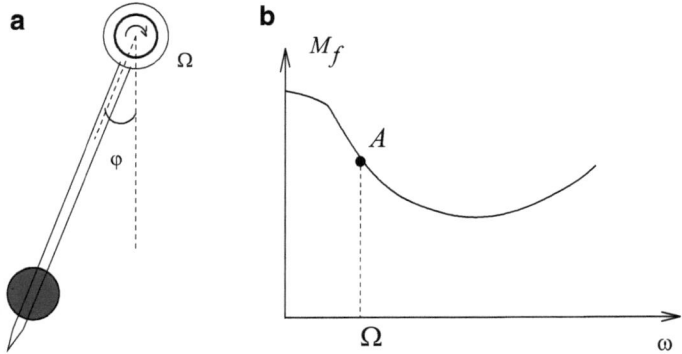

Fig. 4.2 Example of a mechanical self-sustained oscillatory system: (**a**) the Froude pendulum, (**b**) the friction torque as a function of the shaft rotation velocity

4.4 Oscillators with One Degree of Freedom

We now turn our attention to self-sustained oscillatory systems with one degree of freedom and consider several classic examples of such systems.

4.4.1 Froude Pendulum

This is one of the simplest mechanical self-sustained oscillatory systems. It consists of a compound pendulum that is rigidly fastened to a sleeve mounted on a rotating shaft (Fig. 4.2a). The shaft rotates uniformly with angular velocity Ω. The friction force between the sleeve and shaft creates a torque M_f. The angular velocity of the shaft is assumed always to be greater in absolute value than the velocity of the pendulum rotation. Consider the work done by the torque when the pendulum oscillates periodically. During one half-period, the work done by the torque M_f is equal to the energy taken away from the pendulum when the shaft and pendulum rotate in opposite directions. During the other half-period, when the pendulum and shaft move in the same direction, the work done by the torque M_f adds some energy to the pendulum. The friction torque depends on the relative angular velocity of the shaft and sleeve rotation. Assume that the friction force grows with the rate of sliding. The friction torque is greater when the shaft and pendulum rotate in opposite directions than when they rotate in the same direction. In the first case the rate of sliding is greater than in the second. Consequently, the action of friction forces takes energy away from the pendulum during the period and the pendulum oscillations are strongly damped. The energy of pendulum oscillations is spent in the suspension, and the friction of the rotating shaft only increases the damping of oscillations.

This picture can change fundamentally if the friction force decreases as the rate of sliding increases. With a little lubrication, such conditions can be realized

4.4 Oscillators with One Degree of Freedom

in a certain range of the sliding rate variation. A typical plot of the dependence of the friction torque on the shaft rotation velocity is shown in Fig. 4.2b for the motionless pendulum. Let the velocity of shaft rotation correspond to the abscissa of point A. Then the energy of pendulum oscillations will increase over the period. When braking, the relative rotation velocity is greater and the friction force is less. This means that the dissipated energy is less in the declaration phase than in the acceleration phase because the relative rotation velocity is less here, the friction force is greater, and more energy is needed to 'push' the pendulum. The oscillating pendulum will get a certain amount of energy from the shaft over the period, and if it exceeds the energy used by friction against the air, the amplitude of the pendulum oscillations will eventually grow in time.

As the amplitude of pendulum oscillations increases, the amplitude of friction torque against the shaft grows more slowly than the amplitude of friction torque against the air. They become equal at a certain amplitude of pendulum oscillations, at which point the pendulum will make stationary undamped oscillations, i.e., self-sustained oscillations. The rotating shaft of the motor imparts the energy to the pendulum that is needed to cover energy losses in the form of heat in the regime of stationary self-sustained oscillations. The energy is transferred from the motor to the pendulum by the sliding friction force. The frequency of self-sustained oscillations is defined by the basic frequency of the pendulum oscillations.

The equation of motion of the Froude pendulum differs from the classical equation of a pendulum only in that this equation must account for the moment of the friction force of the rotating shaft on the bearing where the pendulum is suspended. Since the friction force depends on the relative velocity of rubbing surfaces, i.e., on the relative angular velocity $\Omega - \dot{\varphi}$ of the shaft and pendulum, the moment of the friction force can be expressed as $M_f(\Omega - \dot{\varphi})$. If we use small angles φ, replace $\sin\varphi$ by φ, and take into account the air resistance by assuming that it is proportional to the angular velocity, then the motion equation of the pendulum can be written in the form

$$I\ddot{\varphi} + h\dot{\varphi} + mga\varphi = M_f(\Omega - \dot{\varphi}), \tag{4.3}$$

where I is the moment of inertia, m is the mass of the pendulum, h is the friction coefficient (excluding friction between the sleeve and the shaft), a is the distance from the center of mass of the pendulum to the axis, Ω is the angular velocity of shaft rotation, and φ is the angle of deviation of the pendulum from the vertical.

We expand the function $M_f(\Omega - \dot{\varphi})$ in a series about the value Ω, viz.,

$$M_f(\Omega-\dot{\varphi}) = M_f(\Omega) + M_f^{(1)}(\Omega)(-\dot{\varphi}) + \frac{1}{2}M_f^{(2)}(\Omega)(-\dot{\varphi})^2 + \frac{1}{6}M_f^{(3)}(\Omega)(-\dot{\varphi})^3 + \ldots, \tag{4.4}$$

where $M_f^{(k)}$ is the kth derivative, and substitute (4.4) into (4.3), omitting the third and higher order terms. This yields

$$\ddot{z} + \frac{h}{I}\dot{z} + \frac{mga}{I}z = -\frac{M_f^{(1)}(\Omega)}{I}\dot{z} + \frac{M_f^{(2)}(\Omega)}{2I}\dot{z}^2 - \frac{M_f^{(3)}(\Omega)}{6I}\dot{z}^3 , \quad (4.5)$$

with

$$z = \varphi - \frac{1}{mga}M_f(\Omega) .$$

Choosing the angular velocity of shaft rotation equal to the abscissa of the point of inflection on the 'falling' section (with negative slope) of the function $M_f(\Omega)$, we then have $M_f^{(1)}(\Omega) < 0$ and $M_f^{(2)}(\Omega) = 0$. In addition, assuming that $M_f^{(3)}(\Omega) > 0$, our Eq. (4.5) takes the form

$$\ddot{z} - (\mu - \beta\dot{z}^2)\dot{z} + \omega_0^2 z = 0 , \quad (4.6)$$

where

$$\mu = -\frac{M_f^{(1)}(\Omega)}{I} - \frac{h}{I} , \quad \beta = \frac{M_f^3(\Omega)}{6I} , \quad \omega_0^2 = \frac{mga}{I} .$$

Changing to the dimensionless time $\tau = \omega_0 t$ and adopting the variables and parameters $\varepsilon = \mu/\omega_0$, $y = \sqrt{\beta\omega_0}z$, we obtain

$$\ddot{y} - (\varepsilon - \dot{y}^2)\dot{y} + y = 0 . \quad (4.7)$$

This is the well-known Rayleigh equation. It reduces to the van der Pol equation under the substitution $x = \sqrt{3}\dot{y}$, whence

$$\ddot{x} - (\varepsilon - x^2)\dot{x} + x = 0 . \quad (4.8)$$

4.4.2 Fastened Weight on a Moving Belt

As another example of a mechanical self-sustained oscillatory system, consider the device shown in Fig. 4.3. The weight of mass m lies on a belt moving uniformly with velocity v_0. This weight is fastened on both sides by springs with elasticity coefficients k_1 and k_2. Let x be the displacement of the weight from its equilibrium position and $\dot{x}$ its velocity. The friction force of the belt on the weight is $F(v_0 - \dot{x})$, a function of the relative velocity $v_0 - \dot{x}$ of the belt and the weight. The 'resulting' coefficient of elasticity of the two springs is denoted by k. All other friction forces acting in the system, such as air resistance or internal friction in the springs, are considered to be proportional to the velocity. The equation of motion for the weight reads

$$m\ddot{x} + b\dot{x} + kx = F(v_0 - \dot{x}) . \quad (4.9)$$

4.4 Oscillators with One Degree of Freedom

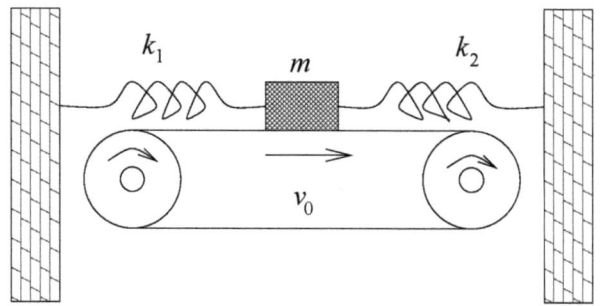

Fig. 4.3 The simplest example of a mechanical self-sustained oscillatory system: a spring-fastened weight on a moving belt

We assume that the constant belt velocity v_0 is much greater than the velocity of the weight's motion. Expanding the function F in a series about the value v_0, we restrict ourselves to the first few terms of the series, viz.,

$$F(v_0 - \dot{x}) = F(v_0) + F^{(1)}(v_0)(-\dot{x}) + \frac{1}{2}F^{(2)}(v_0)(-\dot{x})^2 + \frac{1}{6}F^{(3)}(v_0)(-\dot{x})^3 + \cdots, \quad (4.10)$$

where $F^{(k)}(v_0)$ is the kth derivative at v_0. The equation of motion now becomes

$$m\ddot{x} + b\dot{x} + kx = F(v_0) - F^{(1)}(v_0)\dot{x} + \frac{1}{2}F^{(2)}(v_0)\dot{x}^2 - \frac{1}{6}F^{(3)}(v_0)\dot{x}^3. \quad (4.11)$$

The values of the coefficients $F^{(1)}(v_0)$, $F^{(2)}(v_0)$, and $F^{(3)}(v_0)$ depend on the characteristics of the friction. Let the function $F(v_0 - \dot{x})$ have a negative slope, i.e., $F^{(1)}(v_0) < 0$, the inflection point on the negative slope, and $F^{(3)}(v_0) > 0$. We choose the constant velocity of the belt motion such that its value corresponds to the point of inflection on the negative slope of the function $F(v_0 - \dot{x})$. Then $F^{(2)}(v_0) = 0$ and (4.11) can be rewritten in the form

$$m\ddot{x} + b\dot{x} + kx = F(v_0) + \alpha\dot{x} - \beta\dot{x}^3, \quad (4.12)$$

where

$$\alpha = -F^{(1)}(v_0), \qquad \beta = \frac{1}{6}F^{(3)}(v_0).$$

Using the dimensionless time $\tau = \sqrt{k/m t}$ and changing the variables and parameters in (4.12), we have

$$y = \sqrt{\frac{\beta}{m}}\sqrt{\frac{k}{m}}\left[x - \frac{F(v_0)}{k}\right], \qquad \varepsilon = \frac{\alpha - b}{\sqrt{mk}},$$

we derive the oscillator equation

$$\ddot{y} - \left(\varepsilon - \dot{y}^2\right)\dot{y} + y = 0.$$

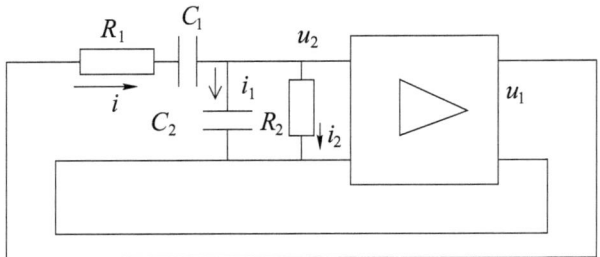

Fig. 4.4 Circuit diagram for an *RC*-oscillator with Wien bridge

Once again, we deduce the Rayleigh equation (4.7), which reduces to the van der Pol equation, as shown earlier.

4.4.3 RC-Oscillator with Wien Bridge

The oscillator with a Wien bridge is the most common type of *RC*-oscillator. It is a series–parallel *RC*-chain connected to the feedback circuit of an amplifier. A simplified scheme for the *RC*-oscillator is shown in Fig. 4.4.

Using Kirchhoff's laws and taking into account the fact that $u_1 = f(u_2)$, we derive the following equation for the voltage u_2:

$$\frac{d^2 u_2}{dt^2} + \left[\frac{1}{R_2 C_2} + \frac{1}{R_1 C_1} + \frac{1}{R_1 C_2} - \frac{1}{R_1 C_2} \frac{df(u_2)}{du_2} \right] \frac{du_2}{dt} + \frac{1}{R_1 C_1 R_2 C_2} u_2 = 0 \,.$$

We consider a symmetric Wien bridge when $R_1 = R_2 = R$ and $C_1 = C_2 = C$. In this case the equation reads

$$\frac{d^2 u_2}{dt^2} + \left[\frac{3}{RC} - \frac{1}{RC} \frac{df(u_2)}{du_2} \right] \frac{du_2}{dt} + \frac{1}{R^2 C^2} u_2 = 0 \,. \tag{4.13}$$

We define $\omega_0 = 1/RC$ and assume that the dependence of the output voltage u_1 of the amplifier on the input voltage u_2 can be approximated by the function

$$f(u_2) = k u_2 - k_1 u_2^3 \,. \tag{4.14}$$

This leads to

$$\frac{d^2 u_2}{dt^2} - \left(\frac{k-3}{RC} - \frac{3k_1}{RC} u_2^2 \right) \frac{du_2}{dt} + \omega_0^2 u_2 = 0 \,. \tag{4.15}$$

4.4 Oscillators with One Degree of Freedom

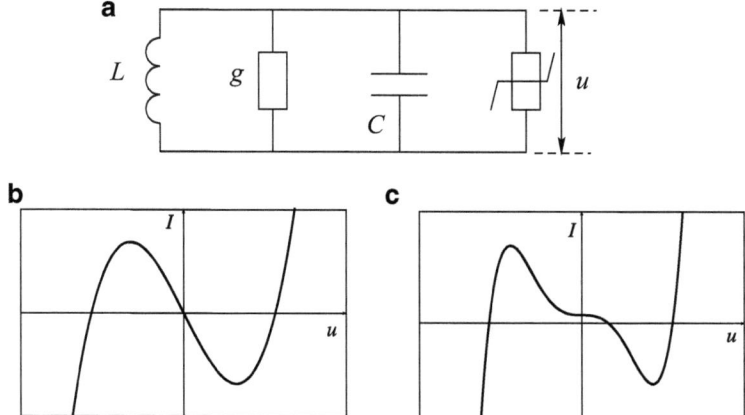

Fig. 4.5 Oscillator with a nonlinear element. (**a**) Circuit diagram of an oscillator. (**b**) and (**c**) Current–voltage characteristics of a nonlinear element

We now introduce the dimensionless time and change the variables and parameters as follows:

$$\tau = \omega_0 t, \qquad x = u_2\sqrt{3k_1}, \qquad \varepsilon = k - 3.$$

As a result, we arrive at the following equation:

$$\ddot{x} - \left(\varepsilon - x^2\right)\dot{x} + x = 0.$$

We thus obtain the van der Pol equation (4.8) for an oscillator with a symmetric Wien bridge, provided that the characteristics of the amplifier can be approximated by the function (4.14).

4.4.4 Oscillatory Circuit with Active Nonlinear Element

A self-sustained oscillatory system can be obtained if a linear oscillatory circuit includes an element which has nonlinear conductivity with a negative slope in the current–voltage characteristic. In radio engineering, self-sustained oscillators of this type are called *oscillators with negative resistance*. As an active nonlinear element, one can use, for example, a tunnel diode, Chua's diode, certain types of vacuum tube, and other devices.

Here we deduce the equation for the oscillator sketched in Fig. 4.5a. Its nonlinear element has a current–voltage characteristic (Fig. 4.5b) that can be approximated by the function

$$I = -\alpha u + \beta u^3. \tag{4.16}$$

Using Kirchhoff's laws, we obtain equations for the current i through the inductor L and for the voltage u across the capacitor C:

$$C\frac{du}{dt} + gu - \alpha u + \beta u^3 + i = 0, \qquad L\frac{di}{dt} = u. \qquad (4.17)$$

The voltage u across the nonlinear element is given by the differential equation

$$\frac{d^2u}{dt^2} - \frac{\alpha - g}{C}\left(1 - \frac{3\beta}{\alpha - g}u^2\right)\frac{du}{dt} + \frac{1}{LC}u = 0.$$

We use the dimensionless variables and parameters

$$\tau = \omega_0 t, \qquad \omega_0^2 = \frac{1}{LC}, \qquad x = \sqrt{\frac{3\beta}{\alpha - g}}u, \qquad \varepsilon = \frac{\alpha - g}{\omega_0 C},$$

and thereby derive the van der Pol equation

$$\ddot{x} - \varepsilon\left(1 - x^2\right)\dot{x} + x = 0.$$

We can also get the Rayleigh equation (4.7) if we write down a differential equation for the current i through the inductor L.

The van der Pol oscillator and the Rayleigh oscillator are two different forms of equation for a self-sustained oscillatory system with soft excitation. As applied to the oscillator considered here (Fig. 4.5a), the excitation of self-sustained oscillations is determined by the the current–voltage characteristic of the nonlinear element. If instead of the cubic polynomial dependence (4.22) we use the nonlinear element with the current–voltage characteristic shown in Fig. 4.5c, we arrive at a self-sustained oscillatory system with hard excitation. Let us deduce an equation for this oscillator.

The current–voltage characteristic of the nonlinear element shown in Fig. 4.5c can be approximated by the function

$$I = -g_0 u - g_1 u^3 + g_2 u^5. \qquad (4.18)$$

The coefficients in (4.18) are positive, i.e., $g_0, g_1, g_2 > 0$. Using Kirchhoff's laws, we get the following dimensionless form of the equation for the self-sustained oscillator:

$$\ddot{x} - \left(\varepsilon_0 + \varepsilon_1 x^2 - x^4\right)\dot{x} + x = 0, \qquad (4.19)$$

where

$$x = u\left(5g_2\sqrt{\frac{L}{C}}\right)^{1/4}, \qquad \varepsilon_0 = (g_0 - g)\sqrt{\frac{L}{C}}, \qquad \varepsilon_1 = \frac{3g_1\sqrt{L/C}}{\sqrt{5g_2\sqrt{L/C}}}.$$

$$(4.20)$$

Equation (4.19) describes the dynamics of the oscillator with hard excitation of self-sustained oscillations. Unlike the van der Pol and Rayleigh oscillators, the behavior of this oscillator with nonlinear dissipation depends on two control parameters ε_0 and ε_1. It can be seen from (4.20) that the excitation parameter ε_0 can take both negative and positive values, while the dissipation parameter ε_1 can only be positive, given that $g_0, g_1, g_2 > 0$.

4.5 Analysis of the van der Pol Equation: Onset of Self-Sustained Oscillations

The van der Pol oscillator equation can be written in different forms depending on how the dynamical variables and parameters of the system are normalized. The van der Pol equation is usually given in the form (4.8). The Rayleigh equation (4.7) can also be reduced to the van der Pol equation.

To define conditions for the appearance of self-sustained oscillations, it is important to know which equilibrium states exist in the system and how the character of their stability depends on the control parameters. Using the material of Chap. 1, we find equilibria and analyze their stability for the van der Pol oscillator

$$\ddot{x} - \left(\varepsilon - x^2\right)\dot{x} + x = 0.$$

We rewrite it as a system of two first-order equations:

$$\dot{x} = f_1(x, y, \varepsilon), \qquad \dot{y} = f_2(x, y, \varepsilon), \qquad (4.21)$$

where

$$f_1(x, y, \varepsilon) = y, \qquad f_2(x, y, \varepsilon) = \left(\varepsilon - x^2\right)y - x;.$$

We can write the equations for equilibrium states by setting the functions f_1 and f_2 equal to zero:

$$y = 0, \qquad \left(\varepsilon - x^2\right)y - x = 0.$$

We thus find that there is a unique singular point located at the coordinate origin $x^0 = 0$, $y^0 = 0$, in the phase plane.

As shown earlier, the stability of an equilibrium can be defined by linearizing the system in the vicinity of this singular point and constructing the linearization matrix (the Jacobi matrix). For the system (4.21), we obtain the matrix

$$\hat{A}\left(x^0, y^0\right) = \begin{bmatrix} \dfrac{\partial f_1\left(x^0, y^0\right)}{\partial x} & \dfrac{\partial f_1\left(x^0, y^0\right)}{\partial y} \\ \dfrac{\partial f_2\left(x^0, y^0\right)}{\partial x} & \dfrac{\partial f_2\left(x^0, y^0\right)}{\partial y} \end{bmatrix} = \begin{bmatrix} 0 & 1 \\ -1 & \varepsilon \end{bmatrix}. \qquad (4.22)$$

The eigenvalues of the matrix (4.22) define the stability of the equilibrium of the van der Pol oscillator. They are easily obtained from the characteristic equation

$$\det \begin{bmatrix} 0-s & 1 \\ -1 & \varepsilon - s \end{bmatrix} = 0$$

or

$$s^2 - \varepsilon s + 1 = 0 . \tag{4.23}$$

As a result, we have two eigenvalues:

$$s_{1,2} = \frac{\varepsilon}{2} \pm \sqrt{\frac{\varepsilon^2}{4} - 1} . \tag{4.24}$$

We now analyze the stability of the equilibrium state ($x^0 = 0, y^0 = 0$) when the control parameter ε is varied:

1. For $\varepsilon < -2$, the eigenvalues s_1 and s_2 are negative real numbers. In this case the equilibrium is a stable node (see Chap. 2).
2. For $-2 < \varepsilon < 0$, the eigenvalues s_1 and s_2 are complex conjugates and Re $s_{1,2} < 0$. The equilibrium is a stable focus.
3. For $0 < \varepsilon < 2$, the eigenvalues s_1 and s_2 are complex conjugates and Re $s_{1,2} > 0$. The equilibrium is an unstable focus.
4. For $\varepsilon > 2$, the eigenvalues s_1 and s_2 are real positive numbers. The equilibrium is an unstable node.
5. $\varepsilon = 0$ is the bifurcation parameter value because the complex conjugate eigenvalues of the linearization matrix at the equilibrium become purely imaginary. As already mentioned in Chap. 3, this situation corresponds to the Andronov–Hopf bifurcation. In order to determine whether this bifurcation is supercritical or subcritical and to find out how the limit cycle which is born from the equilibrium behaves, we use an approximate quasiharmonic description of the self-sustained oscillator in terms of the instantaneous amplitude and phase.

4.5.1 Amplitude and Phase Equations for the Self-Sustained Oscillator

There are several classical methods for finding a quasiharmonic approximate solution to the van der Pol equation. We use the van der Pol averaging method and apply it to the oscillator equation (4.8):

$$\frac{d^2x}{dt^2} - (\varepsilon - x^2) \frac{dx}{dt} + x = 0 .$$

4.5 Analysis of the van der Pol Equation: Onset of Self-Sustained Oscillations

We assume that, for small positive values of ε, the solution of the equations exhibits approximately harmonic oscillations. We write it in the form of a harmonic function with slowly varying parameters (amplitude and phase):

$$x(t) = \text{Re}[a(t)\exp(it)] = \frac{1}{2}[a\exp(it) + a^*\exp(-it)], \qquad (4.25)$$

where $a(t)$ is the complex amplitude, $\text{Re}[\ldots]$ is the real part of the complex value, and a^* is the complex conjugate. Since instead of the real function $x(t)$ we have introduced the new complex function $a(t)$ which is insufficiently defined, we impose an additional condition. The function $a(t)$ is required to obey the condition

$$\frac{da}{dt}\exp(it) + \frac{da^*}{dt}\exp(-it) = 0. \qquad (4.26)$$

Substituting the solution (4.25) into the van der Pol equation, precomputing the first and second derivatives, and taking into account the additional condition (4.26), we arrive at

$$i\frac{da}{dt}\exp(it)$$
$$= \left[\varepsilon - \frac{a^2\exp(2it) + 2|a|^2 + (a^*)^2\exp(-2it)}{4}\right]\frac{ia\exp(it) - ia^*\exp(-it)}{2}.$$

Removing the brackets and dividing both sides of the equation by $i\exp(it)$, we find

$$\frac{da}{dt} = \frac{1}{2}\varepsilon a - \frac{1}{8}|a|^2 a - \frac{1}{2}\varepsilon a^*\exp(-2it) - \frac{1}{8}a^3\exp(2it) \qquad (4.27)$$
$$+ \frac{1}{4}|a|^2 a^*\exp(-2it) - \frac{1}{8}|a|^2 a^*\exp(-2it) + \frac{1}{8}(a^*)^3\exp(-4it).$$

$a(t)$ is assumed to be a slowly varying function in time so its variation over the period can be neglected. The derivative da/dt is also considered to be almost constant over the period. Multiplying both sides of (4.27) by $1/2\pi$ and integrating over time from 0 to 2π, we obtain

$$\frac{da}{dt} = \frac{1}{2}\varepsilon a - \frac{1}{8}|a|^2 a. \qquad (4.28)$$

Equation (4.28) is called the truncated or averaged van der Pol equation for the complex amplitude. To express it in terms of real variables, we write the complex amplitude in polar coordinates:

$$a(t) = \rho(t)\exp(i\varphi(t)),$$

where $\rho(t)$ is the real amplitude and $\varphi(t)$ is the real phase of the oscillations. Equation (4.28) can thus be written as a system of truncated equations for the amplitude and phase:

$$\frac{d\rho}{dt} = \frac{\varepsilon}{2}\rho - \frac{1}{8}\rho^3, \qquad \frac{d\varphi}{dt} = 0. \qquad (4.29)$$

One can see from the system (4.29) that the phase $\varphi(t)$ is in fact time independent, whence its constant value is given by the initial conditions. Thus, the problem of the existence of periodic motions, their stability, and possible bifurcations in the oscillator is reduced in the quasiharmonic approximation to the study of the amplitude equation (4.29).

From (4.29), it is easy to identify the stationary states of the system. We find two possible states by equating the derivatives to zero:

$$\rho^0(t) = 0, \qquad \varphi^0(t) = \varphi_0, \qquad (4.30)$$

and

$$\rho^0(t) = 2\sqrt{\varepsilon}, \qquad \varphi^0(t) = \varphi_0. \qquad (4.31)$$

The first stationary solution corresponds to the absence of oscillations, i.e., to the equilibrium in the van der Pol equation (4.8). The second corresponds to the quasiharmonic self-sustained oscillations in the oscillator:

$$x(t) = 2\sqrt{\varepsilon}\cos(t + \varphi_0), \qquad (4.32)$$

where φ_0 is the initial phase of the oscillations, which can be arbitrary.

Let us analyze the stability of the stationary states. As the two equations of the system (4.29) are independent, they can be considered separately. A small perturbation of the amplitude from its stationary value varies in time according to an exponential law with exponent

$$s_\rho = \frac{\varepsilon}{2} - \frac{3}{8}(\rho^0)^2.$$

For the stationary value $\rho^0 = 2\sqrt{\varepsilon}$, we obtain $s_\rho = -\varepsilon$. It is negative for positive values of ε, so the nonzero stationary state of the equation for the real amplitude is stable. The stationary solution for the phase is neutrally stable, i.e., any perturbation of the initial phase neither increases nor decreases with time.

Thus, the analysis of the truncated equations shows that the Andronov–Hopf bifurcation in the oscillator (4.8) is supercritical (or soft). As the parameter ε goes to positive values, stable periodic oscillations of the form (4.32) arise. Furthermore, the amplitude of the self-sustained oscillations increases from zero in proportion to $\sqrt{\varepsilon}$.

In order to describe a transient process leading to steady-state oscillations, we need to find a solution of the system (4.29) that depends on initial conditions. The solution for the phase equation is obvious. The amplitude equation can also be solved analytically. To do this, we first divide it by ρ^3 to obtain

$$\frac{d}{dt}\left(\frac{1}{\rho^2}\right) = -\varepsilon\frac{1}{\rho^2} + \frac{1}{4}.$$

Making the change of variables $y = 1/\rho^2 - 1/4\varepsilon$, we arrive at the equation

$$\frac{dy}{dt} = -\varepsilon y,$$

which is easily solved. For the original variable $\rho(t)$, the solution can be expressed in the form

$$\rho(t) = \frac{1}{\sqrt{\left[\left(\frac{1}{\rho_0}\right)^2 - \frac{1}{4\varepsilon}\right]\exp(-\varepsilon t) + \frac{1}{4\varepsilon}}}, \qquad (4.33)$$

where $\rho_0 = \rho(0)$ is the initial amplitude of the oscillations ($\rho_0 \neq 0$ according to the solution conditions). The expression (4.33) determines the process of transition to steady-state oscillations in the van der Pol oscillator. As time goes by, the solution (4.33) tends to the stationary state $\rho^0 = 2\sqrt{\varepsilon}$. The value ρ^0 of the stationary amplitude for (4.8) thus depends on ε.

4.6 Oscillator with Hard Excitation of Self-Sustained Oscillations

Consider the oscillator with hard excitation (4.19). We first analyze the stability of its equilibria and derive truncated equations for the amplitude and phase. We then explore bifurcation transitions and finally plot bifurcation diagrams.

4.6.1 Analysis of the Stability of Equilibrium States

We find equilibrium points and analyze their stability in the oscillator with hard excitation described by (4.19):

$$\ddot{x} - \left(\varepsilon_0 + \varepsilon_1 x^2 - x^4\right)\dot{x} + x = 0.$$

Rewriting (4.19) in the form of a system of two first-order equations, viz.,

$$\dot{x} = y, \qquad \dot{y} = \left(\varepsilon_0 + \varepsilon_1 x^2 - x^4\right) y - x, \qquad (4.34)$$

we find that there is one equilibrium with coordinates $x^0 = 0$, $y^0 = 0$ in the phase plane. The linearization matrix in the vicinity of this point has the form

$$\hat{A}\left(x^0, y^0\right) = \begin{bmatrix} \dfrac{\partial f_1\left(x^0, y^0\right)}{\partial x} & \dfrac{\partial f_1\left(x^0, y^0\right)}{\partial y} \\ \dfrac{\partial f_2\left(x^0, y^0\right)}{\partial x} & \dfrac{\partial f_2\left(x^0, y^0\right)}{\partial y} \end{bmatrix} = \begin{bmatrix} 0 & 1 \\ -1 & \varepsilon_0 \end{bmatrix}. \qquad (4.35)$$

The eigenvalues of the matrix (4.34) define the stability of the equilibrium of the oscillator with hard excitation. Considering the characteristic equation

$$s^2 - \varepsilon_0 s + 1 = 0, \qquad (4.36)$$

we obtain the following eigenvalues:

$$s_{1,2} = \frac{\varepsilon_0}{2} \pm \sqrt{\frac{\varepsilon_0^2}{4} - 1}. \qquad (4.37)$$

It can be seen from (4.37) that the character of the equilibrium of the oscillator (4.19) depends on only one control parameter ε_0 and fully coincides with the behavior of the equilibrium depending on the parameter ε in the van der Pol oscillator:

1. For $\varepsilon_0 < 2$, the equilibrium is a stable node.
2. For $-2 < \varepsilon_0 < 0$, the equilibrium is a stable focus.
3. For $0 < \varepsilon_0 < 2$, the equilibrium is an unstable focus.
4. For $\varepsilon > 2$ the equilibrium is an unstable node.

With $\varepsilon_0 = 0$, the equilibrium is characterized by a pair of imaginary eigenvalues, i.e., the condition for the Andronov–Hopf bifurcation is satisfied. We continue to investigate the dynamics of the oscillator with hard excitation in the framework of the quasiharmonic approximation, deriving truncated equations for the amplitude and phase.

4.6.2 Truncated Equations for the Amplitude and Phase for the Oscillator with Hard Excitation

The parameters ε_0 and ε_1 in (4.19) are assumed to be vanishingly small and the system looks like a quasiharmonic oscillator (the right-hand side is treated as a small

4.6 Oscillator with Hard Excitation of Self-Sustained Oscillations

perturbation of a harmonic oscillator). In this case the solution can be sought in the form of a harmonic function with slowly time-varying amplitude and phase.

Carrying out the same transformations as for the van der Pol oscillator with soft excitation, we obtain the truncated equation for the complex amplitude:

$$\dot{a} = \left(\frac{\varepsilon_0}{2} + \frac{\varepsilon_1}{8}|a|^2 - \frac{1}{16}|a|^4\right) a . \quad (4.38)$$

Representing the complex quantity $a(t)$ in the form

$$a(t) = \rho(t) \exp[i\varphi(t)] ,$$

we rewrite (4.38) as a system of equations for the amplitude and phase, viz.,

$$\dot{\rho} = \left(\frac{\varepsilon_0}{2} + \frac{\varepsilon_1}{8}\rho^2 - \frac{1}{16}\rho^4\right)\rho , \qquad \dot{\varphi} = 0 . \quad (4.39)$$

4.6.3 Bifurcation Diagram of the Oscillator with Hard Excitation

As already noted, this problem can be reduced to the study of equilibria, their stability, and bifurcations by varying the control parameters of the system described by the amplitude equation (4.39). From

$$\rho\left(\frac{\varepsilon_0}{2} + \frac{\varepsilon_1}{8}\rho^2 - \frac{1}{16}\rho^4\right) = 0 , \quad (4.40)$$

we obtain three equilibria with coordinates

$$\rho_1^0 = 0 , \qquad \rho_2^0 = \sqrt{\varepsilon_1 + \sqrt{\varepsilon_1^2 + 8\varepsilon_0}} , \qquad \rho_3^0 = \sqrt{\varepsilon_1 - \sqrt{\varepsilon_1^2 + 8\varepsilon_0}} . \quad (4.41)$$

Three characteristic ranges of values can then be distinguished for the control parameter ε_0:

1. If $\varepsilon_0 < -\varepsilon_1^2/8$, the system has one equilibrium ρ_1^0.
2. If $-\varepsilon_1^2/8 < \varepsilon_0 < 0$, there are three equilibria ρ_1^0, ρ_2^0, and ρ_3^0. It should be noted that, for $\varepsilon_0 = -\varepsilon_1^2/8$, the two equilibria merge, i.e., $\rho_2^0 = \rho_3^0 = \sqrt{\varepsilon_1}$, and then disappear below the indicated value of ε_0.
3. If $\varepsilon_0 > 0$, there are two equilibria ρ_1^0 and ρ_2^0. As ε_0 tends to zero from negative values, the fixed point ρ_3^0 moves towards ρ_1^0 and merges with it at $\varepsilon_0 = 0$. There are still two equilibria above this value.

Fig. 4.6 Bifurcation diagram of the oscillator with hard excitation

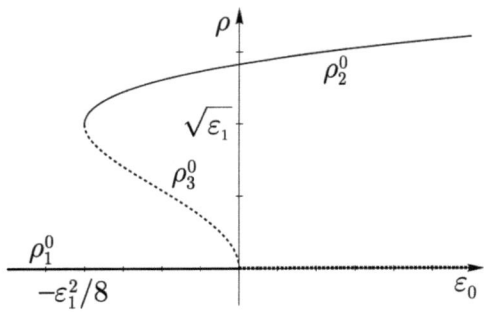

The stability of the equilibria $\rho_1^0, \rho_2^0, \rho_3^0$ is defined by the eigenvalue

$$s_i = \frac{\partial}{\partial \rho}\left(\frac{\varepsilon_0}{2}\rho + \frac{\varepsilon_1}{8}\rho^3 - \frac{1}{16}\rho^5\right)\bigg|_{\rho=\rho_i^0}$$

$$= \frac{\varepsilon_0}{2} + \frac{3\varepsilon_1}{8}\left(\rho_i^0\right)^2 - \frac{5}{16}\left(\rho_i^0\right)^4, \tag{4.42}$$

where the index $i = 1, 2, 3$ indicates one of the three equilibria. We have $s_1 = \varepsilon_0/2$ for ρ_1^0. Hence, this point is stable for $\varepsilon_0 < 0$ and becomes unstable at positive values of ε_0.

Consider the eigenvalue for the equilibrium ρ_2^0:

$$s_2 = -\frac{1}{4}\sqrt{\varepsilon_1^2 + 8\varepsilon_0}\left(\varepsilon_1 + \sqrt{\varepsilon_1^2 + 8\varepsilon_0}\right). \tag{4.43}$$

Before exploring how the stability of ρ_2^0 depends on the parameters, we recall that, in the physical sense, ε_0 can be both positive and negative, while ε_1 can only assume positive values. The equilibrium ρ_2^0 appears when the condition $\varepsilon_0 = -\varepsilon_1^2/8$ is satisfied, and when $s_2 = 0$, that corresponds to the bifurcation point. For $\varepsilon_0 > -\varepsilon_1^2/8$, all three multipliers in (4.43) are positive, and as a consequence, $s_2 < 0$ and ρ_2^0 is stable.

The eigenvalue s_3 of the equilibrium ρ_3^0 is given by

$$s_3 = \frac{1}{4}\sqrt{\varepsilon_1^2 + 8\varepsilon_0}\left(\varepsilon_1 - \sqrt{\varepsilon_1^2 + 8\varepsilon_0}\right). \tag{4.44}$$

The equilibrium ρ_3^0 exists when $-\varepsilon_1^2/8 < \varepsilon_0 < 0$. In this parameter range, the eigenvalue $s_3 > 0$, so ρ_3^0 is unstable.

We can now plot the bifurcation diagram for the oscillator with hard excitation. This is shown in Fig. 4.6. The system dynamics is controlled by the parameter ε_0 whose values are plotted on the horizontal axis. The ordinate is the stationary value of the amplitude $\rho(t)$. For the original equation for the oscillator with hard excitation

(4.19), the points $\rho = 0$ on the bifurcation diagram correspond to the equilibrium at the coordinate origin of the phase plane (x, y). The points with ordinate $\rho > 0$ in Fig. 4.6 correspond to the limit cycle with radius ρ in the phase plane (x, y), centered at the origin.

For the oscillator (4.19) with $\varepsilon_0 < \varepsilon_1^2/8$, there is a unique attractor in the phase plane, namely, the stable equilibrium at the origin. As ε_0 increases and goes beyond the value $\varepsilon_0 = -\varepsilon_1^2/8$, a pair of limit cycles appears: a stable one with radius ρ_2^0 and an unstable one with radius ρ_3^0. When ε_0 increases further, the radius of the stable cycle grows, while that of the unstable cycle decreases. When $\varepsilon_0 = 0$, the unstable limit cycle shrinks to the equilibrium at the origin ($\rho_1^0 = 0$) and the subcritical Andronov–Hopf bifurcation takes place. The equilibrium $\rho_1^0 = 0$ becomes unstable. For $\varepsilon_0 > 0$, the unstable equilibrium and the stable limit cycle exist in the phase plane. For $-\varepsilon_1^2/8 < \varepsilon_0 < 0$, the effect of bistability can be observed in the oscillator with hard excitation. Two attractors coexist in the phase plane: the stable equilibrium point and the stable limit cycle. The unstable limit cycle is the boundary of their basins of attraction. The effect of hysteresis can also be observed in the oscillator. As the control parameter grows, the transition from the equilibrium to the stable limit cycle occurs at $\varepsilon_0 = 0$. Going in the reverse direction of the parameter ε, the transition from the stable limit cycle to the stable equilibrium takes place when $\varepsilon_0 = -\varepsilon_1^2/8$.

4.7 Summary

In this chapter we have considered the two most typical examples of self-sustained oscillatory systems in the phase plane: the van der Pol oscillator and the oscillator with hard excitation. Their dynamics has been explored in the quasiharmonic approximation. We have derived the truncated equations for the amplitude and phase. Using these, we have also plotted the bifurcation diagrams and described the conditions of excitation and the characteristics of self-sustained oscillations. The classic results on the dynamics of dynamical systems with one degree of freedom can be found in [1, 3].

References

1. Andronov, A.A., Vitt, E.A., Khaikin, S.E.: Theory of Oscillations. Pergamon Press, Oxford (1966)
2. Andronov, A.A., Vitt, A.A., Khaikin, S.E.: Theory of Oscillations, p. 222. Nauka Publisher, Moscow (1981) (in Russian)
3. Marsden, L.E., McCraken, V.: The Hopf Bifurcation and Its Applications. Springer, New York (1976)

Chapter 5
Systems with Phase Space Dimension $N \geq 3$: Deterministic Chaos

5.1 Introduction

The transition from the phase plane to a space of higher dimension leads to fundamental qualitative changes. The number of possible bifurcations of equilibrium states and limit cycles increases significantly, and many of them have not yet been studied. Some saddle sets become possible, such as an equilibrium state of the saddle-focus type and a saddle limit cycle. A cycle of the saddle-focus type and a saddle torus can be realized in a phase space with dimension $N \geq 4$. The appearance of multi-dimensional stable and unstable manifolds of saddle sets and new types of doubly asymptotic trajectories such as separatrix loops of saddle foci and Poincaré homoclinic curves, leads in many cases to a complex structure in the phase portrait of a DS. The different kinds of behaviour actually realized are much more complex and varied. Besides periodic oscillations, quasiperiodic and chaotic oscillations can be observed. New types of attractors can emerge, namely, two-dimensional and multi-dimensional tori corresponding to quasiperiodic regimes, and strange chaotic attractors, which are the signature of dynamical chaos. Special types of DS behavior and special 'exotic' attractors can be observed under certain conditions, viz., strange nonchaotic and chaotic nonstrange attractors. Deterministic (dynamical) chaos is the most important and interesting type of system behavior in $\mathbb{R}^N$ for $N \geq 3$.

Chaotic processes in deterministic nonlinear dissipative systems are one of the fundamental problems of the natural sciences today, and an area of intense research activity. It has been convincingly demonstrated that in such systems the cause of complex oscillatory processes, which cannot differ from truly random ones in their physical characteristics, does not lie in the large number of degrees of freedom, nor in the presence of fluctuations as previously believed, but in the exponential instability of modes, which gives rise to a sensitive dependence on the exact choice of the initial state of the system. H. Poincaré understood and foresaw the possibility of such phenomena. In his book *Science and Method* (1908), he pointed out that, in unstable systems "a quite negligible reason which eludes us in its smallness causes a significant effect which we cannot predict. [...] Prediction becomes impossible, we have a random phenomenon." The development of Poincaré's ideas led to today's theory of chaotic dynamics in deterministic systems. As it turned out, the phase space dimension $N \geq 3$ is a necessary condition for the appearance of chaos in differential systems, and the excitation of undamped chaotic pulsations becomes basically possible in oscillators with only one and a half degrees of freedom.

In systems with one degree of freedom whose phase space is the phase plane, the possible dynamical regimes are exhausted by equilibrium states and periodic oscillations (limit cycles). For many years this fact served as a psychological barrier that could not be overcome even by obvious (now!) experimental results. The limitation of 'nonlinear thinking' to the phase plane was understood by many leading experts, but due to the absence of appropriate mathematical tools, a substantiated transition from the plane to a space of three or more dimensions was practically impossible.

5.2 Determinism and Chaos for Beginners

What is the phenomenon of deterministic chaos? Let us try to answer this question. We must first clarify our understanding of the terms 'determinism' and 'chaos', then assess the term 'deterministic chaos'.

5.2.1 Determinism

Determinism is usually understood as an unambiguous relationship between cause and effect (see Chap. 1). As applied to evolution laws, this means that, if some initial state of a system is given at $t = t_0$, it *uniquely* determines the system state at any time $t > t_0$. For example, if a body is uniformly accelerated, then its velocity is determined by the deterministic law

$$v(t) = v(t_0) + at . \tag{5.1}$$

By setting the initial velocity $v(t_0)$, we uniquely determine the velocity $v(t)$ at any time $t > t_0$.

In general, the dependence of the future state $x(t)$ on the initial state $x(t_0)$ can be written as $x(t) = F[x(t_0)]$, where F is a deterministic law (or operator) which performs a unique transformation of the initial state $x(t_0)$ into the future state $x(t)$ for any $t > t_0$. This law can be a function, a differential or integral equation, a simple rule given by a table or a graph, etc. It is important that the law F *uniquely* transforms the initial state (the cause) into the future state (the effect).

5.2.2 Chaos

Now we clarify the notion of chaos. Let us conduct a thought experiment with a Brownian particle. At the initial time $t = t_0$, we put the particle in a liquid solution and the use a microscope to determine its position as time goes by, noting the coordinates of the particle at regular intervals Δt. It is easily seen that, under the influence of random kicks from surrounding molecules, the particle performs an irregular walk, characterized by an intricate trajectory. We now repeat the experiment several times, each time reproducing as far as possible the initial conditions of our experiment. What will be the results? There are two main points. The first is that, in each run, the trajectory of the particle motion will be complex and nonperiodic. The second is that any attempt to unambiguously repeat the experiment will lead to a negative result. Each time we repeat the experiment with the same initial conditions (insofar as this is possible), we will get different trajectories of particle motion, which do not even closely resemble each other!

The classical phenomenon of Brownian particle motion gives us a clear physical idea of chaos as an unpredictable random process. Thus, when we speak about chaos, we mean that the system state varies randomly with time (it cannot be unambiguously predicted) and cannot be reproduced (the process cannot be repeated).

The thoughts described above lead us to the conclusion that the notions of *determinism* and *chaos* are directly opposite in meaning. Determinism is associated with complete and unambiguous predictability and reproducibility, while chaos is associated with complete unpredictability and nonreproducibility. The obvious question is: what could we possibly intend by the term 'deterministic chaos', which combines two such apparently opposite notions? It is not so easy to answer this question, but it is possible and we shall attempt to do so here.

5.2.3 Stability and Instability

We must consider the notion of *stability* (*instability*) of a system's motion. We start with the simplest example and consider a stationary state or an equilibrium of the system. We place a small ball at the lowest point inside a hollow sphere, then push the ball slightly and watch its motion. After making several damping oscillations, the ball will return once again to the bottom of the sphere. In this case the equilibrium is *stable*: small perturbations of the initial state decay with time. But if we place the ball at the top of the sphere (outside), the response to a small perturbation will be different: for any small deviation of the ball from this equilibrium, it will roll off the top. This equilibrium is *unstable*: small perturbations of the initial state grow with time.

The physical meaning of the notion of *stability* (*instability*) considered here for equilibrium is the same for any other regime. The regime in which a dynamical system functions is *stable* if small perturbations in the vicinity of this regime decay with time, tending to zero. If small deviations from this regime increase with time, it is *unstable* (see Chap. 2).

5.2.4 Nonlinearity

We must now discuss another important property of complex systems, namely *nonlinearity*. Consider an unstable regime. If we disturb the regime slightly with a small perturbation, we fix the growth of the perturbation. Will it be infinite? Of course, that could never happen in real life! The deviation will increase only until some nonlinear restriction mechanism begins to act on the perturbation growth. What could this be? We will consider this question from both the physical and the mathematical point of view.

From the physical point of view, the amplitude cannot increase without limit. In the first stage, when the deviation from the initial state is small, it can grow. But what will happen then? Due to the limited amount of energy in the system, this growth must stop or be replaced by a decrease in the deviation amplitude. Any new regime must have a finite amplitude and these processes are governed by nonlinear laws. We speak of nonlinearity when the properties of the system depend directly on its state.

Consider the following example. Let the amplitude of deviation $f(x)$ from the initial state x be determined by the relation

$$f(x) = kx - bx^3, \tag{5.2}$$

where k and b are constant positive coefficients. If $x \ll 1$, then $bx^3 \ll kx$ and

$$f(x) \cong kx. \tag{5.3}$$

In the case of (5.3), $f(x)$ grows linearly as x increases. But if x becomes comparable with unity, the term bx^3 cannot be neglected. In the case of (5.2), the growth of $f(x)$ due to kx will be nonlinearly restricted by the fact of subtracting bx^3. At certain values of x, the deviation (5.2) will again be close to zero and the process can start over again: the deviation will grow, reach its maximum, and then decrease again due to the nonlinear restriction. The system will automatically adjust itself because its properties depend on its current state.

5.2.5 Instability and Nonlinear Restriction

We now consider an unstable deterministic system where deviation growths are nonlinearly restricted. For simplicity we consider an equilibrium state that corresponds

5.2 Determinism and Chaos for Beginners

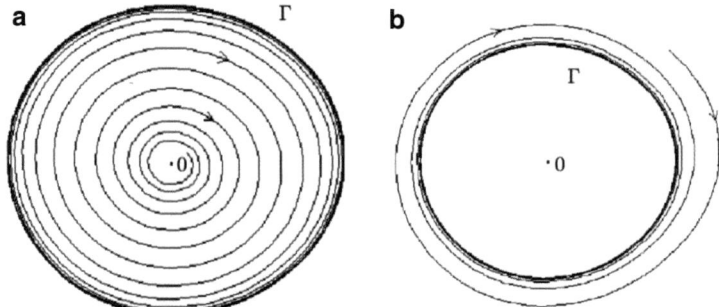

Fig. 5.1 Stable limit cycle Γ born in the vicinity of an unstable equilibrium 0. Behavior of trajectories for (**a**) small deviations and (**b**) large deviations from the equilibrium

to a point in the space of phase coordinates of the system. We remove the system from its equilibrium by a small deviation. This perturbation will increase due to instability. The growth will subsequently slow down because a nonlinear restriction mechanism will start to act. What should we expect in this situation? First, due to the nonlinear restriction, the deviation will decrease strictly to zero. The system will return to its initial equilibrium state. This is theoretically possible, but very unlikely because the initial equilibrium state is unstable. The second situation is more realistic. The system will return to a small neighborhood of the initial state, i.e., it will come very close to the unstable equilibrium state, and will subsequently move away from it again due to instability. This process will continue indefinitely in time! But it can only be realized under certain special conditions.

Suppose we are dealing with a two-dimensional differential dynamical system. Its state space is the phase plane with coordinates x and y. If a small perturbation of an equilibrium in the system will increase and then decrease due to nonlinear restriction, two cases can arise: new stable equilibrium states can appear in the vicinity of the unstable equilibrium or the trajectory can go to a new regime corresponding to periodic oscillations.

The second case is illustrated in Fig. 5.1. For small deviation amplitudes as shown in Fig. 5.1a, the trajectory spirals away from the equilibrium 0. For large deviations (Fig. 5.1b), the trajectory returns. As a result, in the place of the equilibrium which has lost its stability, there appears a new regime, viz., periodic self-sustained oscillations which correspond to the limit cycle Γ in the phase plane.

In the presence of a nonlinear restriction on perturbation growth, the instability of the equilibrium state in the two-dimensional system gives rise to a new regime of stable periodic oscillations. If we imagine a different situation where the deviation from the equilibrium first increases and then tends to zero again due to nonlinearity, we arrive at a contradiction: the phase trajectory must self-intersect (Fig. 5.2). This means that there are initial conditions that evolve to different states! This is impossible, due to the concept of determinism. In this example, it manifests itself through a uniqueness theorem: for given initial conditions, there is a solution and it is *unique*, i.e., there is no other solution.

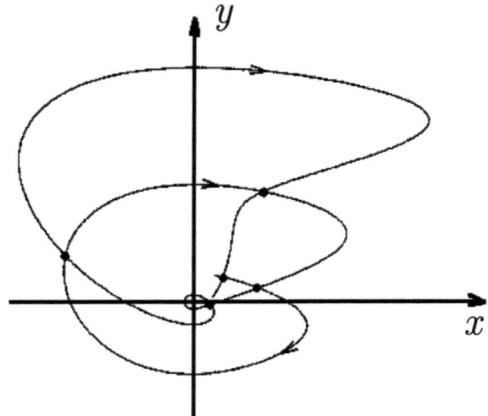

Fig. 5.2 Dynamical system behavior that cannot be realized in the phase plane due to the intersection of phase trajectories. This picture has actually been obtained by projecting a three-dimensional trajectory on the plane of two variables

5.2.6 Deterministic Chaos

The picture changes considerably if we deal with a dynamical system whose state is characterized by three independent variables. So let us repeat our considerations for a three-dimensional space. We can freely realize the situation shown in Fig. 5.2 in the space of three measurements. The trajectory unwinds in the three-dimensional space, moving away from the equilibrium 0 in a spiral. Having reached certain values, and while experiencing the mechanism of nonlinear restriction, the trajectory will return once again to the neighborhood of the initial state. Furthermore, the process will repeat due to instability.

There are two possibilities here. The first goes as follows: having made a few revolutions in the three-dimensional space, after a certain lapse of time, the trajectory closes. This indicates the presence of a complex but periodic oscillatory process in the system. The second option is that, although the trajectory returns to the neighborhood of the origin, it does close but exhibits some aperiodic process which can last forever. The second case corresponds to the regime of deterministic chaos. Indeed, the main principle of determinism works here: the future state is unambiguously determined by the initial one, but the system evolution is nevertheless a complex and non-periodic process. It outwardly resembles a random process.

A more detailed analysis can bring out the important difference between deterministic and random processes. The deterministic process is reproducible! Indeed, repeating the initial state, due to determinism, we can once again unambiguously reproduce the same trajectory, regardless of its complexity. But is this non-periodic process not chaotic in the sense of the above definition of chaos? The answer is affirmative. This is a complex process, similar to a random one, but it is nevertheless deterministic. It is important here that it is unstable. This enables us to understand another fundamentally important property of systems with deterministic chaos, namely *mixing*.

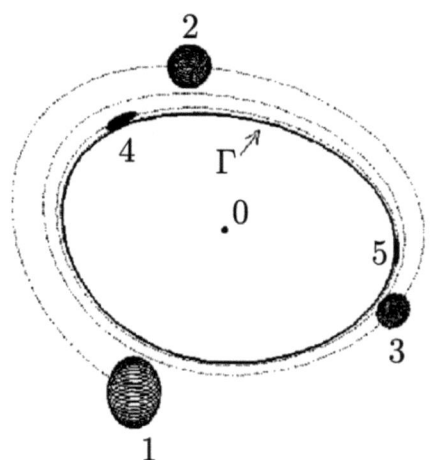

Fig. 5.3 Contraction in time of the initial region of uncertainty 1 in the case when the cycle Γ is a stable limit regime

5.3 Mixing and Probabilistic Properties of Deterministic Systems

We have established that, on a theoretical level, dissipative systems with phase space dimension $N \geq 3$ can exhibit the regime of complex non-periodic pulsations. This kind of motion is deterministic and unstable. What does it lead to? We begin by discussing stable motions in deterministic dissipative dynamical systems.

Suppose the initial state is not a point $\mathbf{x}^0$ with certain coordinates in the state space, but a small sphere with radius $\varepsilon > 0$ surrounding this point. Any point inside the sphere characterizes a small deviation from $\mathbf{x}^0$, so the sphere includes a set of possible deviations from the initial state, not exceeding ε in absolute value. We now apply an evolution operator and track how the sphere is modified. Since the chosen regime is stable, any small deviation must decay in time. This means that, under the deterministic evolution law, the ball of radius ε decreases in time and its radius decreases to zero when $t \to \infty$. This is illustrated in Fig. 5.3. The initial phase volume in dissipative systems decreases in time. This means that small perturbations eventually decay, and the system returns once again to the original stable regime.

Suppose the original regime is unstable. What will happen in this case? The size of the phase space region under consideration can increase indefinitely if the system is linear. But if the system is nonlinear and dissipative, the initial small region of the phase space will evolve in a highly nontrivial way. Let us try to understand this.

Small perturbations grow since the regime is unstable. This is the first factor. The second is that dissipative systems, regardless of their stability type, cause the phase volume element to decrease in time to zero due to energy losses. How can we combine these two factors? There is a unique solution to this dilemma: the phase volume element must stretch in some directions and contract in others. Moreover, on average, the degree of contraction must necessarily prevail over the degree of stretching so that the phase volume eventually decreases in time. This appears to

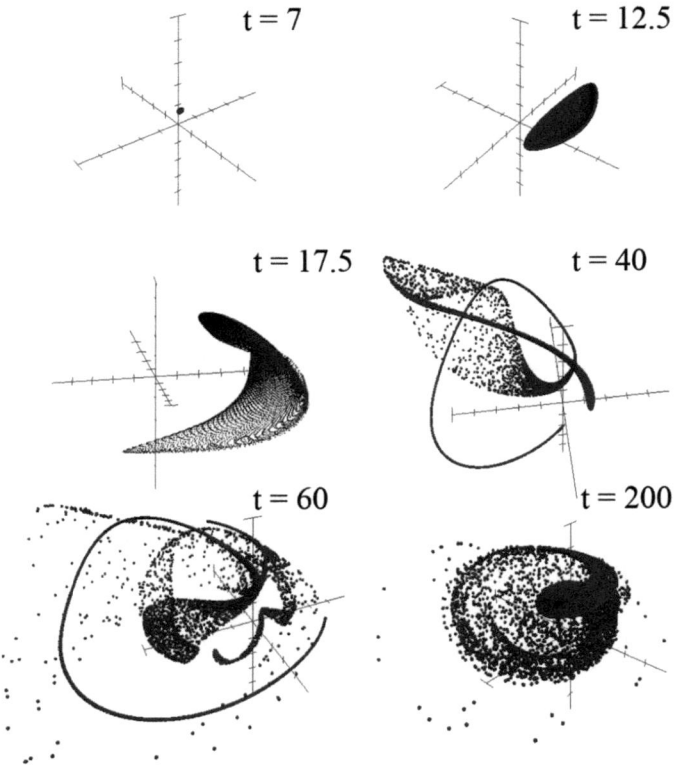

Fig. 5.4 Temporal evolution of a small initial phase volume of size $\varepsilon = 0.1$ containing 200 initial points in the Anishchenko–Astakhov oscillator (1.30) for $m = 1.5$, $g = 0.2$

be possible in nonlinear dissipative systems and is illustrated in Fig. 5.4. Due to the mechanism of nonlinear restriction, the phase trajectory of a complex oscillatory regime is concentrated in a bounded region of the phase space. In this case, any small vicinity of the original initial state evolves as shown in Fig. 5.4 and ends up mixed over the region occupied by the trajectory. It is not easy to picture this process.

Let us conduct a thought experiment. Place a small tealeaf in a glass of water and stir the water with a teaspoon, causing instability. The tealeaf moves along a complex spiralling trajectory caused by the motion of the water in the glass. At any time, we can in theory fix the tealeaf's coordinates $\mathbf{x}(t)$ in the volume of water. Now, instead of the tealeaf, we put a very small drop of ink and again stir the water with the teaspoon. What will happen now? The ink will spread almost evenly throughout the entire volume of water, slightly coloring it. After the mixing time, the ink particles, originally concentrated in a small volume of the droplet, can be found in *any* part of the volume of water in the glass. In everyday life, we usually refer to this process as mixing. In mathematics, this notion also exists and is very close in meaning to the physical interpretation. Indeed, the flow of water in the glass generated by the stirring motion of the teaspoon can be interpreted as the action of the deterministic evolution operator of a deterministic system. The tealeaf moves along a complex but deterministic trajectory. But the ink drop, which can be treated as a small volume in the phase space around the tealeaf, is mixed over the whole volume of water under the action of this same evolution operator.

Thus, in unstable regimes of deterministic nonlinear systems with mixing, the future state can be unambiguously predicted only if the initial conditions are given exactly. However, if an arbitrarily small but finite error is taken into account (i.e., if a drop of ink is considered instead of a tealeaf), deterministic prediction becomes impossible. A small region of the original uncertainty is blurred to a finite region in the phase space due to mixing. Now we are dealing with a process that is associated with true randomness, or true chaos.

The basic property of dynamical systems instantiating deterministic chaos is the sensitive dependence of an operating regime on arbitrarily small changes in initial conditions. This circumstance leads to the loss of deterministic predictability, whereupon probabilistic characteristics must be introduced to describe the dynamics of such systems. In this sense the term 'deterministic chaos' becomes clear. It characterizes the onset of random unpredictable behavior in a system controlled by deterministic laws.

The uncertainty in setting initial conditions is a very real situation from a physical viewpoint. Indeed, since the system state is always registered by measurement devices with a finite accuracy, it is only determined up to a finite (though arbitrarily small) error. This means that we must analyze the temporal evolution, not of an initial point, but of an initial region around this point. Owing to mixing, we are then faced with the process described in detail above.

5.4 Is Deterministic Chaos a Mathematical Oddity or a Typical Property of the Material World?

By simple arguments we have come to the conclusion that the regime of deterministic chaos can be realized in nonlinear dissipative systems. In modern science, this effect is well justified theoretically and reliably confirmed experimentally. However, one might wonder whether this phenomenon is just a mathematical oddity in the sense that its implementation is theoretically possible, but practically unlikely? But the answer is a definite negative. Since deterministic chaos was discovered, its properties have been very clearly understood and diagnostic methods have been developed. Indeed, chaos has been found in almost all fields of modern science, from physics to radio engineering, chemistry, biology, mechanics, economics, and many others. This raises a natural question: Why was this typical regime of operation of dynamical systems not detected and described until very recently? This can be explained as follows.

Although the overwhelming majority of real material systems and processes are theoretically nonlinear, there is a wide class of processes that can be correctly described in a linear or quasi-linear approximation. The linear theory of dynamical systems and processes has been well developed and can often give an exhaustive description that agrees well with experiments. On the other hand, deterministic chaos is a phenomenon that can be realized only in nonlinear systems, and the situation in nonlinear theory is much less encouraging. For example, there is still no general theory for solutions of nonlinear differential equations. Even now analysis of the dynamics of nonlinear systems requires art and creativity, and has to be done on a case-by-case basis. It was precisely the absence of rigorous theoretical results for nonlinear systems that delayed the discovery and understanding of this universal phenomenon. Experimentalists have long been faced with the phenomenon of chaos. However, the limitations of theoretical results due to the influence of the linear and quasi-linear structure of scientific thinking led to errors in interpretation of the observed results. Noise-like oscillations were taken to result either from the influence of fluctuations or from the huge number of degrees of freedom, or indeed from an instrumental malfunction.

Now the situation is different. Modern life more often requires us to quantify factors such as ultra-high densities, temperatures, pressures, speeds, population densities, and so on. And as we know, such considerations necessarily involve a fundamentally nonlinear approach to describe evolutionary processes. These processes are modeled and analyzed by computers, for which the nonlinearity of the model is no obstacle to detailed analysis. It has also been found that a chaotic mode of operation is rather the rule than the exception in such systems!

5.5 Strange Chaotic Attractors

The mathematical image of the steady-state regime of a dissipative dynamical system is associated with an *attractor*, an attracting limit set in phase space, to which all initial trajectories tend. If the steady-state regime is a stable equilibrium state, the attractor of the system is simply a fixed point. If the regime is a stable periodic motion, the attractor is a closed curve called a limit cycle. Attractors used to be thought of as the image of an exceptionally stable regime of system operation. Now we understand that the regime of deterministic chaos is also an attractor in the sense that it specifies a limited trajectory in a bounded region of phase space. However, such attractors display two significant differences: trajectories on it are non-periodic (they do not close) and the operating regime is unstable (small deviations from the regime tend to grow). It was these differences that led to the introduction of a new term. Thanks to the French scientist F. Takens, such attractors were referred to as *strange*. Now the term *chaotic* or *irregular* attractor is more often used.

What is the criterion for strangeness, or rather for chaoticity? Theoreticians have established that the instability of a trajectory is the main criterion for a chaotic attractor. In addition, the instability must be exponential. This means that a small perturbation $D(0)$ of the regime must increase exponentially in time:

$$D(t) = D(0)\exp(\lambda t), \quad \lambda = \lim_{t \to \infty} \frac{1}{t} \ln \frac{D(t)}{D(0)}, \quad (5.4)$$

where λ is the maximal Lyapunov exponent.

It turns out that a positive value of λ indicates not only the exponential instability of the oscillatory regime, but proves the presence of mixing in the system. If it is established that the analyzed regime has $\lambda > 0$, then we know that all the state coordinates are non-periodic in time, the power spectrum is continuous (the spectrum contains all the frequencies in a certain interval), and the autocorrelation function decays in time. Until recently, any behavior with the above characteristics was associated with a truly random process, but we know today that a process generated by deterministic laws may possess similar properties. This is basically why such processes are referred to as deterministic chaos.

Typical chaotic attractors possess geometric 'strangeness' and mixing. In other words, the complex dynamics of a mixing system gives rise to the geometric complexity of the corresponding attractor.

5.6 Strange Nonchaotic and Chaotic Nonstrange Attractors

Chaotic attractors as described above combine two key features: complex geometric structure (and as a consequence, a fractal metric dimension) and exponential instability of individual trajectories. It is these properties that are used as criteria for diagnosing the regimes of deterministic chaos.

However, regimes of complex dynamics are not exhausted by the above-described types of chaotic attractors. It has been shown that chaotic behavior in the sense of mixing and geometric 'strangeness' of an attractor may not be in one-to-one correspondence. Strange attractors with regard to geometry may be nonchaotic due to an absence of exponential instability in the phase trajectories. On the other hand, there are examples of mixing dissipative systems whose attractors are not strictly strange, that is, they are not characterized by a fractal structure and fractal metric dimension (see Chap. 10).

In other words, there are specific examples of dissipative dynamical systems whose attractors have the following properties:

1. The geometric structure of an attractor is regular in the sense of having integer metric dimension, but the individual phase trajectories are exponentially unstable on average.
2. The geometric structure of an attractor is complex, but the trajectories are asymptotically stable. There is no mixing.

The first type is called a *chaotic nonstrange attractor* (CNA). The second is referred to as a *strange nonchaotic attractor* (SNA).

The attractors of dynamical systems can thus be divided into two types: *regular* ones which include differentiable manifolds, i.e., equilibrium points, limit cycles, two-dimensional and multi-dimensional tori, and *irregular* ones which include all attractors having either the property of strangeness or the property of chaoticity, or indeed both, as is most often the case.

Below we give examples of chaotic nonstrange and strange nonchaotic attractors in two-dimensional maps. Strange nonchaotic attractors have also been observed in quasiperiodically driven differential systems. The minimal dimension of such a system reduced to an autonomous form is $N = 3$. Chaotic nonstrange attractors are very rarely discussed in the scientific literature. All the known cases are referred to reversible maps on a two-dimensional torus. Such a map can be considered as the Poincaré map arising in a flow section on a three-dimensional torus. The dimension of a differential system in which a three-dimensional torus can be realized is $N = 4$. However, the existence of a CNA in any specific system defined by differential equations has not yet been strictly determined.

5.6.1 Chaotic Nonstrange Attractors

Chaotic attractors that are nonstrange in terms of their geometry have been known for some time now, but are still not well understood. As an example of a dynamical system with a CNA, we consider the modified Arnold map. This is the well-known 'cat map' with addition of a nonlinear periodic term:

$$\begin{aligned} x_{n+1} &= x_n + y_n + \delta \cos(2\pi y_n) \,, \quad \text{mod } 1 \,, \\ y_{n+1} &= x_n + 2y_n \,, \quad \text{mod } 1 \,. \end{aligned} \tag{5.5}$$

When $\delta < 1/2\pi$, the map (5.5) is a diffeomorphism onto a torus. In other words, the map (5.5) is one-to-one (reversible) and maps a unit square (x_n, y_n) to itself. The map (5.5) is dissipative, i.e., an area element contracts at each iteration. This property is easily proved by calculating the Jacobian of the transformation (5.5), viz.,

$$J = \begin{vmatrix} 1 & 1 - 2\pi\delta \sin 2\pi y_n \\ 1 & 2 \end{vmatrix} \neq 0 \,, \quad \delta < \frac{1}{2\pi} \,. \tag{5.6}$$

The time average is $|J| < 1$. In this case the LCE spectrum signature is $+, -$, i.e., there is mixing.

It would seem that we are dealing with a typical chaotic strange attractor, but this is not so. The distinctive feature of the case considered here is that, despite the

Fig. 5.5 Chaotic nonstrange attractor in the modified Arnold map for $\delta = 0.15$

phase volume contraction, the motion of a representative point of the map (5.5) is ergodic. As $n \to \infty$, the point visits any element of the unit square which represents the full development of a two-dimensional torus. This is evidenced by the fact that the metric dimension of the attractor (Kolmogorov's capacity) is equal to 2. Although the density of the attractor points is nonuniform in the unit square, it vanishes nowhere. Thus, in spite of the contraction, the attractor of the system (5.5) is the whole unit square. In this sense, Arnold's attractor is not strange since its geometry is not fractal.

Despite the fact that the points cover the square almost entirely, as can be seen from the phase portrait of the attractor shown in Fig. 5.5, their density distribution is explicitly inhomogeneous! This inhomogeneity can be quantified using the information dimension $1 < D_I < 2$. For example, for $\delta = 0.05$, $D_I \approx 1.96$, and for $\delta = 0.10$, $D_I \approx 1.84$. In addition, as already mentioned, the capacity $D_C = 2.0$ (this is a rigorous result due to Y. Sinai). The inhomogeneity of the probability density distribution of the points on the attractor results from the fact that the values of all probability-metric dimensions of Arnold's attractor lie in the interval $1 < D < 2$. These dimensions take into account both geometric and dynamical properties of the attractor.

CNAs are found in a number of other maps on a torus. One can assume that ergodic chaotic motions are typical for diffeomorphisms on a torus. The existence of a CNA in such maps suggests that there may be flow (differential) systems in $\mathbb{R}^N$ ($N \geq 4$) which can sustain CNA regimes. However, so far, CNAs have not been detected in differential dynamical systems. In this connection particularly, the possible existence of a chaotic attractor on the surface of a three-dimensional torus embedded in a phase space of dimension $N \geq 4$ remains an open question.

5.6.2 Strange Nonchaotic Attractors

As has been shown above, in the case of chaotic nonstrange attractors, mixing may not lead to the geometric strangeness of an attractor. We now consider the possibility of realizing the opposite situation, when the system exhibits a complex non-periodic oscillatory regime which is asymptotically stable (without mixing), while the attractor is not regular from the standpoint of its geometric structure.

Examples of nonrobust strange nonchaotic attractors (SNAs) are easily identified. In fact, any strange chaotic attractor at the critical point of transition to chaos is an example of an SNA. The Lyapunov exponent is zero (there is no chaos) at the critical point. By definition, this attractor is an SNA. However, it is structurally unstable (nonrobust). From the physical point of view, one is interested in robust attractors which exist on a set of parameter values of nonzero measure and keep their structure in the presence of perturbations. Robust SNAs appear to exist in both differential and discrete dynamical systems.

SNAs are typical for quasiperiodically driven dynamical systems. It is appropriate to specify what we mean by the attractor of a nonautonomous system. Consider an autonomous dynamical system in $\mathbb{R}^N$, subjected to a periodic force with period $T_0 = 2\pi/\omega_0$. We analyze the Poincaré section through the period of the external force. At each time (for any n), we observe a set of points in the secant surface $t = nT_0$. In this case an attractor is the projection of this set obtained in secants for $n \to \infty$ on the initial secant surface for $n = 1$.

An SNA was first found and studied in the following map:

$$\begin{aligned} x_{n+1} &= \lambda \tan(x_n) \cos(2\pi\phi_n) \, , \\ \phi_{n+1} &= \omega + \phi_n \, , \qquad \text{mod } 1 \, . \end{aligned} \quad (5.7)$$

An irrational value of the parameter ω is most often chosen equal to the so-called *golden section*: $\omega = 0.5(\sqrt{5} - 1)$. The existence of an SNA in the map (5.7) has been rigorously proven for $\lambda > 1$ (Fig. 5.6). But SNAs have also been found when introducing a quasiperiodic force in the circle map, the logistic map, the Henon map, etc.

A number of features of SNAs serve as the basis for allocating these objects to a separate class.

5.6.3 Geometric Characteristics of SNAs

An attractor (e.g., in the phase plane) is formed by a curve of infinite length, which is nondifferentiable on a dense set of points. This curve, like the Peano curve, densely covers a part of the phase plane, so the metric dimension (the capacity) of an SNA is strictly equal to 2. But, unlike the map (5.5), we cannot in this case consider the attractor as part of the plane since the total measure of all the points belonging to the

5.6 Strange Nonchaotic and Chaotic Nonstrange Attractors

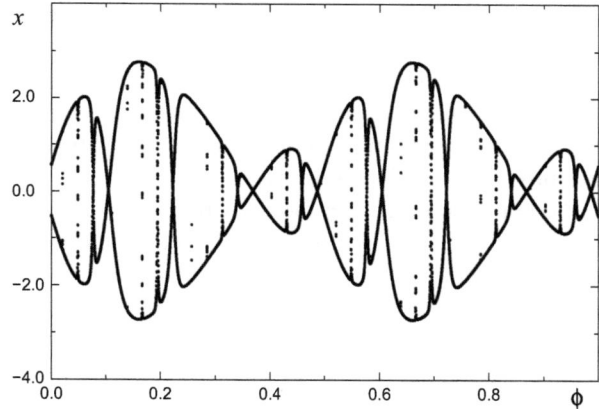

Fig. 5.6 Strange nonchaotic attractor in the map (5.7) for $\lambda = 1.5$

attractor is equal to zero. This fact is reflected in the information dimension $D_I = 1$, which corresponds to the line and not the plane. The LCE spectrum does not contain a positive exponent, so the Lyapunov dimension of an SNA is 1 (different definitions of set dimensions will be given in Chap. 10). Despite the integer metric dimension, an SNA exhibits structural self-similarity as a rule, and consequently also scaling properties. The properties listed above justify speaking of the 'strange' geometry of SNAs.

5.6.4 LCE Spectrum of SNAs

The system dynamics in the SNA regime is not chaotic, due to the absence of mixing. On average, there is no exponential instability of trajectories on the attractor. The LCE spectrum does not contain a positive exponent. The signature of the LCE spectrum of phase trajectories on an SNA is the same as for quasiperiodic motion. However, an SNA is not a quasiperiodic attractor because, in particular, the local (calculated over a finite time interval) maximal exponent of the LCE spectrum is positive. (It has been proven that the probability of the positive local maximal Lyapunov exponent differs from zero.)

5.6.5 Spectrum and Autocorrelation Function

The absence of mixing in the SNA regime leads to the absence, in the strict sense, of a continuous component in the power spectrum. At the same time, the spectrum of a trajectory on an SNA is not discrete. The SNA spectrum occupies an intermediate position between the discrete and continuous cases and has a special name: a singular-continuous spectrum. The peculiarity of the singular-continuous spectrum is that it includes a dense set of self-similarly structured δ-peaks and possesses the properties of fractals.

Since the spectrum of an SNA is not continuous, the autocorrelation function $\Psi(\tau)$ does not tend to zero as $\tau \to \infty$. For trajectories on an SNA, $\Psi(\tau)$ generally decays to a certain limiting nonzero level. In addition, $\Psi(\tau)$, as well as the spectrum, exhibits scale-invariant properties.

It should be noted that the SNA regime is extremely difficult to detect numerically, and this problem requires fine calculations using good modern computer facilities. Otherwise, the SNA regime cannot be distinguished from a quasiperiodic mode with a large number of combination frequencies in its spectrum.

5.7 Summary

In this chapter we have considered the peculiarities of dissipative dynamical systems in a qualitative manner and come to the following fundamental conclusions:

1. Differential dynamical systems with phase space dimension $N \geq 3$ can theoretically exhibit steady-state regimes of oscillations which are neither periodic nor quasiperiodic. They represent dynamical chaos.
2. A fundamental feature of chaotic oscillations is their instability, which leads to a sensitive dependence of the system dynamics on small perturbations.
3. The instability of a nonlinear system together with a limited energy of oscillations may cause mixing.
4. The presence of mixing leads to the need to introduce a statistical description of the dynamics of deterministic systems with chaotic attractors as the most convenient procedure.

The results listed above convince us that the regimes of deterministic nonlinear systems with strange chaotic attractors actually possess a number of specific properties, all of which are included in the notion of deterministic chaos.

Since in the general case the exponential instability of individual trajectories and the 'strange' geometry of an attractor are uniquely related, irregular attractors of a special type can be observed in some cases. There are regimes of chaotic (unstable) self-sustained oscillations which are associated with simple attractors in the geometric sense. These are the so-called chaotic nonstrange attractors (CNA). On the other hand, one can observe non-periodic oscillations which are stable according to Lyapunov and correspond to an attractor which is a strange geometric object. In this case we speak of strange nonchaotic attractors (SNA).

The phenomenon of deterministic chaos is described in many interesting books [1–35]. To get started, we would recommend the monographs [4, 5, 14, 15, 20, 23, 25, 30].

References

1. Afraimovich, V.S., Arnold, V.I., Il'yashenko, Yu.S., Shilnikov, L.P.: In: Dynamical Systems V, Encyclopedia of Mathematical Sciences. Springer, Heidelberg (1989)
2. Afraimovich, V.S., Nekorkin, V.I., Osipov, G.V., Shalfeev, V.D.: Stability, Structures and Chaos in Nonlinear Synchronization Networks. World Scientific, Singapore (1994)
3. Afraimovich, V.S., Shilnikov, L.P.: Strange attractors and quasiattractors. In: Barenblatt, G.I., Iooss, G., Joseph, D.D. (eds.) Nonlinear Dynamics and Turbulence, p. 1. Pitman, Boston (1983)
4. Anishchenko, V.S.: Dynamical Chaos – Models and Experiments. World Scientific, Singapore (1995)
5. Anishchenko, V.S., Astakhov, V.V., Neiman, A.B., Vadivasova, T.E., Schimansky-Geier, L.: Nonlinear Dynamics of Chaotic and Stochastic Systems. Springer, Berlin (2002)
6. Barreira, L., Pesin, Y.: Nonuniform hyperbolicity: dynamics of systems with nonzero Lyapunov exponents. In: Encyclopedia of Mathematics and Its Applications. Cambridge University Press, Cambridge (2007)
7. Berge, P., Pomeau, I., Vidal, C.G.: Order Within Chaos. Wiley, New York (1984)
8. Bohr, T., Jensen, M.H., Paladin, G., Vulpiani, A.: Dynamical System Approach to Turbulence. Cambridge University Press, Cambridge (1998)
9. Chen, G., Dong, X.: From Chaos to Order: Perspectives, Methodologies, and Applications. World Scientific, Singapore (1998)
10. Crownover, R.M.: Introduction to Fractals and Chaos. Jones and Barlett, London (1995)
11. Devaney, R.L.: An Introduction to Chaotic Dynamical Systems. Westview Press, Boulder (1989/2003)
12. Drazin, P.G.: Nonlinear Systems. Cambridge University Press, Cambridge (1992)
13. Eckmann, J.-P., Collet, P.: Iterated Maps as Dynamical Systems. Birkhauser, Basel (1980)
14. Glendinning, P.: Stability, Instability, and Chaos: An Introduction to the Theory of Nonlinear Differential Equations. Cambridge University Press, Cambridge (1994)
15. Guckenheimer, J., Holmes, P.: Nonlinear Oscillations, Dynamical Systems, and Bifurcations of Vector Fields. Springer, New York (1983)
16. Haken, H.: Synergetics: Introduction and Advanced Topics. Springer, Heidelberg (2004)
17. Hilborn, R.C.: Chaos and Nonlinear Dynamics. An Introduction for Scientists and Engineers. Oxford University Press, Oxford (2002/2004)
18. Jackson, E.A.: Perspectives of Nonlinear Dynamics, vols. 1, 2. Cambridge University Press, Cambridge (1989/1990)
19. Katok, A., Hasselblatt, B.: Introduction to the Modern Theory of Dynamical Systems. Cambridge University Press, Cambridge (1995)
20. Kapitaniak, T.: Chaotic Oscillators: Theory and Applications. World Scientific, Singapore (1992)
21. Kapitaniak, T.: Chaos for Engineers: Theory, Applications, and Control. Springer, New York (1998)
22. Lakshmanan, M., Murali, K.: Chaos in Nonlinear Oscillators: Controlling and Synchronization. World Scientific, Singapore (1996)
23. Lichtenberg, A., Lieberman, M.A.: Regular and Stochastic Motion. Springer, New York (1983)
24. Marek, M., Schreiber, I.: Chaotic Behaviour of Deterministic Dissipative Systems. Cambridge University Press, Cambridge (1991/1995)
25. Moon, F.C.M.: Chaotic and Fractal Dynamics: An Introduction for Applied Scientists and Engineers. Wiley, New York (1992)
26. Moon, F.C.M.: Chaotic Vibration: An Introduction for Applied Scientists and Engineers. Wiley, New York (2004)
27. Nicolis, G.: Introduction to Nonlinear Science. Cambridge University Press, Cambridge (1995)
28. Ogorzalek, M.J.: Chaos and Complexity in Nonlinear Electronic Circuits. World Scientific, Singapore (1997)
29. Ott, E.: Chaos in Dynamical Systems. Cambridge University Press, Cambridge (1993/2002)

30. Schuster, H.G.: Deterministic Chaos. Physik-Verlag, Weinheim (1988)
31. Schroeder, M.: Fractals, Chaos, Power Laws. Freeman, New York (1991)
32. Seydel, R.: Practical Bifurcation and Stability Analysis: From Equilibrium to Chaos. Springer, New York (1994/2009)
33. Shilnikov, L.P., Shilnikov, A.L., Turaev, D.V., Chua, L.O.: Methods of Qualitative Theory in Nonlinear Dynamics. World Scientific, Singapore (2001)
34. Thompson, J.M.T., Stewart, H.B.: Nonlinear Dynamics and Chaos. Wiley, New York (1986)
35. Zaslavsky, G.M.: Chaos in Dynamical Systems. Harwood Academic, New York (1985)

Chapter 6
From Order to Chaos: Bifurcation Scenarios (Part I)

6.1 Introduction

A dynamical system (DS) displays its nonlinear properties in different ways when system control parameters are varied. But as a rule, when the influence of nonlinearity increases, the dynamical regime becomes more complicated. Simple attractors in the phase space of a dissipative system are replaced by more complicated ones. Under certain conditions, nonlinearity can lead to the onset of dynamical chaos. Moving along a relevant direction in the parameter space, a sequence of bifurcations can be observed, resulting in the appearance of a chaotic attractor. Such typical bifurcation sequences are called *bifurcation mechanisms*, or *scenarios of the transition to chaos*.

The first scenario of transition to nonregular behavior was proposed by L.D. Landau in 1944 and independently by E. Hopf when attempting to explain the onset of turbulent behavior in fluid flow with increasing Reynolds number. The corresponding bifurcation mechanism was called the *Landau–Hopf scenario*. The latter involves a sequence of bifurcations (of the Neimark–Saker bifurcation type), each giving rise to a new incommensurate frequency, i.e., $\omega_1 \to \omega_1, \omega_2 \to \ldots \to \omega_1, \omega_2, \ldots, \omega_k$. As a result, a multi-frequency quasiperiodic regime appears, following a multi-dimensional torus in the phase space of a DS. If the number k of bifurcations is large enough and if fluctuations inevitably present in real systems are taken into account, the power spectrum of the process becomes broadband and similar to the spectrum of chaotic oscillations. However, multi-frequency oscillations, even in the presence of noise, can remain stable according to Lyapunov. Mixing in such a system is connected only with noise and not with the deterministic evolution operator. Thus, the Landau–Hopf scenario does not necessarily presuppose the transition to chaotic dynamics and, strictly speaking, is not a scenario for chaos development. Besides, this scenario cannot explain the onset of oscillations with a continuous spectrum in low-dimensional systems.

In the early 1970s, the idea of developing turbulence through quasiperiodic oscillations was revised by D. Ruelle, F. Takens, and S. Newhouse. They connected the turbulent behavior with dynamical chaos and were the first to introduce the notion of a *strange attractor* (SA) as the mathematical image of chaos in a DS. It was also shown that a strange attractor can arise in low-dimensional systems ($N \geq 3$).

By now, three typical bifurcation scenarios of chaos development in dissipative systems had been revealed and studied. They can already be realized in a three-dimensional phase space. In addition, each of them has *universality* properties, i.e., certain general features, independent of the particular form of the evolution operator. These scenarios will be discussed in this and the next chapter. Furthermore, in the quasiperiodic framework, we shall focus on the peculiarities of the development of nonregular dynamics in systems with a robust ergodic two-dimensional torus.

6.2 Transition to Chaos via a Cascade of Period-Doubling Bifurcations: Feigenbaum Universality

Many different dynamical systems, ranging from the simplest maps to distributed media, demonstrate the transition to chaos through a *cascade of period-doubling bifurcations*, in both numerical and full-scale experiments. Since period-doubling is a codimension-one bifurcation, this route assumes a one-parameter analysis and works as follows. Let α be the control parameter of a DS, and suppose that, at some $\alpha = \alpha_0$, the system has a stable limit cycle C with period T_0. As α is increased to α_1, the supercritical period-doubling bifurcation takes place, resulting in the birth of a stable limit cycle $2C$ with period $2T_0$. At $\alpha = \alpha_k, k = 1, 2, 3, \ldots$, the system exhibits an infinite sequence of period-doubling bifurcations of cycles $2^k C$. The power spectrum has new components that appear at subharmonics of the fundamental frequency $\omega_0 = 2\pi/T_0$, whence the period-doubling bifurcation sequence is sometimes called the *subharmonic cascade*. When $k \to \infty$, the bifurcation points α_k accumulate to some critical value $\alpha = \alpha_{cr}$, at which the period becomes infinite and the power spectrum continuous. For $\alpha > \alpha_{cr}$, there occur aperiodic oscillations that are unstable according to Lyapunov. These oscillations follow a chaotic attractor in the system phase space.

Figure 6.1 exemplifies the changes occurring in the oscillator with inertial nonlinearity (the Anishchenko–Astakhov oscillator, see Chap. 11) during the period-doubling route to chaos. The oscillator is governed by the following system of equations:

$$\dot{x} = mx + y - xz ,$$
$$\dot{y} = -x , \quad\quad\quad (6.1)$$
$$\dot{z} = -gz - g\Phi(x) ,$$

where $\Phi(x) = x^2$ for $x \geq 0$ and $\Phi(x) \equiv 0$ for $x < 0$.

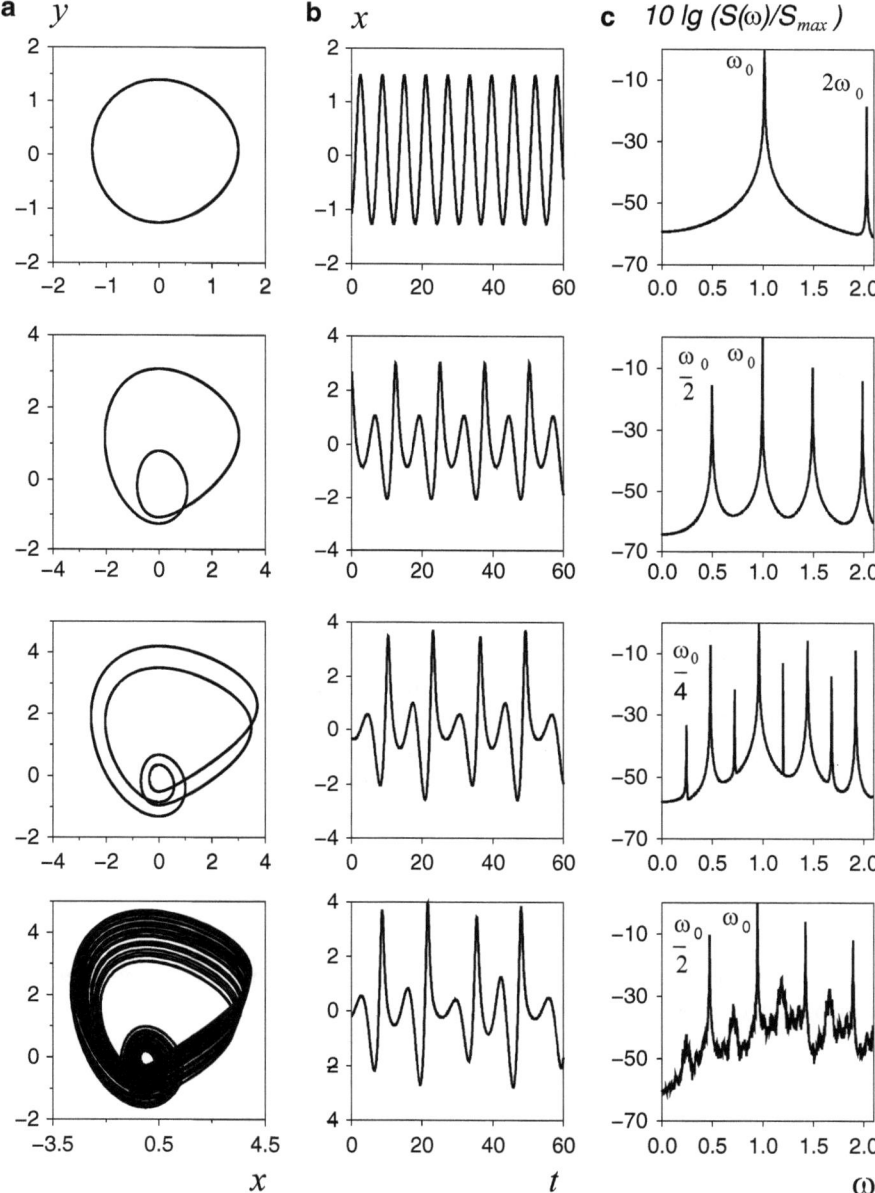

Fig. 6.1 Cascade of period-doubling bifurcations in (6.1). (**a**) Projections of phase trajectories, (**b**) waveforms $x(t)$, and (**c**) power spectra for period $2^k T_0$ cycles, $k = 1, 2, 3$, and for a chaotic attractor

It has been established experimentally that, in all continuous-time systems, a chaotic attractor arising through the period-doubling route has the fractal dimension $2 < d < 3$, and its section resembles a horseshoe in shape. In this case the Poincaré map arising in the secant surface can be modelled by the simple and well-known Henon map:

$$x_{n+1} = 1 + y_n - ax_n^2 ,$$
$$y_{n+1} = bx_n .$$
(6.2)

The map (6.2) is reversible and contracts a square element for $b < 1$. As one of the system control parameters a or b is varied, the Henon map illustrates the period-doubling route to chaos. If the square element is strongly contracted, then the transverse Cantor structure of the horseshoe[1] can be neglected, and points in the map can be considered to lie on one smooth bent curve. Introducing a new coordinate along this curve one can arrive at an irreversible model map of an interval to itself. This is defined by a smooth first return function with a single extremum. The map stretches an interval element and then 'folds' it into the same interval. Since the first return function is assumed to be smooth everywhere, then, as with the horseshoe, there exists a region near the extremum for which stretching is absent. The existence of such a region causes the birth of stable periodic orbits inside a chaotic zone, the so-called *stability* or *periodic windows*. Periodic windows can be eliminated, provided that the map stretches everywhere. An example is the tent map with a break at the extremum point. However, in such models, periodic windows disappear together with a period-doubling cascade. The theory of the period-doubling route to chaos was developed by M. Feigenbaum on the basis of one-dimensional model maps, so this bifurcation mechanism is known as the *Feigenbaum scenario*.

The simplest model for studying the Feigenbaum scenario is the *logistic map*

$$x_{n+1} = f(x_n) = r - x_n^2 ,$$
(6.3)

where r is a parameter. The logistic map may also be rewritten in another form which reduces to (6.3) by a linear change of variables, e.g.,

$$x_{n+1} = rx_n(1 - x_n) , \qquad x_{n+1} = 1 - rx_n^2 .$$

Consider how the map (6.3) evolves as the parameter r increases. It has a fixed point x_0, also called a period-1 cycle, or 1-cycle, with coordinate $x_0 = -1/2 + \sqrt{r + 1/4}$. The fixed point x_0 is stable in the interval $r \in [-1/4, 3/4]$, and its multiplier μ_1 is equal to $-2x_0$. At $r = r_1 = 3/4$, we have $\mu_1 = -1$, and the first period-doubling bifurcation takes place, giving rise to a stable period-2 cycle. The latter consists of two points $x_{1,2} = 1/2 \pm \sqrt{r - 3/4}$. The 2-cycle has a multiplier

---

[1] The Cantor set and Smale's horseshoe are described in detail in Chap. 10.

6.2 Transition to Chaos via a Cascade of Period-Doubling Bifurcations:...

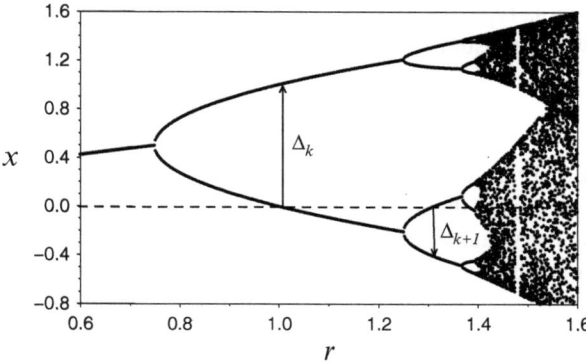

Fig. 6.2 Phase-parameter diagram of dynamical regimes for the map (6.3)

$$\mu_2 = f'_x(x_1) f'_x(x_2) = 4(1-r)$$

and is stable in the range $r \in [3/4, 5/4]$. At $r = r_2 = 5/4$, the second period-doubling bifurcation occurs, and a period-4 cycle is created, etc. By this means, the map (6.3) generates the sequence of period-doubling bifurcation values of the parameter r, namely, $r_1 = 3/4$, $r_2 = 5/4$, $r_3 \approx 1.368\,099$, $r_4 \approx 1.394\,046$, $r_5 \approx 1.399\,637, \ldots$, with the accumulation point $r_{cr} \approx 1.401\,15\ldots$. When $k \to \infty$, the convergence rate of the bifurcation values tends to a finite limit:

$$\delta = \lim_{k \to \infty} \frac{r_{k+1} - r_k}{r_{k+2} - r_{k+1}} = 4.669\,201\ldots. \tag{6.4}$$

Figure 6.2 shows a phase-parameter diagram for the logistic map (6.3), which appears to be typical for systems exhibiting the period-doubling route to chaos. Such a diagram is called *Feigenbaum's tree*. The diagram readily illustrates the scale splitting of the dynamical variable as well as *scaling* properties, i.e., *scale invariance*, when the same image element reproduces itself under an arbitrary small change in scale. Denoting the distances between similar points of the tree branches by Δ_k (see Fig. 6.2), we can introduce the scaling factors $a_k = \Delta_k / \Delta_{k+1}$, which in the limit as $k \to \infty$ converge to

$$a = \lim_{k \to \infty} \frac{\Delta_k}{\Delta_{k+1}} = -2.502\,9\ldots. \tag{6.5}$$

It has been shown numerically that the values of δ and a do not depend on the particular form of the map. However, the map must be unimodal, i.e., it must have a single extremum, and the extremum must be quadratic.

M. Feigenbaum explained the universal character of the quantitative regularities of the period-doubling route and created the *universality theory*. To analyze maps of the logistic parabola type, he applied a *renormalization group (RG) method*, which works as follows. We assume that, at the critical point $r = r_{cr}$, we have the map

$$x_{n+1} = f_0(x_n), \qquad (6.6)$$

where f_0 is an arbitrary unimodal function with a quadratic extremum at $x_n = 0$, and $f_0(0) = 1$. Applying the map (6.6) twice, we obtain a map $x_{n+1} = f_0(f_0(x_n))$. We then rescale the variable $x \to x/a_0$ so that the new map is also normalized to unity at the origin, i.e., $a_0 = 1/f_0(f_0(0))$. Denoting the new map by

$$x_{n+1} = f_1(x_n) = a_0 f_0(f_0(x_n/a_0)),$$

we repeat this procedure many times and hence derive the *renormalization group equation*

$$f_{i+1}(x) = a_i f_i(f_i(x/a_i)), \qquad (6.7)$$

where $a_i = 1/f_i(f_i(0))$. Due to self-similarity, the following limits exist at the critical point:

$$\lim_{i \to \infty} f_i(x) = g(x), \qquad \lim_{i \to \infty} a_i = a. \qquad (6.8)$$

The function $g(x)$ is a fixed point of the functional *Feigenbaum–Cvitanovic equation*:

$$\hat{T} g(x) = a g(g(x/a)) = g(x), \qquad (6.9)$$

where $\hat{T}$ is a doubling operator and $a = 1/g(g(0))$.

For the critical point corresponding to the period-doubling transition to chaos, the boundary conditions for (6.9) are $g(0) = 1$ and $g'_x(0) = 0$. The function $g(x)$ is *universal*, because it does not depend on a particular form of the original map and is defined by the order of the extremum alone. Taking into account the renormalization of the variable x, this function yields an asymptotic form of the map at the critical point, which is given by 2^i applications of the evolution operator as $i \to \infty$. The constant a in the fixed point equation is also universal. The fixed point solution of (6.9) was found numerically by Feigenbaum under the assumption of a quadratic extremum and the above-stated boundary conditions. It is defined by

$$g(x) \approx 1 - 1.527\,633\,0 x^2 + 0.104\,815\,2 x^4 + 0.026\,705\,7 x^6 - 0.003\,527\,4 x^8$$
$$+ 0.000\,081\,6 x^{10} + 0.000\,025\,4 x^{12} - 0.000\,002\,7 x^{14}. \qquad (6.10)$$

The *universal Feigenbaum constant a* is thus equal to $-2.502\,907\,876\ldots$

If the evolution operator $f(x_n)$ is slightly perturbed due to a small deviation of the parameter from its critical value, the doubling operator $\hat{T}$ and the function $g(x)$

6.2 Transition to Chaos via a Cascade of Period-Doubling Bifurcations:...

turn out to be perturbed as well. Having linearized $\hat{T}$ at the point $g(x)$ for $r = r_{cr}$, one can obtain an operator $\hat{L}_g$ which defines the behavior of perturbations as well as the equation for eigenfunctions $h(x)$ and eigenvalues ρ of the linearized operator:

$$\hat{L}_g h(x) = a\Big[g'(g(x/a))h(x/a) + h(g(x/a))\Big] = \rho h(x) \,. \quad (6.11)$$

The behavior of perturbations will be mainly determined by the eigenvalues exceeding unity in modulus. In the case of the quadratic extremum, there is one such value corresponding to an unremovable component of the perturbations. This value defines a second *universal Feigenbaum constant*, $\rho_1 = \delta = 4.669\,201\,609\,1\ldots$

Maps with a nonquadratic extremum are characterized by different values of the universal constants. However, computer and physical experiments performed for a variety of continuous-time systems, including distributed ones, with the Feigenbaum scenario to chaos have shown that the scaling factor a and the convergence rate of the bifurcation sequence δ coincide within experimental error with the theoretical values found for maps with a quadratic extremum. Evidently, the typical case is when a map generated by the evolution operator of a continuous-time system in the neighborhood of the critical point may be approximated by a one-dimensional map with a quadratic extremum. Other cases are considered to be atypical.

The Feigenbaum scenario is universal and its universality manifests itself in the behavior of spectral amplitudes of subharmonics which appear with each period doubling. The ratio of amplitudes of subharmonics ω_0/k and $\omega_0/(k-1)$ in the limit as $k \to \infty$ is a universal constant ($\approx -13.5\,\text{dB}$).

At $r = r_{cr}$, the logistic map (6.3) generates a limit set of points called the *Feigenbaum attractor*. It is strange since its capacity dimension is fractal,[2] but nonchaotic because the Lyapunov exponent λ is zero at the critical point. The universal properties also hold in a supercritical region $r > r_{cr}$, for both model maps and continuous-time systems. The Lyapunov exponent, having become positive beyond the critical point, grows according to the universal law

$$\lambda \sim \varepsilon^\gamma, \qquad \gamma = \frac{\ln 2}{\ln \delta} \approx 0.449\,8\,, \quad (6.12)$$

where $\varepsilon = r - r_{cr}$ is the *supercriticality parameter*. By analogy with the theory of second-order phase transitions, the coefficient γ is called the *critical index* of the transition to chaos. Figure 6.3 shows numerically computed Lyapunov exponents as a function of the control parameter for the logistic map and for the Anishchenko–Astakhov oscillator.

Beyond the critical point, systems with period doubling demonstrate a cascade of *merging* bifurcations. This kind of bifurcation consists in merging parts, or bands, of a chaotic attractor, which are visited by a phase point in a certain order. Each attractor-band-merging bifurcation is accompanied by the disappearance of

---

[2] Fractal dimensions of sets with a complex geometric structure will be discussed in Chap. 10.

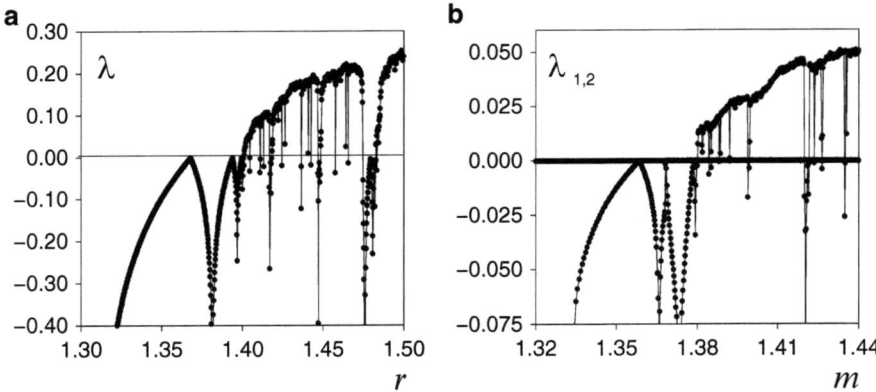

Fig. 6.3 Lyapunov exponents versus the control parameter (**a**) for the logistic map and (**b**) for the Anishchenko–Astakhov oscillator with $g = 0.3$

appropriate subharmonics in the power spectrum. Phase portraits and corresponding power spectra are shown in Fig. 6.4 for the band-merging bifurcation cascade observed in the system (6.1). For a one-dimensional map, the band-merging bifurcation looks like the merging of neighboring intervals which are filled with points of a chaotic set. Let $\bar{r}_k$ denote the parameter values corresponding to the merging bifurcations, where the index $k = 1, 2, \ldots$ increases as one approaches the critical point from right to left. 2^i-periodic orbits exist on intervals to the left of the critical point, whereas 2^i-band chaotic attractors are realized in the range of the parameter r to the right of r_{cr}. Thus, the r-axis is divided into two ranges which appear to be symmetric with respect to the critical point. Segments of multi-band chaotic sets at relevant points of each interval possess a property of similarity, with scaling factors tending to the universal constant a. The values $\bar{r}_k$ accumulate to the critical point at a rate equal to the universal constant δ.

Apart from chaotic trajectories, in a supercritical region, the logistic map has a set of periodic orbits with different periods. In 1964, A.N. Sharkovsky established a hierarchy for cycles of a smooth noninvertible map of an interval. A cycle of period M is considered to be more complicated than a cycle of period N if the existence of the N-cycle follows from the existence of the M-cycle. Their periods are said to be in a ratio of order $M \to N$. In accordance with Sharkovsky's theorem, this ratio arranges cycles in a certain order, the so-called *Sharkovsky series*:

$$3 \to 5 \to 7 \to 9 \to \ldots \to 3 \cdot 2 \to 5 \cdot 2 \to 7 \cdot 2 \to 7 \cdot 2 \to \ldots \to 3 \cdot 2^2$$
$$\to 5 \cdot 2^2 \to 7 \cdot 2^2 \to 9 \cdot 2^2 \to \ldots \to 2^3 \to 2^2 \to 2 \to 1. \qquad (6.13)$$

A 3-cycle appears to be the most complicated in the sense of Sharkovsky. Its existence implies the existence of cycles of arbitrary period. Independently, T. Li and J.A. Yorke proved an analogous result in 1975. In their paper entitled *Period*

6.2 Transition to Chaos via a Cascade of Period-Doubling Bifurcations:... 101

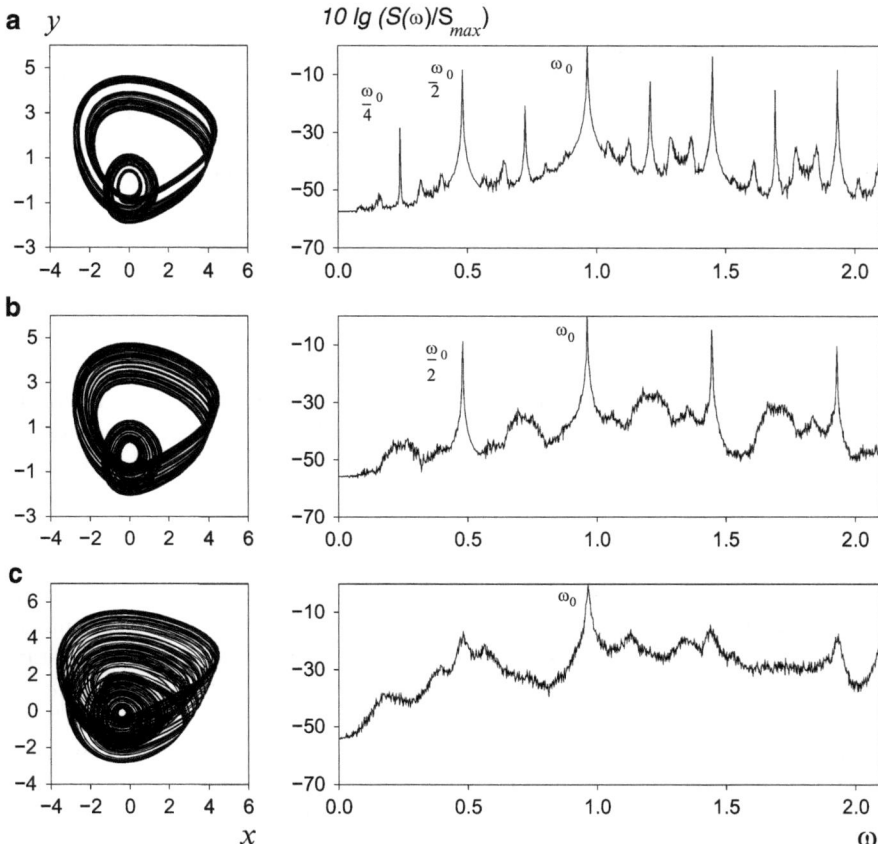

Fig. 6.4 Phase portraits and power spectra under the attractor-band-merging bifurcations in the Anishchenko–Astakhov oscillator (6.1). (**a**) A four-band chaotic attractor, (**b**) a two-band chaotic attractor, and (**c**) a single-band chaotic attractor. ω_0 is the basic spectrum frequency corresponding to the period T_0 of the generating cycle

three implies chaos, they also proved that a map with a 3-cycle can be characterized by the presence of chaotic sets.

However, neither Sharkovsky's theorem nor Li and Yorke's theorem tells us about the stability of cycles. Stability regions, or *periodic windows*, of cycles with different periods are arranged in a supercritical region according to the sequence: 6, 5, 3, 6, 5, 6, 4, 6, 5, 6, ... As a rule, the window's width and bifurcation parameter values corresponding to the window boundaries are different for different dynamical systems. However, the regularity of stability window emergence with increasing supercriticality is so universal that it does not even depend on the order of the extremum of the first return function. The widest periodic window corresponds to a 3-cycle generated by a tangent bifurcation. As the system parameter increases, the doubling process of the 3-cycle occurs, resulting in the onset of chaos. Cycles of higher periodicity arise and evolve in periodic windows in a similar manner. Generally speaking, in the supercritical region one can find a stability window of a certain cycle in an arbitrarily small neighborhood of any parameter value. The cycle can have a very high period, and its stability window can be so narrow that the cycle cannot be detected even in numerical experiments. However, this fact implies that a chaotic attractor arising via the period-doubling cascade is nonrobust towards small perturbations.

6.3 Crisis and Intermittency

As our knowledge of dynamical chaos has developed, it has been established that the transition from regular oscillations to chaos may occur suddenly as a result of the bifurcation alone. Such a mechanism for chaos onset is said to be *hard* and is accompanied by the phenomenon of *intermittency*. The latter is a regime in which chaotic behavior (the *turbulent phase*), appearing immediately after crossing the border into chaos, is interspersed in an intermittent fashion with a periodic type of behavior (the *laminar phase*). Figure 6.5 shows a typical oscillation process $x(t)$ for intermittency. The abrupt transition to chaos and the intermittency phenomenon were first considered in the work of I. Pomeau and P. Manneville. The corresponding bifurcation mechanism for chaos onset was thus called the *Pomeau–Manneville scenario*.

In the intermittency route, the sole bifurcation of a periodic regime may cause drastic qualitative changes to the structure of phase space, but also to the structure of the basin of attraction of an attractor. Such attractor bifurcations are called *crises*. Typical crises of periodic regimes (limit cycles) are related to certain kinds of local codimension-one bifurcations, namely, tangent (saddle-node), subcritical period-doubling, and subcritical torus birth bifurcations (the Andronov–Hopf bifurcation in the Poincaré map). In the case of tangent bifurcation, a stable limit cycle disappears by merging with a saddle cycle. In the two other situations, the limit cycle still exists after a bifurcation, but becomes unstable, i.e., a saddle cycle.

Suppose that, for $\alpha < \alpha_{cr}$, a system has an attractor in the form of a limit cycle C. When any of the aforementioned bifurcations takes place at $\alpha = \alpha_{cr}$, the attractor C disappears. For $\alpha > \alpha_{cr}$, phase trajectories coming from the close vicinity of the former cycle C must fall on another attractor which either already exists in the system for $\alpha < \alpha_{cr}$ or is generated by the bifurcation. Let us assume that the system already possesses one more attractor. In this case the bifurcation results in a simple transition from one attractor to another. Intermittency does not arise, even if the new regime is chaotic. This fact can be explained as follows. The chaotic attractor is not created through a limit cycle crisis and does not capture the local neighborhood of the cycle C. Phase trajectories leave this region and never return.

The question is: what are the conditions under which the limit cycle crisis results in the appearance of intermittent chaos? Evidently, this situation may happen when, at the bifurcation point $\alpha = \alpha_{cr}$, there already exists a chaotic set which becomes

6.3 Crisis and Intermittency

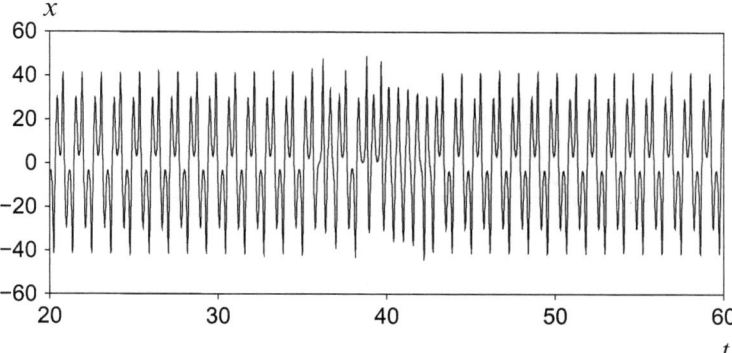

Fig. 6.5 Intermittency in the Lorenz system $\dot{x} = \sigma(y - x)$, $\dot{y} = -xy + rx - y$, $\dot{z} = xy - bz$ when $r = 166.1$, $\sigma = 10$, $b = 8/3$

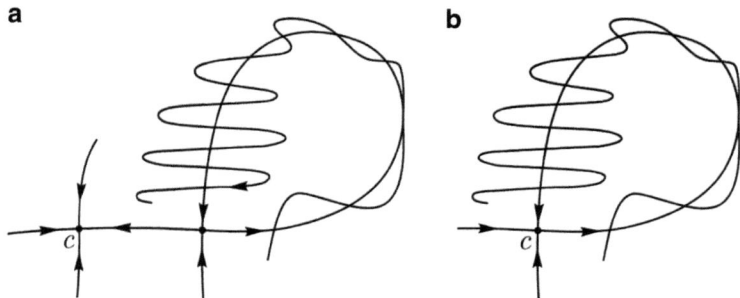

Fig. 6.6 Qualitative picture of the Poincaré section for the tangent bifurcation of stable and saddle cycles, leading to the emergence of chaos via intermittency: (**a**) before the bifurcation and (**b**) at the bifurcation point

attracting for $\alpha > \alpha_{cr}$ and encloses the local vicinity of the cycle C, so that a phase trajectory moving on the chaotic attractor returns to this vicinity from time to time. This kind of system behavior is realized whenever a saddle limit set involved in the cycle C crisis possesses a homoclinic structure.

As an example, Fig. 6.6 illustrates the tangent bifurcation of cycles, leading to chaotic intermittency. The saddle cycle has a pair of robust homoclinic orbits. At the bifurcation point $\alpha = \alpha_{cr}$, a nonrobust saddle-node orbit is created with a homoclinic structure in its vicinity. Phase trajectories move away from it and approach it along double-asymptotic homoclinic curves (they correspond to the intersection points of the manifolds in the section shown in Fig. 6.6). For $\alpha > \alpha_{cr}$, the nonrobust closed orbit disappears and the non-attracting homoclinic structure becomes attracting. As a result, a chaotic attractor emerges in the system phase space. Trajectories lying on this attractor are concentrated in the region of the former saddle-node orbit and repeat the motion on it for a long period of time. This motion testifies to a laminar phase of intermittent chaos.

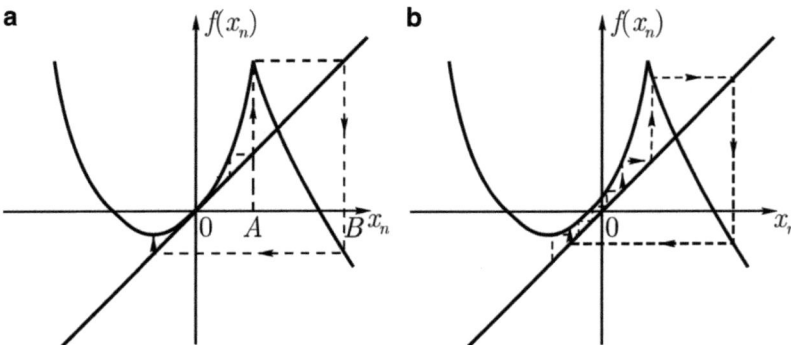

Fig. 6.7 The map modeling type-I intermittency (**a**) at the bifurcation point and (**b**) immediately after the bifurcation

The most typical intermittency for a wide variety of dynamical systems relates to the tangent bifurcation of cycles. This phenomenon was discovered and studied much earlier than the other types of intermittency and is called *type-I intermittency*. To analyze its properties, the following one-dimensional model map is usually used:

$$x_{n+1} = f(x_n) = \varepsilon + x_n + \beta |x_n|^p + \text{'return'}. \tag{6.14}$$

Here, the parameter ε denotes the supercriticality $(\alpha - \alpha_{\text{cr}})$ of the system, since in (6.14) a tangent bifurcation occurs at $\varepsilon = 0$. Here p is an integer which defines the order of an extremum of the first return function. The return of a phase point to a bounded interval of x values can proceed in different ways. For example, for the map presented in Fig. 6.7, the branch of the return function graph on the interval AB serves to return the phase point. Figure 6.7a shows the map at the moment of tangent bifurcation, $\varepsilon = 0$. The dashed lines constructed using Lamerei's diagram indicate a double-asymptotic trajectory of a saddle node. The map displayed in Fig. 6.7b corresponds to the case $\varepsilon > 0$. In the neighborhood of the former fixed point, a so-called *channel* opens up, along which the phase point moves for a long period of time. This motion reflects the laminar phase of intermittency. When the point leaves the channel, a turbulent phase is observed, during the course of which the point must reach the interval AB, with the latter providing its return to the channel.

The study of maps of the form (6.14) reveals certain quantitative features of type-I intermittency, such as the way the average duration of the laminar phase depends on the supercriticality. These regularities are universal in the sense that they do not depend on the particular form of the map and are determined by the order of an extremum p. For a typical case $p = 2$, these features are in good agreement with the results of numerical and experimental investigations of type-I intermittency in continuous-time systems.

The RG method has been applied to study this type of intermittency. Let us consider a map at the critical point and restrict ourselves to the interval $x_n \in [0, 1]$, where the map is defined by a monotonic function of the form $x_{n+1} = f_0(x_n)$,

6.3 Crisis and Intermittency

with $f_0(0) = 0$ and $f'_0(0) = 1$. Repeating all the arguments as for the Feigenbaum scenario, one can derive the same Feigenbaum–Cvitanovic equation (6.9), but with different boundary conditions: $g(0) = 0$ and $g'(0) = 1$. The RG method enables us to determine theoretically the asymptotic behavior of the average duration of the laminar phase, which reads

$$T_1 \sim \varepsilon^{-\nu}, \qquad \nu = \frac{p-1}{p}. \tag{6.15}$$

For $p = 2$, we have $T_1 \sim 1/\sqrt{\varepsilon}$, which concurs well with numerical results.

As already mentioned, other types of intermittency are associated with a subcritical Andronov–Hopf bifurcation in the Poincaré section and with subcritical period doubling. These are called *type-II* and *type-III* intermittency, respectively. Type-II intermittency is modeled by the following map of the plane, defined in polar coordinates:

$$\begin{aligned} r_{n+1} &= (1+\varepsilon)r_n + \beta r_n^3 + \text{'return'}, \\ \phi_{n+1} &= \phi_n + \Omega, \quad \text{mod } 1. \end{aligned} \tag{6.16}$$

For $\varepsilon = 0$, the map displays the subcritical Andronov–Hopf bifurcation in which an unstable invariant circle and a stable focus merge. The unstable invariant circle corresponds to a saddle torus in a continuous-time system with dimension $N \geq 4$. For type-II intermittency the asymptotic behavior of the average duration of the laminar phase is defined by

$$T_1 \sim \frac{1}{\varepsilon}, \tag{6.17}$$

where $\varepsilon = \alpha - \alpha_{\text{cr}}$ is the supercriticality parameter.

Type-III intermittency can be described by a one-dimensional model map in the form

$$x_{n+1} = -(1+\varepsilon)x_n - \beta x_n^2 + \text{'return'}. \tag{6.18}$$

For $\varepsilon = 0$, the subcritical period-doubling bifurcation of a period-1 cycle occurs in the map. The average duration of the laminar phase is evaluated approximately by the same ratio as in the case of type-II intermittency.

Chapter 7
From Order to Chaos: Bifurcation Scenarios (Part II)

7.1 Route to Chaos via Two-Dimensional Torus Destruction

According to the Ruelle–Takens–Newhouse scenario, the transition from quasiperiodicity to chaos occurs after the third frequency birth, when Lyapunov-unstable chaotic trajectories appear on a three-dimensional torus. However, the study of particular dynamical systems has shown that the appearance of chaos following the destruction of two-frequency quasiperiodic motion is also a typical scenario of the transition to chaos. According to this route, a two-dimensional (2D) torus T^2 in phase space is destroyed and trajectories fall in a set with fractal dimension $2 + d$, $d \in [0, 1]$. This set is created in the vicinity of T^2 and is thus called *torus chaos*. Such a route may be thought of as a special case of the quasiperiodic transition to chaos.

In contrast to the Feigenbaum scenario, the transition $T^2 \rightarrow$ strange attractor (SA) requires a two-parameter analysis. This is associated with the fact that the character of quasiperiodic motion depends on a winding number θ, which determines the ratio of the basic oscillation frequencies. If the frequencies are rationally related, i.e., the winding number θ has a rational value, resonance takes place on a torus, and consequently, periodic oscillations are realized. If the frequencies become irrationally related, the motion on the torus will be ergodic. The transition from the torus T^2 to chaos at a fixed winding number can be realized only by simultaneously controlling at least two system parameters. On the torus birth line, specified by the bifurcation condition $\mu_{1,2} = \exp(\pm i\phi)$, where $\mu_{1,2}$ is a pair of complex conjugate multipliers of a limit cycle, the winding number is defined as $\theta = \phi/2\pi$. Resonance regions on the plane of two control parameters have a tongue-like shape and originate from the relevant points on the torus birth line. Such regions are called *Arnold tongues* (in honor of V.I. Arnold, who studied the structure of resonance regions). Each direction of the paths on the parameter plane is characterized by its own sequence of bifurcations associated with the appearance and disappearance of different resonances on the torus.

7.1.1 Two-Dimensional Torus Breakdown Theorem

The results obtained by mathematicians in the framework of the qualitative theory of dynamical systems play an important role in understanding the mechanisms which lead to the destruction of T^2 and to the birth of torus chaos. L.P. Shilnikov and V.S. Afraimovich have proved the theorem on two-dimensional torus T^2 breakdown and indicated possible routes to chaotic dynamics.

Consider an N-dimensional DS ($N \geq 3$):

$$\dot{\mathbf{x}} = \mathbf{F}(\mathbf{x}, \boldsymbol{\alpha}), \qquad (7.1)$$

where components of the vector function $\mathbf{F}$, $j = 1, 2, \ldots, N$ belong to the smoothness class C^k, $k \geq 3$, and $\boldsymbol{\alpha}$ is the system parameter vector. We make the following assumptions:

- Suppose that, at $\boldsymbol{\alpha} = \boldsymbol{\alpha}_0$, a smooth attracting torus $T^2(\boldsymbol{\alpha}_0)$ exists in some region of the phase space of the system (7.1) and has a robust structure consisting of an even number of stable and saddle cycles. The torus T^2 is formed by unstable manifolds of saddle cycles. Assume for simplicity that there is a stable cycle $C^{\mathrm{st}}(\boldsymbol{\alpha}_0)$ and a saddle cycle $C^{\mathrm{sd}}(\boldsymbol{\alpha}_0)$. Then we have $T^2(\boldsymbol{\alpha}_0) = W^{\mathrm{u}}(\boldsymbol{\alpha}_0) \bigcup C^{\mathrm{st}}(\boldsymbol{\alpha}_0)$, where $W^{\mathrm{u}}(\boldsymbol{\alpha}_0)$ is the unstable manifold of the saddle cycle. The Poincaré section of a resonance torus is shown in Fig. 7.1a. If we assume that the invariant torus does not exist at $\boldsymbol{\alpha} = \boldsymbol{\alpha}_1$, then for the continuous curve $\boldsymbol{\alpha}(s)$, where $s \in [0, 1]$, $\boldsymbol{\alpha}(0) = \boldsymbol{\alpha}_0$, $\boldsymbol{\alpha}(1) = \boldsymbol{\alpha}_1$, there exists a value $s = s^*$ such that at $\boldsymbol{\alpha}(s^*)$ the torus T^2 is destroyed and no longer exists in the system (7.1), at least for some values of s arbitrarily close to s^*, $s > s^*$.
- For all $0 \leq s < s^*$, let the attracting set of (7.1) coincide with the torus $T^2(\boldsymbol{\alpha}(s))$.
- For $s > s^*$, assume that the unstable manifold $W^{\mathrm{u}}(\boldsymbol{\alpha}(s))$ of the saddle cycle does not contain periodic orbits different from C^{st} and C^{sd}.

These assumptions imply the *theorem on two-dimensional torus breakdown* which asserts the following three possible mechanisms for T^2 destruction:

- Loss of cycle C^{st} stability,
- Emergence of a homoclinic tangency between the unstable and stable manifolds W^{u} and W^{s}, respectively, of the saddle cycle C^{sd}, or
- A tangent bifurcation of C^{st} and C^{sd} on T^2.

Before being destroyed, T^2 loses its smoothness for $s > s^{**}$, i.e., $T^2(\boldsymbol{\alpha}(s > s^{**}))$ is *homeomorphic*, but not *diffeomorphic* to the torus.

Figure 7.1b shows a sketch of the qualitative bifurcation diagram in the two parameter plane (α_1, α_2). The directions indicated in Fig. 7.1b by A, B, C correspond to the three routes to destruction of the resonance torus. In the diagram, the curve l_0 corresponds to the bifurcation of T^2 birth. The phase-locking region is bounded by the two bifurcation curves l_1 related to a tangent bifurcation of C^{st} and C^{sd} on T^2. On the curve l_2, the resonance cycle C^{st} loses its stability inside the synchronization

7.1 Route to Chaos via Two-Dimensional Torus Destruction

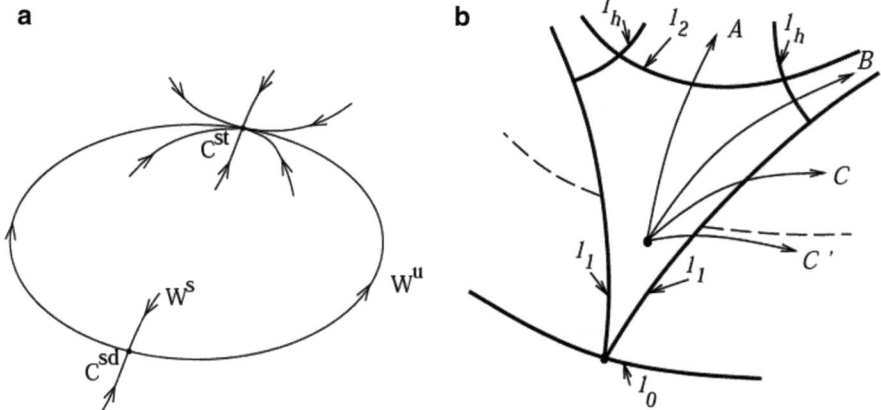

Fig. 7.1 (a) Section of the resonance torus T^2 and (b) qualitative bifurcation diagram for the breakdown of T^2. Routes of torus destruction are indicated by A, B and C

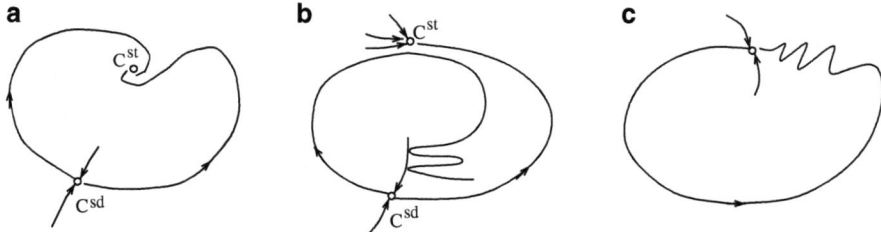

Fig. 7.2 Qualitative illustration of the invariant curve in the Poincaré section at the moment of torus T^2 destruction as one moves along the routes (**a**) A, (**b**) B, and (**c**) C indicated in Fig. 7.1b

region. The bifurcation lines l_h correspond to a tangency between the manifolds W^s and W^u of the saddle cycle C^{sd}. Outside the resonance region, T^2 is destroyed on the curves shown in Fig. 7.1b as dashed lines. As a matter of fact, this boundary has a complicated fractal structure. Moving along the route C' results in a tangent bifurcation on the curve l_1 without torus destruction. In this case there is a transition from the resonant T^2 to an ergodic T^2.

As one moves along the direction A, C^{st} on T^2 becomes unstable on the bifurcation curve l_2 due to either period doubling or a 2D torus birth from C^{st}. The resonance torus T^2 loses its smoothness when a pair of multipliers of the cycle become complex conjugates or one of the multipliers is negative. At the moment of bifurcation, the length of an invariant curve in the Poincaré map becomes infinite (Fig. 7.2a), which implies torus destruction. The transition to chaos along route A can come either from a period-doubling bifurcation cascade or via the breakdown of the torus born on the curve l_2.

As one moves along route B, the manifold W^u of C^{sd} which forms the torus surface is distorted, and on the bifurcation curve l_h, a tangency occurs between W^u and W^s. This is illustrated in Fig. 7.2b. At this moment ($s = s^*$), a structurally unstable homoclinic trajectory Γ_0 arises and T^2 is destroyed. For $s > s^*$, two robust homoclinic orbits appear, and a homoclinic structure of the cycles and of chaotic trajectories forms in their vicinity. However, C^{st} is still stable and remains an attractor. A chaotic attractor may arise either through the disappearance of C^{st} or when C^{st} loses its stability. When we cross the synchronization region above the curve l_h, a transition to chaos appears, accompanied by type-I intermittency.

Along route C, W^u is also distorted as one approaches C^{st}. T^2 is destroyed as our route intersects the curve l_1 corresponding to a tangent bifurcation. Suppose an invariant curve in the Poincaré section of the torus becomes unsmooth at the bifurcation point (Fig. 7.2c). This means that, when we apply the Poincaré map repeatedly, the image of a small segment of the unstable separatrix of a saddle node bends into the shape of a horseshoe. The disappearance of the saddle node leads to the emergence of a Smale horseshoe map in its vicinity. This map generates a countable set of saddle cycles and a continuous set of aperiodic hyperbolic trajectories. Under certain additional conditions, these trajectories can be transformed into a chaotic attractor.

The mechanisms of resonance torus destruction thus lead to the emergence of a chaotic set in the vicinity of the torus, and this set may become attracting. The chaotic attractor originates from a horseshoe-type map with a smooth bend and is nonhyperbolic. The bifurcation scenarios for the transition to chaos described here concern bifurcations of resonance cycles on T^2. They do not cause the absorbing area to change abruptly, so they represent the bifurcation mechanism of a soft transition to chaos. All the conclusions of the torus breakdown theorem are generic, as has been verified by physical experiments and computer simulations for a variety of discrete and continuous-time systems.

If we consider the evolution of an invariant curve in the Poincaré section of a torus by changing system parameters in such a way that the winding number is kept irrational, the following phenomena can be observed. First, the invariant curve in the Poincaré map of the ergodic torus is distorted in shape and repeats the unstable manifold of a resonance saddle cycle. Then the ergodic torus loses its smoothness and is destroyed. However, chaotic motion does not yet appear, since in the neighborhood (in parameter space) of the former torus with an irrational winding number, there always exist other resonance tori, remaining as attractors of the system. Hence, a resonance on T^2 always precedes the transition to chaos. The torus destruction line in the plane of two control parameters is characterized by a complex structure. It consists of a countable set of intervals on which the resonance torus is destroyed according to the scenarios indicated by the theorem, together with a set of points corresponding to the breakdown of an ergodic torus and having a joint zero measure.

7.1.2 Circle Map: Universal Regularities of Soft Transition from Quasiperiodicity to Chaos

In the general case, the motion on a two-dimensional torus can be modeled by an isomorphic dissipative map of the ring Q into itself. For different parameter values of the map, the following cases are possible:

- Inside the ring there is a closed contour L (Fig. 7.3a) which is transformed into itself. In other words, there exists a closed invariant curve corresponding to a two-dimensional torus in a continuous-time system. On the contour L, a new map can be defined which will be one-dimensional and homeomorphic to the circle map

$$\phi_{n+1} = \Phi(\phi_n, \alpha) \quad (\text{mod } 1), \tag{7.2}$$

where α is the parameter vector of the circle map.
- A horseshoe-type map arises, generating a countable set of periodic orbits and a continuous set of aperiodic hyperbolic trajectories. Such a structure is created when some part of the region Q, denoted by σ, is transformed into $\tilde{\sigma}$ as shown in Fig. 7.3b. In this case the closed contour L no longer exists and the model map (7.2) becomes noninvertible.

The circle map is often defined by

$$\phi_{n+1} = \Phi(\phi_n, \Omega, K) = \phi_n + \Omega - \frac{K}{2\pi} \sin(2\pi\phi_n) \quad (\text{mod } 1). \tag{7.3}$$

The angle ϕ is determined in the interval $[0, 1]$, while $K \geq 0$ and $\Omega \in [0, 1]$ are considered as parameters of the map. Generally speaking, the form of the function $\Phi(\phi)$ is not so important (as in the case of the logistic map), but the following conditions must be satisfied:

- $\Phi(\phi + 1) = 1 + \Phi(\phi)$,
- For $K < K_{\text{cr}}$, $\Phi(\phi)$ and its inverse function $\Phi^{-1}(\phi)$ exist and are differentiable, i.e., the map is a diffeomorphism of the circle, and
- At $K = K_{\text{cr}}$, the inverse function $\Phi^{-1}(\phi)$ becomes nondifferentiable at the point $\phi = 0$, and the single-valued inverse function no longer exists for $K > K_{\text{cr}}$.

All these conditions are fulfilled for (7.3) with $K_{\text{cr}} = 1$.

The dynamics of a point in the circle map is characterized by the winding number θ, which is given by

$$\theta = \lim_{n \to \infty} \frac{\Phi^n(\phi_0) - \phi_0}{n}. \tag{7.4}$$

This number represents the mean angle through which a phase point rotates on a circle in one iteration. For a smooth one-to-one map, i.e., when $0 \leq K < 1$, the limit

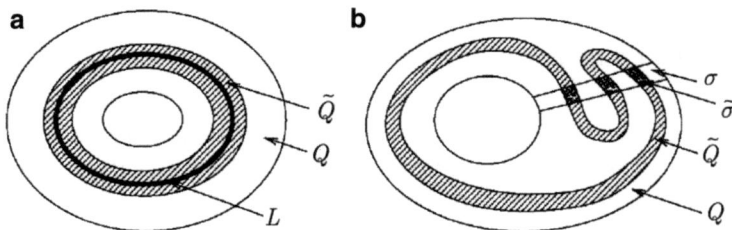

Fig. 7.3 Ring map (**a**) when there is an invariant closed curve L and (**b**) when a local horseshoe-type map arises

(7.4) exists and does not depend on the initial point ϕ_0. From the above statement, it follows that, for $K < 1$, the map (7.3) has no fixed points if the winding number is irrational. When θ takes a rational value $\theta = p/q$ with p, q integers, the circle map possesses an even number of stable and unstable fixed points of multiplicity q, i.e., the map has at least one stable and one unstable q-cycle. The numerator p determines the number of full turns around a circle in q iterations.

The resonance structure corresponding to a rational value of the winding number is robust. Each rational value of θ remains unchanged within a certain range of parameter variation, the so-called *Arnold tongue*. The dependence of the winding number on the parameter Ω is known as the *devil's staircase* and represents a fractal curve consisting of an infinite number of 'steps' corresponding to rational values of θ and a set of isolated points for which θ is irrational. At $K = 0$, the winding number for (7.3) coincides with Ω and has a set of rational values of measure zero. When $0 < K < 1$, there are both rational and irrational values of the winding number, and their number is not equal to 0. With increasing K, the set of rational values grows, while the set of irrational values decreases and vanishes at the critical line $K = 1$ (the sum of all step lengths becomes equal to 1). However, a countable set of points with irrational winding numbers still exists at $K = 1$.

For $K > 1$, the circle map does not exhibit quasiperiodic motion. The dependence $\theta(\Omega)$ that corresponds to the overlapping of Arnold tongues becomes ambiguous. In a supercritical region, the circle map describes resonances on a torus as well as chaotic motions in the vicinity of the former torus T^2. The map demonstrates the scenarios of torus breakdown and appearance of chaos, indicated in the Afraimovich–Shilnikov theorem. Inside the Arnold tongues, the stable resonance cycle loses its stability in a period-doubling bifurcation, and a transition to chaos via the Feigenbaum scenario occurs. In the regions where the resonance tongues overlap, crises may take place, leading to the merging of chaotic attractors which appear from different resonance cycles. As a result, torus chaos is generated. Figure 7.4 shows the dynamical regimes for (7.3) in the (Ω, K) plane and reflects the complicated self-similar structure of Arnold tongues.

The map (7.3) exhibits certain quantitative universal properties, independent of any particular form of $\Phi(\phi)$, provided that $\Phi(\phi)$ satisfies the above-mentioned conditions. However, these features depend on θ.

7.1 Route to Chaos via Two-Dimensional Torus Destruction

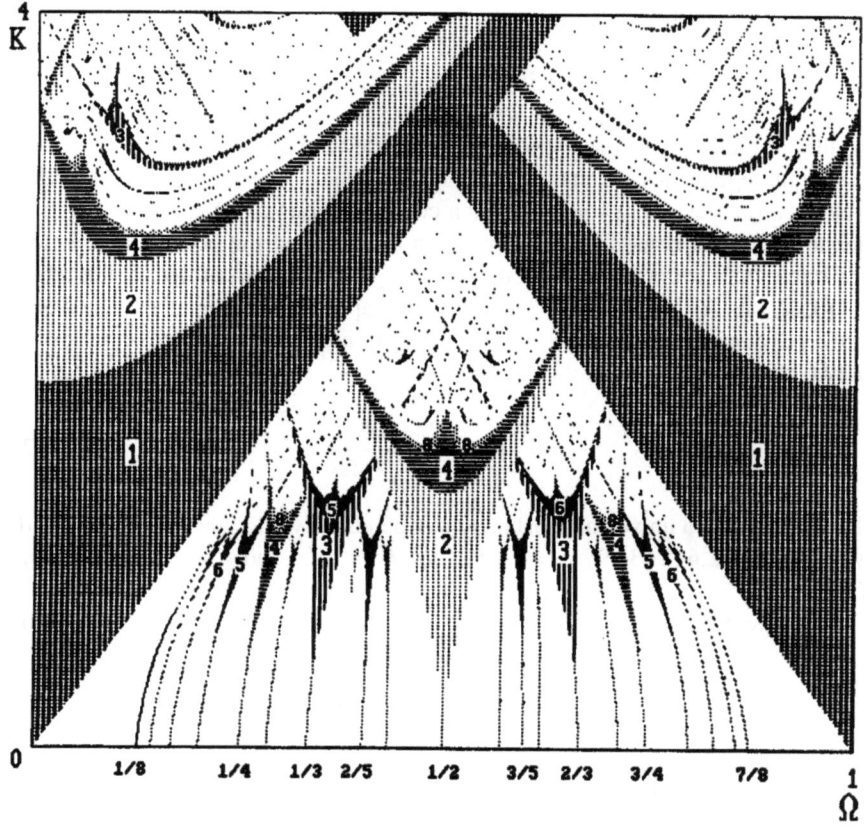

Fig. 7.4 Dynamical regimes of the circle map. Periodic regimes are the *shaded regions* and their periods are indicated by numbers (Taken from the paper Kuznetsov et al. [8])

An irrational number can be represented in the form of a *continued fraction*:

$$\theta = \cfrac{1}{m_1 + \cfrac{1}{m_2 + \cfrac{1}{\cdots}}} = \langle m_1, m_2, \ldots, m_k, \ldots \rangle. \tag{7.5}$$

If only the first k expansion terms are taken into consideration, one obtains a rational number $\theta_k = p_k/q_k$, which is called the *rational approximation* of θ of order k. In this case the irrational number can be defined as the limit of a sequence of rational numbers:

$$\theta = \lim_{k \to \infty} \theta_k. \tag{7.6}$$

The irrational number called the *golden mean*,[1] viz.,

$$\theta_g = 0.5(\sqrt{5} - 1) = \langle 1, 1, 1, \ldots \rangle ,$$

has the simplest expansion as a periodic continued fraction.

Certain regularities are characteristic for irrational values of θ, having a periodic expansion (at least from some m_k) as a continued fraction. Assume $\Omega_k(K)$ to be the value of Ω at some fixed K, for which $\theta = \theta_k$ and the point $\phi = 0$ belongs to a stable limit cycle with period q_k. In other words, Ω_k is defined by the relation $\Phi^{q_k}(0, \Omega_k, K) = p_k$, where Φ^{q_k} is a function applied q_k times. The values of Ω_k converge to some value $\Omega_\infty(\theta, K)$ according to a geometric progression law with rate δ given by

$$\delta = \lim_{n \to \infty} \frac{\Omega_k - \Omega_{k-1}}{\Omega_{k+1} - \Omega_k} . \tag{7.7}$$

The quantity δ is a universal constant and depends only on θ and K. For the golden mean, $\delta = -2.6180339\ldots = -\theta_g^{-2}$ for $K < K_{cr}$ and $\delta = -2.83362\ldots$ at $K = K_{cr}$.

The scale given by $d_k = \Phi_{\Omega_k}^{q_k-1}(0, \Omega_k, K) - p_{k-1}$ is characterized by the limit

$$\lim_{k \to \infty} \frac{d_k}{d_{k+1}} = a , \tag{7.8}$$

where a is a universal constant. For $\theta = \theta_g$, $a = -1.618\ldots = -\theta_g^{-1}$ for $K < K_{cr}$ and $a = -1.28857\ldots$ at $K = K_{cr}$.

The spectrum of the circle map at $K = K_{cr}$ also exhibits a number of universal properties. If θ can be presented in the form of a periodic continued fraction, the spectrum has the property of *scaling*. For $\theta = \theta_g$, the frequencies of spectral components, reduced to the interval $[0, 1]$, satisfy the following relation:

$$\nu = |F_{k+1}\theta_g - F_k| , \tag{7.9}$$

where F_k, F_{k+1} are sequential terms of one of the Fibonacci series. The spectral series, arranged in decreasing order with respect to their amplitudes, correspond to the Fibonacci series with the following bases: the main series $(0,1)$, the second $(2,2)$, the third $(1,3)$, the fourth $(3,3)$, the fifth $(1,4)$, the sixth $(2,5)$, and so on. The reduced spectral power for the lines of each series has a limit as $j \to \infty$:

---

[1] For the golden mean, p_k and q_k are the sequential terms of the main Fibonacci series, namely, $p_k = F_k$, $q_k = F_{k+1}$, and consequently $\theta_g = \lim_{k \to \infty} F_k/F_{k+1}$. The Fibonacci series is determined by the recurrence formula $F_{k+1} = F_{k-1} + F_k$, where (F_0, F_1) are the base of the series. The main series has base $(0, 1)$.

$$S_i = \lim_{j \to \infty} \frac{S_i^j}{v^2(j)} = \text{const.} \quad (7.10)$$

The normalized spectrum $a_i^j = S_i^j/(S_1^1 v^2(j))$, represented in the coordinates $\log a_i^j$, $\log v$, is divided into identical intervals located between the relevant lines of each series.

The above-mentioned and other quantitative regularities of two-frequency quasi-periodicity destruction appear to be typical, not only for model one-dimensional maps, but also for invertible maps and continuous-time systems. These properties have been observed within the given accuracy in full-scale experiments and in computer simulation of various dynamical systems.

The RG method has been applied to analyze maps in the form of (7.3). When $\theta = \theta_g$, one can derive a functional equation for the fixed point, which reads

$$\Phi^*(\phi) = a\Phi^*\big(a\Phi^*(\phi/a^2)\big), \quad (7.11)$$

where $\Phi^*(\phi + 1) = \Phi^*(\phi) + 1$. Its solution $\Phi^*(\phi)$ is a universal function and the scaling multiplier a represents a universal constant. Equation (7.11) has the linear solution $\Phi^*(\phi) = \phi - 1$, for which $a_{1,2} = 0.5(\pm\sqrt{5} - 1)$. The value of a, found numerically for $K < K_{\text{cr}}$, coincides with the solution

$$a_2 = 0.5(-\sqrt{5} - 1) = -\theta_g^{-1} \approx -1.618 \;.$$

At $K = K_{\text{cr}}$, the linear solution does not obey (7.11). Since $\Phi(\phi)$ has a cubic inflection point at zero, the universal function $\Phi^*(\phi)$ must contain a cubic term ϕ^3. A nontrivial function of this sort has been deduced numerically in the form

$$\Phi^*(\phi) = 1 + c_1\phi^3 + c_2\phi^6 + \cdots . \quad (7.12)$$

The value of a agrees with numerically obtained results. One of the eigenvalues of the linearized equation at the fixed point coincides with the constant derived numerically from (7.7). This fact explains the universal character of the constant δ. Results obtained for $\theta = \theta_g$ by means of the RG method have been generalized to the case of an arbitrary irrational winding number, which can be presented in the form of a periodic continued fraction. With this, the form of $\Phi^*(\phi)$ together with the values of a and δ naturally depend on θ.

7.2 Route to Chaos via Ergodic Torus Destruction: Chaotic Nonstrange Attractors

As mentioned above, the resonance phenomena always preceding the onset of chaos play an important role in the quasiperiodic scenario for transition to chaos. But there is a whole class of systems for which ergodic quasiperiodic motion is robust and the transition to chaos is not accompanied by the emergence of resonance periodic motions. This class encompasses quasiperiodically forced systems. Quasiperiodic forcing with a fixed irrational relation between frequencies imposes an irrational winding number upon the system, independently of its internal properties.

Consider the simplest case when one of the forcing frequencies coincides with the basic frequency of periodic oscillations of the system. With this, a robust two-frequency quasiperiodic regime with a fixed winding number can be observed in a synchronization region of basic oscillations. This winding number is defined from the outside and assumed to be irrational. By varying the system parameters and the amplitude of the external forcing, one can achieve T^2 destruction and the transition to chaos. The transition from an ergodic torus to chaos is characterized by its own peculiarities as compared with the case considered above, where the destroyed torus had an arbitrarily varying winding number. Numerous studies of flow systems and maps with quasiperiodic forcing have shown that the appearance of a special class of attractors, the so-called strange nonchaotic attractors (SNA), is typical for such systems. An SNA is defined as an attracting limit set of a DS which is not a manifold. The peculiarity of this set is that its phase trajectories do not diverge, while the set is fractal.

In order to study the mechanisms of transition from an ergodic torus to chaos, it is convenient to use quasiperiodically forced maps presented in the following autonomous form:

$$\mathbf{x}_{n+1} = \mathbf{F}(\mathbf{x}_n, \phi_n, \boldsymbol{\alpha}),$$
$$\phi_{n+1} = \phi_n + \theta \pmod{1}. \tag{7.13}$$

Here, $\mathbf{x} \in \mathbb{R}^N$ is the state vector of the system, ϕ denotes the phase of the forcing, $\mathbf{F} \in \mathbb{R}^N$ is periodic in the argument ϕ with period 1, $\boldsymbol{\alpha}$ is the parameter vector of the system, and θ is the winding number. The forcing is quasiperiodic if θ is irrational. The winding number is usually chosen to be the golden mean, $\theta = 0.5(\sqrt{5} - 1)$. One-dimensional maps ($N = 1$) are easiest to use. A two-dimensional torus in the map (7.13) corresponds to an invariant closed curve. As θ is irrational, this curve is densely covered everywhere by points of phase trajectories. Since no resonance structure arises on the torus and there is no need to control the winding number, the bifurcation mechanisms of T^2 destruction and the appearance of chaos allow a one-dimensional analysis in this case. Hence, we consider α to be a scalar.

Suppose that, at $\alpha = \alpha_0$, there exists an ergodic T^2 with a strictly fixed irrational winding number, whereas at $\alpha = \alpha_1$, a chaotic attractor (CA) arises. What is the scenario of transition to chaos in this case? Studies have shown that the destruction of a robust ergodic T^2 leads initially to the appearance of an SNA that is then transformed into a CA. In the map, the invariant curve is first deformed and then loses its smoothness. According to the Afraimovich–Shilnikov theorem, a resonance torus also loses its smoothness before it is destroyed. This happens on a finite set of fixed points of the invariant curve, corresponding to the points of the stable resonant cycle. Such a 'nonsmooth torus' can exist for some time in the system phase space before being destroyed.

7.2 Route to Chaos via Ergodic Torus Destruction: Chaotic Nonstrange Attractors

In the case of an ergodic torus, the invariant curve has no fixed points, and, at some $\alpha = \alpha_{cr1}$, it loses its smoothness simultaneously on an everywhere dense set of points. As a result, the invariant curve is destroyed and a set that is not a manifold appears. However, the torus destruction does not automatically lead to the emergence of exponential instability of the motion. The dynamics becomes chaotic later, when $\alpha = \alpha_{cr2} > \alpha_{cr1}$. Thus, there exists a finite range of values of the parameter α, viz., $\alpha_{cr1} < \alpha < \alpha_{cr2}$, where an SNA is observed. The SNA regime has the property of being intermediate between quasiperiodicity and chaos. The problem when exploring SNAs is the following. While an SNA is easily distinguished numerically from a chaotic attractor (there is no positive Lyapunov exponent), it is harder to reveal its difference from a quasiperiodic regime. The numerical criteria available do not allow one to detect with any confidence whether the set under study is an SNA or a strongly deformed but still smooth torus.

The most reliable numerical methods for diagnosing the SNA regime have been proposed in work by Pikovsky et al.[2] The first approach is related to the rational approximation of the winding number. The second method is based on the phase sensitivity property of the SNA. The last concerns the criterion provided by local Lyapunov exponents. Consider (7.13) for the case of $N = 1$. The map attains the following form:

$$\begin{aligned} x_{n+1} &= f(x_n, \phi_n, \alpha), \\ \phi_{n+1} &= \phi_n + \theta \pmod{1}. \end{aligned} \quad (7.14)$$

Using this map, we can explore the destruction of an invariant curve in a phase plane. The *rational approximation method* is based on the bifurcational analysis of cycles arising in the map under the rational approximation of the winding number $\theta_k = p_k/q_k$, $\lim_{k\to\infty} \theta_k = \theta$. In this case the behavior of the map strongly depends on the choice of the initial phase ϕ_0. If the cycles exhibit bifurcations as ϕ_0 changes, for sufficiently large (theoretically for arbitrarily large) k, one can conclude that an SNA exists in (7.14). When using this method, a phase-parameter diagram $x(\phi_0)$, $\phi_0 \in [0, 1/q_k]$, is constructed for $\theta = \theta_k$. This diagram is called the *approximating attracting set*. Indeed, at $\theta = \theta_k$, the ϕ_0 dependence of the attractor coordinates of the map approximates a small segment of an ergodic attractor. The fact that the approximating set is smooth testifies to the quasiperiodic regime of the map, whereas the presence of points of nonsmoothness corresponds to an SNA.

Another approach is based on *the sensitivity of the dynamical variable to the phase of the external forcing*. The map is now considered for an irrational value of the winding number. The derivative $\partial x_n / \partial \phi_0$ is calculated along a trajectory and its maximum is estimated. Writing $f_\phi = \partial f(x,\phi)/\partial \phi$ and $f_x = \partial f(x,\phi)/\partial x$, for (7.14) it is easy to obtain

---

[2]For example, Pikovsky and Feudel [12].

$$\frac{\partial x_n}{\partial \phi_0} = \sum_{k=1}^{n} f_\phi \mu_{n-k}(x_k, \phi_k) + \mu_n(x_0, \phi_0) \frac{\partial x_0}{\partial \phi_0} , \qquad (7.15)$$

where

$$\mu_m(x_k, \phi_k) = \prod_{i=0}^{m-1} f_x(x_{k+i}, \phi_{k+i}) , \qquad \mu_0 = 1 , \qquad (7.16)$$

is a 'local multiplier' of the phase trajectory. The map (7.14) has one Lyapunov exponent

$$\lambda = \lim_{n \to \infty} \frac{1}{n} \ln |\mu_n| , \qquad (7.17)$$

which is not equal to zero. Since it is negative in the SNA regime, the local multiplier must tend to zero with n. Taking this into account, the derivative (7.15) may be rewritten as

$$\frac{\partial x_n}{\partial \phi_0} = \sum_{k=1}^{n} f_\phi \mu_{n-k}(x_k, \phi_k) . \qquad (7.18)$$

We introduce the quantity Γ_n by

$$\Gamma_n = \min_{x_0, \phi_0} \max_{0 \le i \le n} \left| \frac{\partial x_i}{\partial \phi_0} \right| , \qquad (7.19)$$

where the maximum is sought at all points of a single trajectory and the minimum is determined with respect to randomly chosen initial points. The fact that the value of Γ_n grows infinitely with n means that the derivative $\partial x_n / \partial \phi_0$ does not exist, i.e., the invariant curve loses its smoothness and an SNA appears. Γ_n may be represented in the form

$$\Gamma_n \sim n^\eta . \qquad (7.20)$$

Here, η is called a *phase sensitivity exponent*. In the SNA regime, η is very close to unity.

Phase sensitivity is connected with the existence of a nonzero measure of positive local Lyapunov exponents. The *local Lyapunov exponent* is a Lyapunov exponent of a trajectory which is calculated on a finite time interval. For (7.14), the local Lyapunov exponent reads

$$\Lambda_n(x, \phi) = \frac{1}{n} \ln |\mu_n(x, \phi)| , \qquad (7.21)$$

7.2 Route to Chaos via Ergodic Torus Destruction: Chaotic Nonstrange Attractors

where $\mu_n(x, \phi)$ is defined by (7.16). It is clear that both the value and the sign of the local Lyapunov exponent depend on the initial point (x_0, ϕ_0). Since the SNA is nonsmooth and the local multiplier μ_n is unbounded, a measure of positive values Λ_n must differ from zero, even for sufficiently large n. With this, $\lim_{n \to \infty} \Lambda_n = \lambda < 0$, because the attractor is nonchaotic.

Unfortunately, the criteria considered can only usually be applied to 1D discrete models with quasiperiodic forcing and are little suited to continuous-time systems. The exception may be the criterion of positive local Lyapunov exponents, but its application is associated with a problem concerning the minimal time interval for which the local exponents of a quasiperiodic regime of the given system cannot yet be positive. Other numerical criteria for the existence of SNAs have also been proposed. They involve the properties of the spectrum and autocorrelation function. However, in most cases, they too do not provide unambiguous results. Therefore, the mechanism of ergodic torus destruction should be explored with caution, using as many available SNA criteria as possible.

The reasons why an invariant curve in a model map might lose its smoothness and then be destroyed have been analyzed, and two mechanisms leading to the appearance of SNAs have been discovered:

- A crisis of an ergodic torus at $\alpha = \alpha_{cr1}$, through a nonlocal bifurcation.
- A gradual evolution of the torus, leading to its loss of smoothness and destruction at $\alpha = \alpha_{cr1}$.

The ergodic torus crisis occurs when a stable torus touches an unstable torus or its stable manifold, the latter playing the role of a separatrix surface. In 1D models with quasiperiodic forcing, viz., (7.14), a saddle torus corresponds to an unstable invariant curve (repeller). The crisis may be connected with the merging of bands of a quasiperiodic attractor.[3] The point is that, as a result of torus doubling bifurcation in one of the periods, the invariant curve in the torus section will consist of 2^k bands. These bands are visited by a phase point in a strictly defined order and separated by a separatrix surface. The latter can be represented by a surface in the section of the stable manifold of a saddle torus. In terms of one-dimensional maps, a separatrix corresponds to an unstable invariant curve. Figure 7.5 illustrates the band-merging crisis of an invariant curve in the logistic map with quasiperiodic parameter modulation:

$$\begin{aligned} x_{n+1} &= \alpha(1 + \varepsilon \cos 2\pi \phi_n) x_n (1 - x_n) \,, \\ \phi_{n+1} &= \phi_n + \theta \quad (\text{mod } 1) \,. \end{aligned} \quad (7.22)$$

The invariant curve exists for $\alpha < \alpha_{cr1}$, and consists of two bands separated by a repeller. A representative point visits each band in one iteration of the map. The distortion of the invariant curve leads to the situation that, at $\alpha = \alpha_{cr1}$, the

---

[3] Strictly speaking, such a merger is not a crisis since it does not cause the absorbing area to change. But in this case the term 'crisis' is commonly accepted.

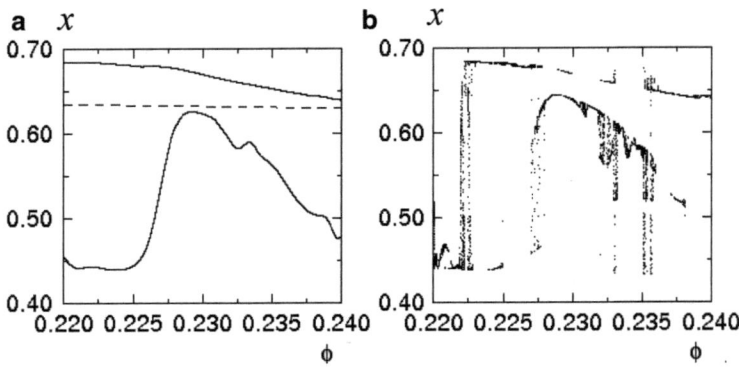

Fig. 7.5 Band merging crisis of the invariant curve in the map (7.22) for $\varepsilon = 0.1$, $\theta = \theta_g$. A segment of the invariant curve before tangency ($\alpha = 3.271$) (**a**) and after tangency ($\alpha = 3.272$) (**b**) with the repeller (*dashed line*) (Taken from the paper Heagy and Hammel [7])

invariant curve touches the repeller separating the attractor bands. At the moment of tangency, the stable invariant curve loses its smoothness and is destroyed. For $\alpha > \alpha_{cr1}$, there exists an SNA uniting both bands of the destroyed curve. These are now visited randomly by the representative point. This mechanism of torus destruction restricts the sequence of ergodic torus doublings, which is typically finite.

Torus crisis can also be related to the merging of different quasiperiodic attractors. This situation is possible when the crisis occurs without collision of quasiperiodic attractors or bands of a single attractor. This case is especially typical for a resonance on T^3 for which one winding number has a fixed irrational value and the other is varied arbitrarily. The distortion of the shape of invariant curves in the section of stable and unstable two-dimensional tori on T^3 results in a crisis of tangency at separate points, instead of a tangency bifurcation of the tori on the boundary of the synchronization region.

Smoothness loss and destruction of an ergodic torus can also take place without bifurcations related to any separatrix tangency. In this case, as the control parameter α is varied, the shape of an invariant curve in a torus section is gradually distorted, and this causes the phase sensitivity to increase. The derivative with respect to the initial phase is no longer bounded at the critical point $\alpha = \alpha_{cr1}$. If the winding number is replaced by rational approximations, the dynamics of the map turns out to be chaotic at some ϕ_0 values, but remains regular at other ϕ_0 values. However, the Lyapunov exponent still converges to a negative value when $n \to \infty$. In this case the approximating set cannot be smooth and an SNA is observed within some interval $\alpha_{cr1} \leq \alpha \leq \alpha_{cr2}$.

It is assumed that an SNA arising from a crisis or by evolution exists on a set of nonzero measure in parameter space, but this statement is difficult to prove. It is known that, on the interval $\alpha \in [\alpha_{cr1}, \alpha_{cr2}]$, an SNA can degenerate into an invariant curve which possesses a finite number of points of discontinuity and then reappear once again. The bifurcation mechanism for transition from SNA to chaos is still not understood. The well-studied scenarios of chaos formation assume the presence of homoclinic trajectories of saddle cycles or dangerous separatrix loops of a saddle focus. In their neighborhood, a horseshoe-type map arises and a chaotic set of trajectories can be created. When the winding number θ is kept constant and has an irrational value, a system possesses neither equilibrium states nor limit cycles. In this case a homoclinic structure must be connected with the tangency and intersection of manifolds of saddle tori. At the present time, homoclinics of this sort have not been adequately explored.

It is worth noting that the quasiperiodic route to chaos considered above does not assume the emergence of a third independent oscillation frequency. Consequently, the described route to chaos may already occur in continuous-time systems with phase space dimension $N = 3$. The quasiperiodic route to chaos in multi-dimensional systems assumes the appearance of fourth, fifth, etc., independent frequencies. The Ruelle–Takens–Newhouse scenario can be reliably realized in systems with dimension $N \geq 4$. However, the theorems proved in the work of Ruelle, Takens, and Newhouse do not explain the bifurcation sequences which lead to chaos. They only provide evidence that the quasiperiodic motion on a T^3 can be converted into chaotic motion if the system is slightly perturbed. Apparently, there are many different specific scenarios of chaos development in this case. Generally speaking, they have not been sufficiently well studied yet. We note only that, due to the nonlinear interaction of oscillatory modes of the system, the measure of ergodic multi-frequency regimes is very small and the transition to chaos is usually associated with the emergence of resonant structures on multi-dimensional tori.

7.3 Summary

In Chaps. 6 and 7, we have described three typical routes to chaos:

- The period-doubling cascade route (Feigenbaum scenario),
- The crisis of periodic oscillations and the intermittency transition route (Pomeau–Manneville scenario), and
- Different kinds of quasiperiodic routes to chaos (the Ruelle–Takens–Newhouse scenario).

It is worth noting that, for the same system, different routes to chaos can be observed corresponding to different regions and directions in parameter space. Moreover, the observed bifurcational sequences may be combined in a complex way. Thus, to imagine the full picture of the appearance of chaotic motion, one should not restrict oneself to a one-parameter analysis. One must have some idea, at least in general terms, of the bifurcation diagram of dynamical regimes of the system in parameter space, what happens on its different 'leaves', where they are 'sewn' together, etc.

The scenarios considered are typical in the sense that they are observed for a wide class of dynamical systems with both low- and high-dimensional phase spaces, as well as for distributed systems with arbitrarily chosen control parameters and different types of variation of those parameters. Evidently, other scenarios for chaos formation may also be possible, but they are not typical. They may be related to some peculiarities (degeneracy) of the DS or to a special choice of directions of motion in parameter space which pass through high-codimension critical points.

The features of bifurcation transitions to chaotic dynamics are described in [1–5, 9–11, 13–16]. The classic description of this problem is given in [13]. The destruction of quasiperiodic oscillations is described in more detail in [1, 2, 6].

References

1. Anishchenko, V.S.: Dynamical Chaos – Models and Experiments. World Scientific, Singapore (1995)
2. Anishchenko, V.S., Astakhov, V.V., Neiman, A.B., Vadivasova, T.E., Schimansky-Geier, L.: Nonlinear Dynamics of Chaotic and Stochastic Systems. Springer, Berlin (2002)
3. Berge, P., Pomeau, I., Vidal, C.G.: Order Within Chaos. Wiley, New York (1984)
4. Binney, J.: The Theory of Critical Phenomena: An Introduction to the Renormalization Group. Oxford University Press, Oxford (1992)/World Scientific, Singapore (2010)
5. Collins, J.: Renormalization: An Introduction to Renormalization. Cambridge University Press, Cambridge (1984, 1986)
6. Feudel, U., Kuznetsov, S., Pikovsky, A.: Strange Nonchaotic Attractors: Dynamics Between Order and Chaos in Quasiperiodically Forced Systems. World Scientific, Singapore (2006)
7. Heagy, J.F., Hammel, S.M.: The birth of strange nonchaotic attractors. Physica D **70**, 140 (1994)
8. Kuznetsov, A.P., Kuznetsov, S.P., Sataev, I.P.: Codimension and typicality in the context of description of the transition to chaos through period-doubling in dissipative dynamical systems. Regular and Chaotic Dyn. **2**, 90–105 (1997) (in Russian)
9. Marek, M., Schreiber, I.: Chaotic Behaviour of Deterministic Dissipative Systems. Cambridge University Press, Cambridge (1991, 1995)
10. Nicolis, G.: Introduction to Nonlinear Science. Cambridge University Press, Cambridge (1995)
11. Ogorzalek, M.J.: Chaos and Complexity in Nonlinear Electronic Circuits. World Scientific, Singapore (1997)
12. Pikovsky, A.S., Feudel, U.: Characterizing strange nonchaotic attractors. Chaos **5**, 253 (1995).
13. Schuster, H.G.: Deterministic Chaos. Physik-Verlag, Weinheim (1988)
14. Seydel, R.: Practical Bifurcation and Stability Analysis: From Equilibrium to Chaos. Springer, New York (1994, 2009)
15. Shilnikov, L.P., Shilnikov, A.L., Turaev, D.V., Chua, L.O.: Methods of Qualitative Theory in Nonlinear Dynamics. World Scientific, Singapore (2001)
16. Thompson, J.M.T., Stewart, H.B.: Nonlinear Dynamics and Chaos. Wiley, New York (1986)

Chapter 8
Robust and Nonrobust Dynamical Systems: Classification of Attractor Types

8.1 Introduction

We consider a class of autonomous continuous-time dynamical systems with phase space dimension $N \geq 3$. Besides robust systems similar to Andronov–Pontryagin systems on the plane, there appears a class of robust systems with nontrivial hyperbolicity, i.e., systems with chaotic dynamics. Chaotic attractors of robust hyperbolic systems are, in the rigorous mathematical sense, strange attractors. They usually represent some mathematical idealization and are not as a rule observed in experiments. In most cases systems with irregular dynamics are nonrobust. Mathematicians have proven that robust hyperbolic systems are not everywhere dense on the set of dynamical systems with $N \geq 3$. Structural instability (nonrobustness) is associated with the emergence of nonrobust double-asymptotic trajectories, such as separatrix loops, homoclinic curves, and heteroclinic curves, which are formed when manifolds of saddle cycles and another saddle sets intersect non-transversally.

The rigorous mathematical description of nonrobust systems with complex dynamics causes great difficulties because many bifurcations can be observed when system parameters are even slightly perturbed. The number of bifurcations in any interval of control parameter values can be theoretically infinite and cannot be fully described. In addition, nonrobust systems are very sensitive to the accuracy in setting initial conditions and to the presence of noise (even if this is round-off noise in numerical experiments). Under such conditions, we are generally no longer interested in particular bifurcations and details of the phase portrait. The important thing is now the general character of behavior in the presence of fluctuations and statistical characteristics of the regime, which can be measured experimentally.

Nonrobust dissipative dynamical systems can be divided into two classes: nearly hyperbolic (quasihyperbolic) DS and nonhyperbolic DS. Quasihyperbolic systems have attractors which are most similar in their characteristics to robust hyperbolic attractors. The main feature distinguishing them from nonhyperbolic attractors is that they do not include stable cycles and tori. Nonhyperbolic attractors always contain stable cycles or tori, and these significantly affect their properties.

The notion of attractor has recently become important, and not only due to the discovery of deterministic chaos. This concept in the nonlinear theory of oscillations is associated with self-sustained oscillations in the dynamical system. If in contrast to Andronov, we base the definition of self-sustained oscillations on the main fact, i.e., that the steady-state regime is independent of initial conditions, then any type of attractor of an autonomous DS is the mathematical image of self-sustained oscillations of the system in its phase space. Using this concept, the properties of self-sustained oscillatory regimes will be determined by the structure and properties of the relevant attractors. For example, stable regular self-sustained oscillations will be associated with limit cycles and tori, and chaotic self-sustained oscillations with corresponding chaotic attractors.

In this chapter we consider robust hyperbolic attractors, quasihyperbolic attractors which also include the so-called Lorenz-type attractors, and nonhyperbolic attractors which are the ones most often realized in experiments.

8.2 Homoclinic and Heteroclinic Curves

Manifolds of saddle sets and double-asymptotic trajectories play a major role in understanding the dynamics of a nonlinear system. We start by giving a more general definition of the manifolds of a saddle equilibrium point and a saddle cycle.

Consider the following DS:

$$\dot{\mathbf{x}} = \mathbf{F}(\mathbf{x}), \qquad \mathbf{x} \in \mathbb{R}^N. \tag{8.1}$$

Let $\mathbf{x}_0$ be an equilibrium. Its stability is characterized by the eigenvalues s_i of the linearization matrix

$$\hat{A}(\mathbf{x}) = \left.\frac{\partial \mathbf{F}}{\partial \mathbf{x}}\right|_{\mathbf{x}_0}. \tag{8.2}$$

Suppose there are m eigenvalues s_i such that $\operatorname{Re} s_i < 0$, $i = 1, 2, \ldots, m \ne 0$, and k eigenvalues s'_j such that $\operatorname{Re} s'_j > 0$, $j = 1, 2, \ldots, k \ne 0$ $(m + k = N)$. This means that $\mathbf{x}_0$ is a saddle-type equilibrium (a saddle or a saddle focus). The set of points $\mathbf{x} \in \mathbb{R}^N$ such that $\mathbf{x} \to \mathbf{x}_0$ as $t \to \infty$ is called the *stable manifold* $W^s_{\mathbf{x}_0}$ *of the equilibrium* $\mathbf{x}_0$. It has dimension m. The set of points $\mathbf{x} \in \mathbb{R}^N$ such that $\mathbf{x} \to \mathbf{x}_0$ as $t \to -\infty$ is called the *unstable manifold* $W^u_{\mathbf{x}_0}$ *of the equilibrium* $\mathbf{x}_0$. It has dimension k.

Let L be a periodic trajectory (limit cycle) with period T. Its stability is defined by multipliers which are eigenvalues of the monodromy matrix M_T. Suppose there are m multipliers μ_i such that $|\mu_i| < 1$, $i = 1, 2, \ldots, m \ne 0$, and k multipliers μ'_j

8.2 Homoclinic and Heteroclinic Curves

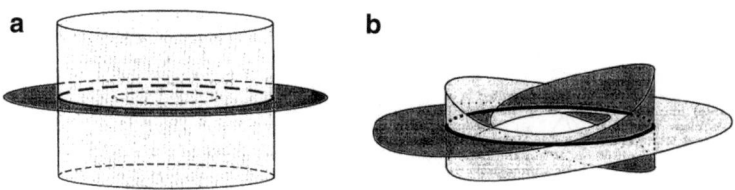

Fig. 8.1 Two cases of intersection of manifolds of a saddle limit cycle: (**a**) a cylinder and a plane and (**b**) two Möbius bands

such that $|\mu'_j| > 1$, $j = 1, 2, \ldots, k \neq 0$ ($m + k + 1 = N$). This means that the periodic trajectory L is of saddle type. The set of points $\mathbf{x} \in \mathbb{R}^N$ such that $\mathbf{x} \to L$ as $t \to \infty$ is called the *stable manifold* W_L^s *of the periodic trajectory* L. It has dimension $m + 1$. The set of points $\mathbf{x} \in \mathbb{R}^N$ such that $\mathbf{x} \to L$ as $t \to -\infty$ is the *unstable manifold* W_L^u *of the periodic trajectory* L. It has dimension $k + 1$.

Let W^s and W^u be the stable and unstable manifolds, respectively, of saddle periodic orbits or saddle equilibria. We now define the notion of *transversality* for the intersection of these manifolds. To do this, we introduce tangent planes (or tangent surfaces for $N > 3$) $E_\mathbf{x}(W^s)$ and $E_\mathbf{x}(W^u)$ to the manifolds W^s and W^u at points $\mathbf{x} \in W^s \cap W^u$. The manifolds W^s and W^u are said to intersect transversally if

$$\dim\big(E_\mathbf{x}(W^s)\big) + \dim\big(E_\mathbf{x}(W^u)\big) - \dim\big(E_\mathbf{x}(W^s) \cap E_\mathbf{x}(W^u)\big) = N , \quad (8.3)$$

where dim denotes the dimension of the set. The property of transversality of these manifolds is preserved in the presence of small perturbations, i.e., it is structurally stable.

A saddle limit set itself belongs to the transverse intersection of its manifolds. There are two types of intersection of manifolds at points of a saddle limit cycle, as shown in Fig. 8.1. Besides the saddle limit set, the intersection of its manifolds may include double-asymptotic trajectories. A trajectory that tends to a saddle (a saddle equilibrium or a saddle cycle) both forward and backward is called a *double-asymptotic trajectory of the saddle*. This trajectory can be exemplified in the plane by a separatrix loop of a saddle fixed point. In this case the manifolds do not intersect transversally and the loop is always nonrobust.

Double-asymptotic trajectories of saddle limit cycles are called *Poincaré homoclinic curves* (Fig. 8.2a). Double-asymptotic trajectories belonging to the intersection of manifolds of two different saddle cycles are called *Poincaré heteroclinic curves*. A heteroclinic curve is wound on one saddle cycle in forward time (along its stable manifold) and on another one in reverse time (along its unstable manifold).

It should be noted that in $\mathbb{R}^N$, $N > 3$, double-asymptotic trajectories similar to Poincaré homoclinic and heteroclinic curves may arise when the manifolds, not only of saddle cycles, but also of saddle equilibria intersect.

In contrast to separatrix loops in the phase plane, double-asymptotic curves in a phase space of dimension $N \geq 3$ can be robust. A necessary and sufficient condition

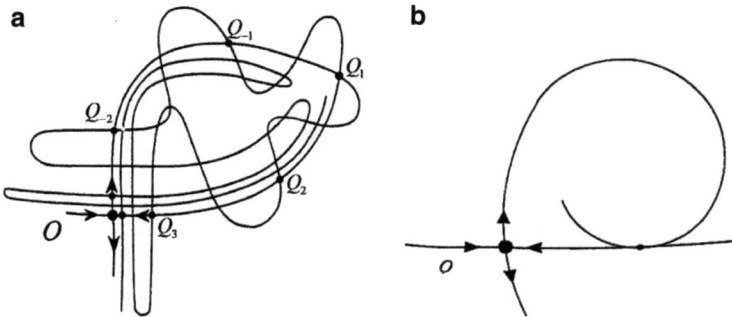

Fig. 8.2 Transverse intersection of manifolds (**a**) and tangency of manifolds (**b**) of a saddle cycle (in a secant plane). The sequence of points $Q_{\pm 1}$, $Q_{\pm 2}$,... belongs to the cross-section of one of the transverse homoclinic curves

for robustness of homoclinic and heteroclinic curves is the condition of transverse intersection of the stable and unstable manifolds which these curves belong to. When the manifolds are tangent, the transversality condition is violated and the double-asymptotic trajectory existing at the moment of tangency is nonrobust (Fig. 8.2b).

8.3 Structurally Stable Systems in $\mathbb{R}^N$, $N \geq 3$: Hyperbolicity

For the space $\mathbb{R}^N$, $N \geq 3$, three types of structurally stable (robust) systems can be distinguished: *Morse–Smale systems*, *Anosov systems*, and *Smale systems with nontrivial hyperbolicity*.

8.3.1 Morse–Smale Systems

These form a class of DS satisfying the following requirements:

- The phase space of a DS contains a finite number of structurally stable limit sets which are equilibria and periodic orbits.
- Stable and unstable manifolds of equilibria and periodic trajectories intersect transversally.

Morse–Smale (MS) systems are the ones with the simplest behavior and constitute a generalization of the class of Andronov–Pontryagin systems to phase space dimensions $N > 2$. There are only a finite number of regular attractors in the phase space of MS systems. A peculiarity of MS systems is the absence of separatrix

8.3 Structurally Stable Systems in $\mathbb{R}^N$, $N \geq 3$: Hyperbolicity

loops, contours, and nontransverse intersections of manifolds of saddle orbits. The notion of MS systems can also be extended to discrete-time DS (return maps).

8.3.2 Hyperbolic Sets

Anosov systems or *C-systems*, together with Smale systems with nontrivial hyperbolicity, are a more complicated type of structurally stable system in $\mathbb{R}^N$, $N \geq 3$. But before defining C-systems and Smale systems, we must start by explaining the notion of *hyperbolic set*. Generally speaking, the property of hyperbolicity plays a very important role in nonlinear dynamics. For greater simplicity, we start by defining a *hyperbolic equilibrium* and a *hyperbolic limit cycle*.

An equilibrium of a DS is said to be hyperbolic if it has no eigenvalues on the imaginary axis. Thus, a node (stable or unstable), a focus (stable or unstable), a saddle, and a saddle focus are hyperbolic equilibria. A center-type equilibrium point is an example of a nonhyperbolic equilibrium. A limit cycle of a continuous-time DS is hyperbolic if it has only one multiplier whose absolute value is equal to 1. A fixed point or cycle of a return map is hyperbolic if it has no multiplier whose absolute value is equal to one. As can be seen from the definition, hyperbolic equilibria and hyperbolic limit cycles are structurally stable points and cycles. All equilibria and limit cycles in Anosov and MS systems are hyperbolic.

We generalize the definition of hyperbolicity to an arbitrary invariant set of the continuous-time autonomous system (8.1). Let $T^\tau(\mathbf{x})$ be the evolution operator in the time interval τ, acting on the state $\mathbf{x}$. The space $E_\mathbf{x}$ spanned by eigenvectors of the linearized evolution operator $DT_\mathbf{x}^\tau$ is a tangent space of the DS at the point $\mathbf{x}$. The invariant set $M \in \mathbb{R}^N$ of the system (8.1) is hyperbolic if at any point $\mathbf{x} \in M$, the tangent space $E_\mathbf{x}$ can be represented as a direct sum of three subspaces, viz.,

$$E_\mathbf{x} = E_\mathbf{x}^s \oplus E_\mathbf{x}^u \oplus E_\mathbf{x}^c .$$

The space $E_\mathbf{x}^s$ is the *subspace of contracting vectors*, or the *stable subspace*, $E_\mathbf{x}^u$ is the *subspace of stretching vectors*, or the *unstable subspace*, and $E_\mathbf{x}^c$ is the *central subspace* generated by the phase velocity vector. The central subspace has dimension equal to one. For any vectors $\boldsymbol{\xi} \in E_\mathbf{x}^s$, $\boldsymbol{\eta} \in E_\mathbf{x}^u$, $\boldsymbol{\kappa} \in E_\mathbf{x}^c$ and at any $\tau > 0$ the following inequalities hold:

$$\|DT_\mathbf{x}^\tau \boldsymbol{\xi}\| < a|\boldsymbol{\xi}|e^{-C\tau} , \qquad \|DT_\mathbf{x}^{-\tau} \boldsymbol{\eta}\| < b|\boldsymbol{\eta}|e^{-C\tau} , \qquad \|DT_\mathbf{x}^{\pm\tau} \boldsymbol{\kappa}\| < k|\boldsymbol{\kappa}| .$$

Here a, b, k, and C are positive constants independent of $\mathbf{x}$.

The above definitions of a hyperbolic equilibrium and a hyperbolic limit cycle follow from this general definition. It also implies that double-asymptotic trajectories belonging to the transverse intersection of manifolds are also hyperbolic sets. Thus, all invariant limit sets of Morse–Smale systems are hyperbolic. This is the underlying reason for the structural stability of MS systems.

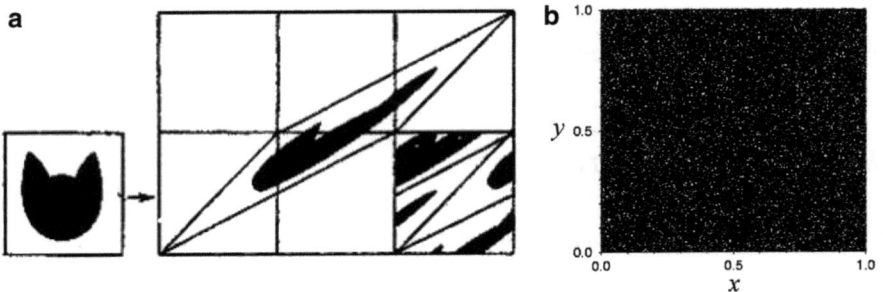

Fig. 8.3 The cat map action: (**a**) in one iteration, and (**b**) in 1,000 iterations (Taken from the book Arnold [5])

8.3.3 Anosov Systems

Time-reversible dynamical systems for which the hyperbolicity condition is fulfilled at each point of the phase space are called *Anosov systems* or *C-systems*. They can be *C-flows* or *C-cascades*. The latter are also called *Anosov diffeomorphisms*. Mathematicians have rigorously proven that C-systems are robust (structurally stable).

As an example of the Anosov diffeomorphism, consider the following map on an N-dimensional torus:

$$\Theta_{n+1} = \hat{A}\Theta_n + \mathbf{F}(\Theta_n) \pmod{1}, \tag{8.4}$$

where $\hat{A}$ is a constant matrix with integer elements, different from the identity matrix and such that $|\det \hat{A}| = 1$, and $\mathbf{F}(\Theta)$ is a periodic function with period 1. A special case of the map (8.4) can be represented by the linear automorphism of a torus

$$\begin{aligned} x_{n+1} &= 2x_n + y_n \pmod{1}, \\ y_{n+1} &= x_n + y_n \pmod{1}, \end{aligned} \tag{8.5}$$

known as *Arnold's map* or the *cat map*. This map is area-preserving (conservative). An area element contracts in one direction and is stretched in the other. The action of the map is illustrated in Fig. 8.3.

The system (8.5) can exhibit chaotic dynamics. However, since the area element is preserved, the map (8.5) has no attractor. Contracting Anosov diffeomorphisms on a torus are also possible. The modified cat map (5.5) considered in Chap. 5 exemplifies a contracting torus diffeomorphism. This map contracts the area. In addition, the system has a chaotic nonstrange attractor which is the whole torus, since any trajectory is everywhere dense in it.

8.3.4 Smale Systems with Nontrivial Hyperbolicity: Strange Attractors

Systems for which a hyperbolic set does not occupy the whole phase space can also be structurally stable. Morse–Smale systems constitute the simplest example of such systems. In general, structurally stable systems are those that satisfy certain conditions called *Smale conditions*.

A system satisfies axiom A and is called an A-system if the nonwandering set[1] Ω of the DS is hyperbolic (consists of hyperbolic sets). The nonwandering set Ω is the closure of the periodic trajectories.[2] In addition, the system satisfies the strong transversality conditions for intersection of the manifolds of all saddle limit sets.

It has been proven that, if the DS satisfies axiom A and the condition of strong transversality, it is structurally stable. The necessity and sufficiency of Smale conditions for structural stability have been proven for discrete-time systems. In the case of continuous-time systems, the Smale conditions are apparently necessary, but this has not been rigorously proven.

Morse–Smale systems satisfy axiom A and the condition of strong transversality. However, there are cascades with $N \geq 2$ and flows with $N \geq 3$ which satisfy Smale conditions, but are not Morse–Smale systems. They are called *Smale systems with nontrivial hyperbolicity*. Nontrivial hyperbolicity means that the system contains a nontrivial hyperbolic set. This set has an everywhere dense transverse homoclinic curve. Nontrivial hyperbolic sets can be attracting (attractors) or non-attracting. A classic example of a system with a nontrivial hyperbolic set is the *Smale horseshoe*, which will be considered in Chap. 10.

A *strange attractor* (SA) is, in the strict sense, an attracting nontrivial hyperbolic set. Strange attractors are robust hyperbolic objects. They possess the following properties:

- Saddle periodic orbits, robust homoclinic and heteroclinic trajectories, and also nonclosed Poisson-stable hyperbolic trajectories are everywhere dense in strange attractors.
- A strange attractor has the property of transitivity, i.e., there is at least one phase trajectory that is everywhere dense in the attractor.

---

[1]The *nonwandering set* of a DS is the set of all nonwandering points in phase space. A point is *nonwandering* if, for any specified time interval θ, under the action of the evolution operator, any neighborhood of this point crosses its original position in a time $\tau > \theta$. Points of limit sets are nonwandering. However, the notion of nonwandering sets, unlike limit sets, is applicable to conservative systems. It should also be noted that the notion of a nonwandering point is somewhat different from the notion of a point that is stable according to Poisson. Obviously, Poisson-stable points are nonwandering, but the converse statement is not always true. For example, separatrix loops in the plane consist of nonwandering points but double-asymptotic trajectories that are components of them are not Poisson-stable.

[2]This implies that periodic trajectories are everywhere dense in Ω.

Fig. 8.4 Construction of the Smale–Williams attractor

- Unstable manifolds of all periodic orbits of a strange attractor are located in the attractor.
- The local structure at any point of the cross-section of an attractor is homeomorphic to the Cantor set.

In mathematics, very few examples of robust hyperbolic strange attractors are known. They include the Smale–Williams attractor, the Plykin attractor, and some others, including the attractor in the torus map (5.5), even though we referred to it as nonstrange. Consider, for example, the construction of the Smale–Williams attractor. When a two-dimensional torus is stretched in length, its cross-section contracts so that the volume of the transformed torus is less than the volume of the original one. Afterwards, the transformed torus is wrapped twice and placed inside the original one as shown in Fig. 8.4. The same procedure is repeated with the new torus and so on an infinite number of times. In the limit, we obtain a hyperbolic set with a complex structure, called the Smale–Williams attractor.

Recently, a number of mathematical models and physical devices have been proposed which can realize attractors very similar to robust hyperbolic attractors.[3] However, these models and devices were specially constructed to obtain hyperbolic chaos. Hyperbolic attractors have not yet been found in systems of natural origin.

8.4 Structurally Unstable Dynamical Systems

Most dynamical systems modeling real systems and processes are nonrobust hyperbolic systems. The structural instability of dynamical systems satisfying axiom A is usually related to the violation of the strong transversality condition, i.e., to the appearance of nonrobust homoclinic and heteroclinic curves.

Systems with structurally unstable homoclinic trajectories were first studied by L.P. Shilnikov and N.K. Gavrilov. In their work, they showed that, under certain conditions in systems possessing a homoclinic saddle-focus loop (see Sect. 3.7) and systems possessing a saddle periodic orbit with quadratic tangency of manifolds (as in Fig. 8.2b) in the neighborhood of a nonrobust homoclinic trajectory, a countable

---

[3] See, for example, Kuznetsov [12].

set of nonrobust homoclinic trajectories can arise, together with a countable set of periodic orbits undergoing bifurcations.

Smooth two-dimensional diffeomorphisms with nonrobust homoclinic tangencies were studied by S.E. Newhouse. It was shown that, besides the bifurcations established by Gavrilov and Shilnikov, in the neighborhood of the quadratic tangency of manifolds, there are parameter regions with very complex behavior. We give two basic theorems proved by Newhouse.

Theorem 8.1. *In any neighborhood of a C^r-smooth ($r \geq 2$) two-dimensional diffeomorphism having a saddle fixed point with a structurally unstable homoclinic trajectory, there exist regions where systems with structurally unstable homoclinic trajectories are everywhere dense. These regions are called Newhouse regions.*

Theorem 8.2. *Any arbitrarily small interval $(-\alpha, \alpha)$, where a system parameter α is varied, contains a countable set of Newhouse regions.*

We note that $\alpha = 0$ corresponds to the case of homoclinic tangency.

The Newhouse results were generalized to a multi-dimensional case in work by S.V. Gonchenko, L.P. Shilnikov, and D.V. Turaev. From the theoretical results, it follows that, besides nonrobust homoclinic trajectories, saddle periodic orbits, and chaotic trajectories, a set of stable periodic orbits can also arise in systems with a quadratic homoclinic tangency of manifolds (similar to the one shown in Fig. 8.2b). These orbits may have a very long period and very narrow basins of attraction, so that only a few of them can be detected in numerical simulation.

The results obtained by Shilnikov, Gavrilov, and Newhouse are especially important for solving many problems of nonlinear dynamics. In particular, from these results it follows that chaotic attractors of structurally unstable systems are not, in the strict sense, strange attractors. They are characterized by a more complex structure and properties that we shall now discuss.

8.5 Quasihyperbolic Attractors: Lorenz-Type Attractors

The hyperbolicity conditions stated above are not usually fulfilled for real dynamical systems. Nevertheless, there are dynamical systems whose attractors are close to hyperbolic ones. Such attractors are chaotic, do not enclose stable regular attractors, and preserve these properties under perturbations. From the mathematical point of view, the strong transversality condition is violated for these systems, and nonrobust double-asymptotic curves are formed. However, the formation of these curves does not lead to the appearance of stable periodic orbits.

We shall call nearly (almost) hyperbolic attractors *quasihyperbolic*. There are quasihyperbolic Lozi, Belykh, and Lorenz-type attractors. It has been rigorously proven that these attractors are quasihyperbolic in the above-stated sense. It is advisable to identify and classify the distinctive experimental characteristics of quasihyperbolic attractors that can be used to diagnose them in computer simulations.

8.5.1 Quasihyperbolic Attractor in the Lozi Map

The existence of a quasihyperbolic attractor in a DS can be justified by proving the following two statements:

- All phase trajectories in the attractor are unstable, and
- When parameters of the system are varied, no stable trajectories arise.

This purely mathematical problem cannot be solved in a general way due to the nonlinearity of the DS. Fortunately, this challenge has a solution for some specific dynamical systems.

Consider the simplest example, namely, the *Lozi attractor* in the two-dimensional discrete-time system

$$x_{n+1} = 1 - a|x_n| + y_n ,$$
$$y_{n+1} = bx_n .$$
(8.6)

The map (8.6) is nonlinear, one-to-one, and dissipative (for $b < 1$). By virtue of the diffeomorphism, it is in the strict sense the Poincaré map of some differential system with phase space dimension $N = 3$. Thus, the properties revealed and proven for this map can be reliably applied to a flow in $\mathbb{R}^3$.

It has been established theoretically that, in the parameter range $1.3 < a < 1.8$, the system (8.6) has a single chaotic attractor. In the literature this attractor is known as the *quasihyperbolic Lozi attractor*. It includes nonrobust homoclinic curves resulting from the tangency of manifolds of saddle cycles. However, the break in the unstable manifold at the tangency point leads to the fact that the Newhouse theorems are not applicable. Moreover, such tangencies do not generate stable periodic orbits.

The Lozi attractor and the region (basin) of its attraction are depicted in Fig. 8.5. The attractor G_0 is the only attracting set in the interval $1.3 < a < 1.8$ for fixed $b = 0.3$. It has the homogeneous basin of attraction G_1. Any initial point (x_0, y_0) belonging to G_1 tends to the Lozi attractor in time.

The dependence of the maximal Lyapunov exponent on the parameter a is quite typical. At fixed $b = 0.3$, the Lozi attractor appears hard at $a_{cr} = 1.3$ and remains chaotic in the whole region of its existence $1.3 < a < 1.8$. The dependence $\lambda_1(a)$ has no zero values and is a smooth positive function. This result indicates the absence of stable fixed points (stability windows) in the region where the Lozi attractor exists.

The power spectrum $S(\omega)$ calculated for the x_n coordinate in the region where the Lozi attractor exists is a smooth function and does not include explicit maxima

8.5 Quasihyperbolic Attractors: Lorenz-Type Attractors

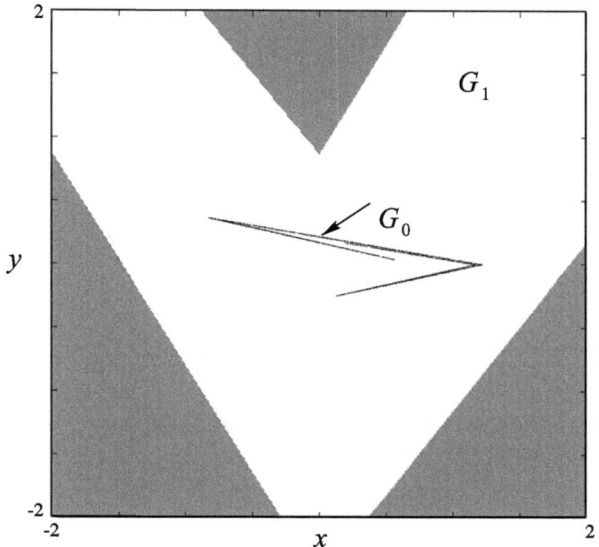

Fig. 8.5 Lozi attractor G_0 and the basin of its attraction G_1 for $a = 1.5$ and $b = 0.3$ (trajectories from the *gray region* go to infinity)

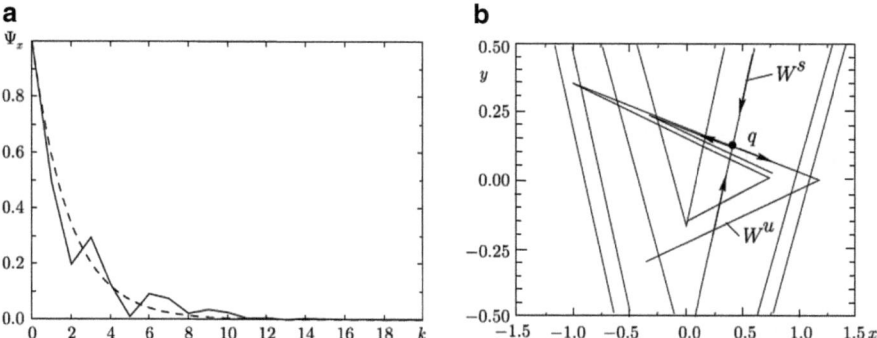

Fig. 8.6 (a) Autocorrelation function for the map (8.6) at $a = 1.75$. The exponential approximation is shown by the *dashed line* ($\lambda_1 = 0.53$ is the maximal Lyapunov exponent for $a = 1.75$). (b) Stable and unstable manifolds of a saddle equilibrium q in (8.6) for $a = 1.7, b = 0.3$

at any characteristic frequencies. Consequently, the autocorrelation function (ACF) of the x_n process decreases nearly exponentially (Fig. 8.6a).

Comparison of Figs. 8.5 and 8.6b shows that the stable manifolds of saddle cycles define the boundaries of the basin of attraction, and the chaotic attractor itself is located along the unstable separatrices repeating their form. As can be seen from Fig. 8.6b, the tangency of the manifolds cannot be quadratic, so the appearance of homoclinic trajectories does not lead to the birth of stable periodic orbits. The

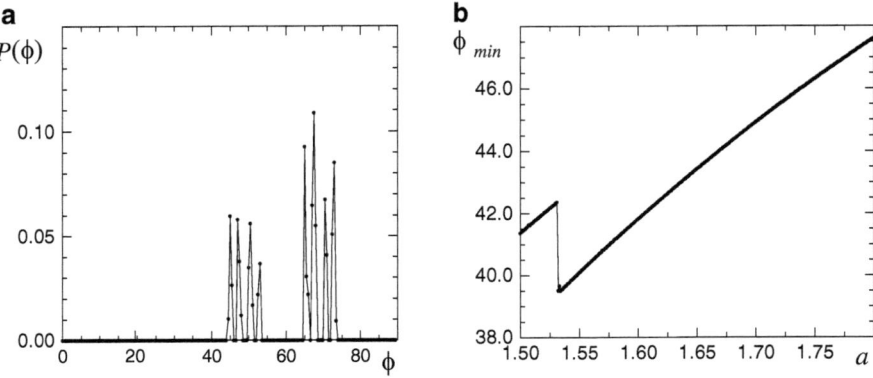

Fig. 8.7 (a) Angle probability distribution between the stable and unstable manifolds of a chaotic trajectory on the Lozi attractor for $a = 1.7$, $b = 0.3$ and (b) ϕ_{min} as a function of the parameter a for $b = 0.3$

hyperbolic chaotic set is the only attracting limit set in the phase space of the system (8.6).

The Lozi map is one of the simplest systems for which the hyperbolicity condition can be checked numerically. For this purpose, a special computer program has been developed to calculate the probability $P(\phi)$ of the angle ϕ between the stable and unstable manifolds of a saddle trajectory (x_n, y_n) as $n \to \infty$. The numerical results are shown in Fig. 8.7a for 18,000 points on the attractor. It can be seen that there is a minimal nonzero angle value ϕ_{min} that depends on the system parameters (Fig. 8.7b). In the whole interval of values of the parameter a where the chaotic attractor exists, the minimal angle between the stable and unstable manifolds of the phase trajectory is greater than 39° and never vanishes.

8.5.2 The Lorenz Attractor

Consider the differential Lorenz system that has a quasihyperbolic attractor, known as the *Lorenz attractor*. One of the hyperbolicity requirements (the condition of strong transversality) is violated both for the Lorenz attractor and the Lozi attractor. The Lorenz attractor includes a saddle equilibrium point at the origin and homoclinic loops arising from this point. However, these nonrobust trajectories do not lead to the emergence of stable limit cycles and do not affect the observed characteristics of chaos. Lorenz-type attractors are found in a number of systems and are a typical example of quasihyperbolic attractors. The Lorenz attractor is qualitatively characterized by the same properties as the Lozi attractor and is considered to be a classic example of quasihyperbolic chaos.

8.5 Quasihyperbolic Attractors: Lorenz-Type Attractors

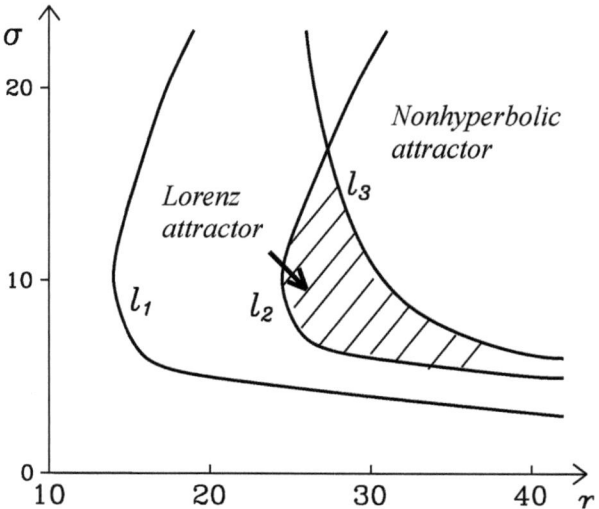

Fig. 8.8 Bifurcation diagram of the Lorenz system in the parameter plane (r, σ) for $b = 8/3$

The Lorenz equations were obtained from the Navier–Stokes equations for convective fluid flow and have the form

$$\dot{x} = -\sigma(x - y), \qquad \dot{y} = rx - y - xz, \qquad \dot{z} = xy - bz, \qquad (8.7)$$

where σ, b, and r are the control parameters. Certain laser models as well as the disc dynamo model can be reduced to equations of type (8.7).

We note that the quasihyperbolic attractor regime in the system (8.7) is realized in a finite region of its control parameter values. The bifurcation diagram of the system is shown in Fig. 8.8. The Lorenz attractor exists in the shaded region in the parameter space. The phase portrait of the Lorenz attractor is presented in Fig. 8.9a. Outside the indicated region, the properties of the chaotic attractor are different: the Lorenz attractor is transformed into a nonhyperbolic attractor.

The Lorenz attractor has the following typical properties. The LCE spectrum does not change when the initial conditions are varied because the Lorenz attractor is the only one and the basin of its attraction is the whole phase space. The LCE spectrum remains practically unchanged if the system control parameters are varied in the region where the Lorenz attractor exists. These properties clearly illustrate the robustness of the Lorenz attractor from an experimental standpoint: the attractor preserves its structure and undergoes no bifurcations when the parameters and initial conditions are varied.

The autocorrelation function of the Lorenz attractor decreases exponentially and almost monotonically with time, as illustrated in Fig. 8.9b. Comparison of the ACFs for the Lorenz and Lozi attractors indicates a qualitative equivalence of dynamical processes in nearly hyperbolic systems.

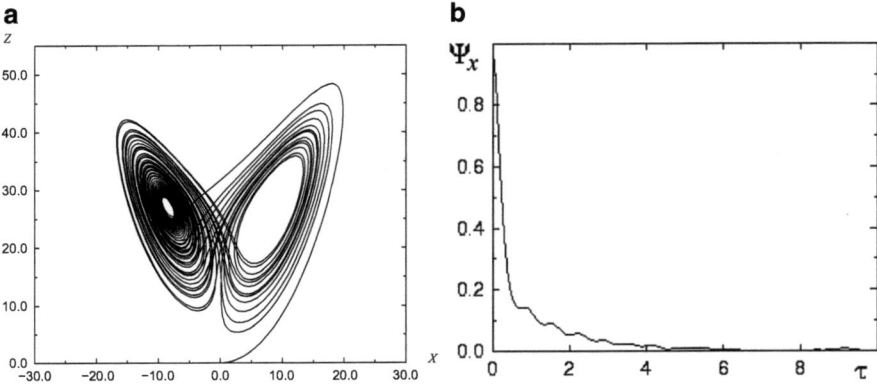

Fig. 8.9 Phase portrait (**a**) and autocorrelation function (**b**) of the Lorenz attractor for $\sigma = 10$, $r = 28$, and $b = 8/3$

8.6 Nonhyperbolic Attractors and Their Properties

Systems with nonhyperbolic attractors exhibit regimes of deterministic chaos which are characterized by exponential instability of trajectories and a fractal structure of the attractor. From this point of view, the characteristics of the indicated regimes are similar to those of robust hyperbolic and quasihyperbolic attractors. However, there is a crucial difference which must be understood in order to avoid misinterpretation of experimental results. A distinctive feature of nonhyperbolic attractors is the simultaneous coexistence of a countable set of different chaotic and regular attracting subsets in a bounded element of the system phase space volume for fixed system parameters. This collection of all coexisting limit subsets of trajectories in the bounded region G_0 of phase space, which are approached by all or almost all trajectories from the region G_1 including G_0, is called a *nonhyperbolic attractor* of the DS. But the complexity is actually greater than this. When the system parameters are varied in some finite range, both regular and chaotic attractors exhibit cascades of different bifurcations. Accordingly, their basins of attraction are also rebuilt. The complexity of nonhyperbolic attractors is caused by the effects of homoclinic tangencies of stable and unstable manifolds of saddle trajectories or by the appearance of a saddle-focus separatrix loop. All these effects are realized on a set of parameter values of nonzero measure. In other words, the condition of transversality of manifolds is violated.

If in this case we take into account the fact that basins of attraction of coexisting limit sets may have a fractal structure and occupy extremely narrow regions in phase space, it becomes clear that the accuracy of computer calculations and the effect of fluctuations are crucially important.

8.6 Nonhyperbolic Attractors and Their Properties

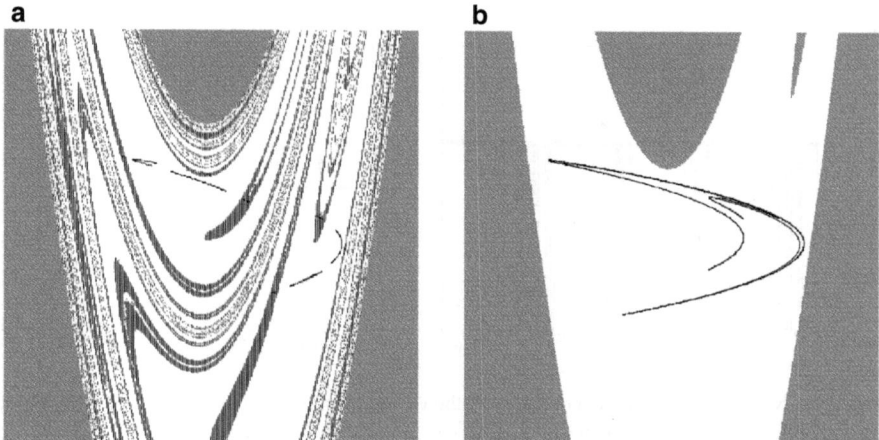

Fig. 8.10 Attractors in the Henon map and their basins of attraction in the phase plane (x_n, y_n) at $b = 0.3$: (**a**) for $a = 1.078$, and (**b**) for $a = 1.32$. The *white region* in (**a**) is the basin of attraction of the four-band chaotic attractor, the *dark region* in (**a**) corresponds to the basin of attraction of the six-band chaotic attractor, and trajectories from the *gray regions* in (**a**) and (**b**) go to infinity

8.6.1 Nonhyperbolic Attractor in the Henon Map

Consider the well-known Henon map

$$\begin{aligned} x_{n+1} &= 1 - ax_n^2 + y_n , \\ y_{n+1} &= bx_n . \end{aligned} \tag{8.8}$$

This map is dissipative for $0 < b < 1$ and possesses a nonhyperbolic attractor. Furthermore, it is one-to-one, i.e., a diffeomorphism. The Henon map differs from the Lozi map by a smooth quadratic function that specifies the nonlinearity.

Consider phase portraits of attracting sets in the system (8.8), their basins of attraction, and how they evolve when the parameter a is varied for fixed $b = 0.3$. The regime of coexistence of two attracting subsets in the phase plane is illustrated in Fig. 8.10a. When the initial conditions change, a sharp alternation of two regimes is observed. Figure 8.11a shows calculation results for the maximal Lyapunov exponent when $a = 1.078$ and for different initial values of the x coordinate at a fixed y. It can be seen that λ_1 'jumps' randomly between $\lambda_1 = 0.126$ and $\lambda_1 = 0.062$, indicating that the system goes from one chaotic attractor to another. If these results are compared with the structure of the basins of attraction (Fig. 8.10a), it becomes clear why this is happening. The change in the initial conditions leads to the boundary crossing of the corresponding basins of attraction.

The basin of the attractor for $a = 1.32$ (Fig. 8.10b) looks homogeneous, and this indicates the presence of a single attractor only. But it is known that the system

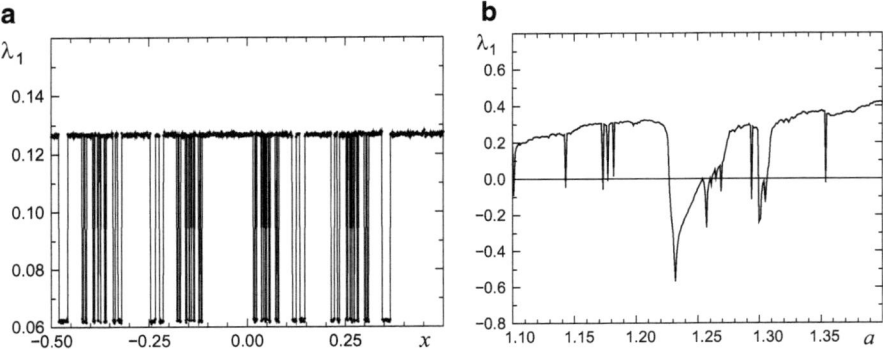

Fig. 8.11 Maximal Lyapunov exponent λ_1 of the Henon map as a function (**a**) of initial values of the x coordinate at $y = 0.5$ when $a = 1.078$, $b = 0.3$, and (**b**) of the parameter a for $b = 0.3$

can possess coexisting stable cycles with long periods and very narrow basins of attraction, which cannot be detected numerically.

When the control parameter a is varied in the nonhyperbolic attractor region $1.1 < a < 1.4$, an alternating picture of change between regular and chaotic attractors is observed. The calculation results for the maximal Lyapunov exponent λ_1 as a function of a are presented in Fig. 8.11b. The dependence $\lambda_1(a)$ shows both positive and negative values of λ_1, suggesting an irregular alternation of chaotic and periodic attractors in the system when the parameter a is varied.

Chaotic attractors of the Henon model also possess other typical features: their power spectra have δ-peaks at multiple frequencies, depending on a number of bands of chaotic attractors, and the ACF may not even decrease to zero if there are 2 or more bands.

The Henon attractor does not satisfy the transversality condition for the intersection of stable and unstable manifolds of saddle cycles on a countable set of the parameter values for which there is a tangency of manifolds.

The calculation results for stable and unstable separatrices of a saddle point are presented in Fig. 8.12. At certain points, the stable W^s and unstable W^u separatrices touch each other and the angle between them is equal to zero. There is then a quadratic tangency of the manifolds.

Following the evolution of the angle ϕ between the manifolds along a chaotic trajectory, one can calculate the angle probability distribution $P(\phi)$. The calculation results are shown in Fig. 8.13a and testify to the fact that the probability of zero angles is reliably greater than zero, i.e., the tangency of manifolds has nonzero probability.

The probability $P^{\delta\phi}$ of angles close to zero ($\delta\phi \leq 1°$) can be calculated for different values of the system parameter a. The numerical results are presented in Fig. 8.13b. We see that there is a countable set of parameter values at which the probability $P^{\delta\phi}$ is zero. These values of the parameter a clearly correspond to the presence of periodic attractors in the map (compare with the data in Fig. 8.11b).

8.6 Nonhyperbolic Attractors and Their Properties

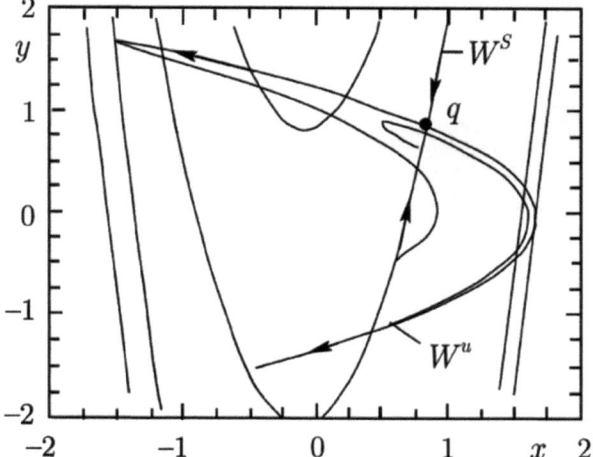

Fig. 8.12 Stable and unstable manifolds of a saddle point q of the Henon map for $a = 1.3$ and $b = 0.3$

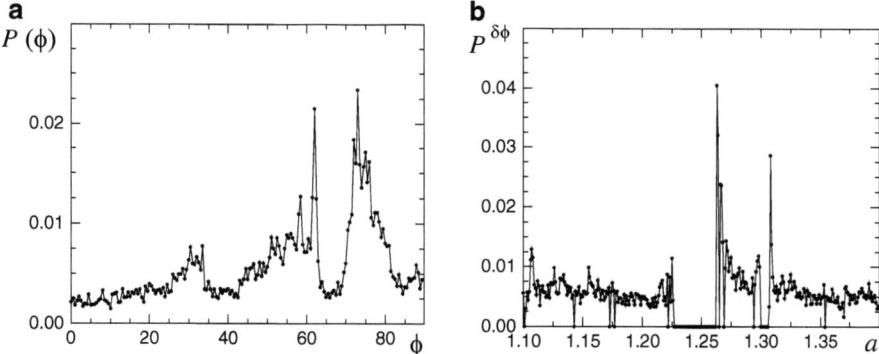

Fig. 8.13 (a) Angle probability distribution between the directions of stable and unstable manifolds of a chaotic trajectory of the Henon attractor for $a = 1.179$ and $b = 0.3$, and (b) probability $P^{\delta\phi}$ of angles ϕ close to zero ($\delta\phi \leq 1°$) as a function of the control parameter a for $b = 0.3$

Generally speaking, the equality $P^{\delta\phi}=0$ may indicate the regime of hyperbolic chaos. However, in the experiments actually conducted, this condition corresponds to the regimes of regular cycles in the Henon map.[4]

Finally, we discuss the properties of nonhyperbolic attractors, which are quite important in the analysis and interpretation of experimental results. We return to Fig. 8.10. The structure of basins of attraction shows that the system is highly sensitive to the accuracy in setting initial conditions. Let us conduct the following

---

[4]Due to certain specifics of the calculation algorithm, the probability is set to zero if the phase trajectory has no unstable manifold.

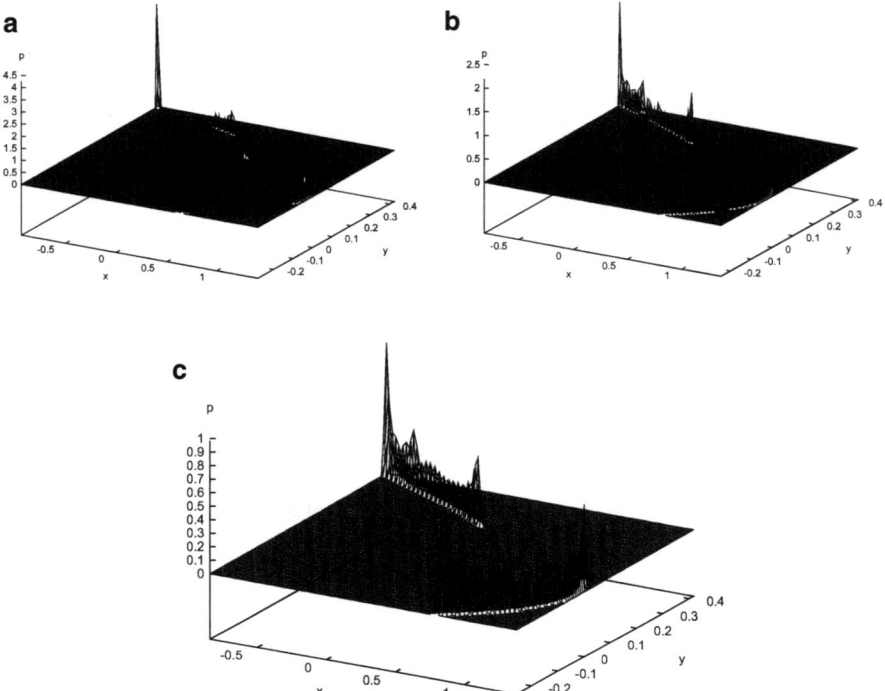

Fig. 8.14 (a) and (b) Probability distribution densities $p(x, y)$ on the chaotic attractors coexisting in the Henon map for $a = 1.08$ and $b = 0.3$ without noise, and (c) the effect of noise with intensity $D = 5 \times 10^{-6}$ on these distributions

experiment. We choose the parameter $a = 1.078$ at which two chaotic attractors coexist in (8.8). Figure 8.14a, b show probability distribution densities $p(x, y)$ for these attractors. We now add nearly white noise of intensity $D = 5 \times 10^{-6}$ to both equations of the map (8.8). The influence of weak noise causes the two coexisting attractors to merge into one (Fig. 8.14c). The resulting regime does not depend on which basin of attraction the initial conditions are chosen from.

We note that, as $b \to 0$, the Henon model (8.8) reduces to the well-known one-dimensional logistic map:

$$x_{n+1} = 1 - ax_n^2 . \tag{8.9}$$

The map (8.9) is irreversible and not a diffeomorphism. Nevertheless, the properties of attractors in the model (8.9) will be qualitatively the same as the above-mentioned properties of the Henon map.

8.6 Nonhyperbolic Attractors and Their Properties

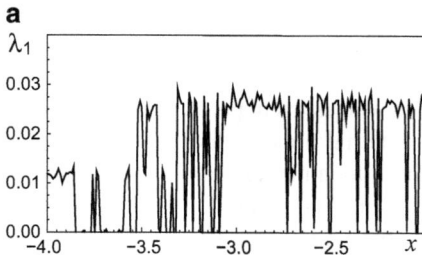
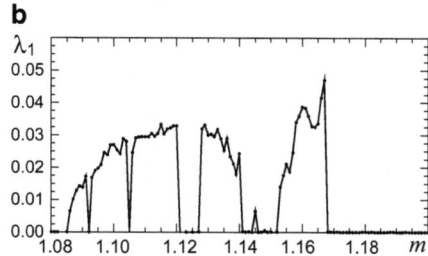

Fig. 8.15 Maximal Lyapunov exponent of the system (8.10) as a function (**a**) of initial values of the x coordinate for $m = 2.412$ and $g = 0.097$, and (**b**) of the parameter m for $g = 0.3$

8.6.2 Nonhyperbolic Attractor in the Oscillator with Inertial Nonlinearity

Consider the Anishchenko–Astakhov oscillator which illustrates the mechanism of birth and properties of a nonhyperbolic attractor in systems with a saddle-focus separatrix loop. This model is described by the three-dimensional two-parameter differential system

$$\dot{x} = mx + y - xz,$$
$$\dot{y} = -x, \qquad (8.10)$$
$$\dot{z} = -gz + gI(x)x^2, \quad I(x) = \begin{cases} 1, & x > 0, \\ 0, & x \leq 0. \end{cases}$$

We fix $m = 2.412$ and $g = 0.097$ and calculate the dependence of the maximal Lyapunov exponent on the initial conditions. Numerical results are shown in Fig. 8.15a and suggest that two chaotic and one periodic regime of oscillations coexist in the system.

When the system parameters are varied, its limit sets undergo bifurcations. This can be illustrated by the dependence of the maximal Lyapunov exponent on the parameter m, as shown in Fig. 8.15b. Zero values of λ_1 indicate the birth of one of the sets of limit cycles undergoing cascades of period-doubling bifurcations. The bifurcations of attractors are accompanied by a change in the structure of their basins of attraction, which assume fractal properties.

The presence of stable and saddle cycles as well as chaotic limit sets in the nonhyperbolic attractor manifests itself in the structure of the autocorrelation function and the power spectrum. The ACF decreases on average exponentially in time, but contains periodic components. The power spectrum is continuous, but possesses sharp peaks at certain characteristic frequencies. These features of the ACF and the power spectrum of the chaotic regime are typical and clearly distinguish a nonhyperbolic attractor from the Lorenz attractor, as well as the nonhyperbolic Henon attractor from the quasihyperbolic Lozi attractor.

The fractal boundaries of the basins of attraction and the fact that the system exhibits a set of bifurcations of various regimes when system parameters are slightly varied make the system highly sensitive to external noise influences, similarly to what is observed for the Henon attractor. The fundamental reason for the above-mentioned properties of a nonhyperbolic attractor is undoubtedly the countable set of homoclinic tangencies of stable and unstable manifolds. This is why systems of the type (8.10) are nonrobust. In the Henon map, the effects of homoclinic tangency can be observed numerically, while this problem is still unsolved for three-dimensional systems. However, the data presented in Fig. 8.15 are undoubtedly a consequence of the effect of homoclinic tangency, and can be considered to be typical features of a nonhyperbolic attractor.

8.7 Summary

Examination of the structure and properties of attractors of nonlinear dissipative systems, viewed as the images of complex nonperiodic self-sustained oscillations, leads to the following conclusions:

1. Robust hyperbolic systems and Lorenz-type systems exhibit classic properties of deterministic chaos as nonperiodic exponentially unstable solutions of the corresponding dynamical systems. Strange (or nearly strange) attractors are their mathematical images. Their distinctive feature is a fractal geometrical structure of the attractor, a fractional metric dimension, and the presence of at least one positive exponent in the LCE spectrum as a consequence of mixing. Robust hyperbolic attractors and Lorenz-type attractors are barely sensitive to the effect of noise. Basins of attraction of such attractors are smooth and homogeneous. The properties of these attractors are not sensitive to changes in initial conditions.
2. Nonhyperbolic attractors are more complex objects. They include a finite or infinite set of regular and chaotic attracting subsets that coexist simultaneously at fixed values of the system parameters. When these parameters are varied in a finite interval, such subsets can undergo an infinite number of different bifurcations. Basins of attraction of coexisting attractors represent an extremely complex structure of embedded regions which have a fractal geometry. As a result, nonhyperbolic attractors are highly sensitive to changes in initial conditions and to the effect of noise.

A detailed mathematical analysis of the problem considered here is given in [1–4, 6–11, 13, 14].

References

1. Afraimovich, V.S., Arnold, V.I., Ilyashenko, Yu.S., Shilnikov, L.P.: Dynamical Systems V. Encyclopedia of Mathematical Sciences. Springer, Heidelberg (1989)
2. Afraimovich, V.S., Shilnikov, L.P.: Strange attractors and quasiattractors. In: Barenblatt, G.I., Iooss, G., Joseph, D.D. (eds.) Nonlinear Dynamics and Turbulence, p. 1. Pitman, Boston (1983)
3. Anishchenko, V.S.: Dynamical Chaos – Models and Experiments. World Scientific, Singapore (1995)
4. Anishchenko, V.S., Astakhov, V.V., Neiman, A.B., Vadivasova, T.E., Schimansky-Geier, L.: Nonlinear Dynamics of Chaotic and Stochastic Systems. Springer, Berlin (2002)
5. Arnold, V.I.: Additional Chapters of the Theory of Differential Equations. Nauka, Moscow (1978) (in Russian)
6. Barreira, L., Pesin, Y.: Nonuniform Hyperbolicity: Dynamics of Systems with Nonzero Lyapunov Exponents. Encyclopedia of Mathematics and Its Applications. Cambridge University Press, Cambridge (2007)
7. Bowen, R.: Equilibrium States and the Ergodic Theory of Anosov Diffeomorphisms. Lecture Notes in Mathematics, vol. 470. Springer, Berlin (1975)
8. Devaney, R.L.: An Introduction to Chaotic Dynamical Systems. Westview, Boulder (1989, 2003)
9. Feudel, U., Kuznetsov, S., Pikovsky, A.: Strange Nonchaotic Attractors: Dynamics Between Order and Chaos in Quasiperiodically Forced Systems. World Scientific, Singapore (2006)
10. Guckenheimer, J., Holmes, P.: Nonlinear Oscillations, Dynamical Systems, and Bifurcations of Vector Fields. Springer, New York (1983)
11. Katok, A., Hasselblatt, B.: Introduction to the Modern Theory of Dynamical Systems. Cambridge University Press, Cambridge (1995)
12. Kuznetsov, S.P.: Dynamical chaos and uniformly hyperbolic attractors: from mathematics to physics. Phys. Uspekhi **181**(2), 121–149 (2011)
13. Lichtenberg, A., Lieberman, M.A.: Regular and Stochastic Motion. Springer, New York (1983)
14. Zaslavsky, G.M.: Chaos in Dynamical Systems. Harwood, New York (1985)

Chapter 9
Characteristics of Poincaré Recurrences

9.1 Introduction

The analysis of Poincaré recurrences is one of the fundamental problems in the theory of dynamical systems. Poincaré recurrence means that practically any phase trajectory starting from some point of the system phase space passes arbitrarily close to the initial state an infinite number of times. H. Poincaré called these phase trajectories stable according to Poisson. Since Poincaré's day, the analysis of the dynamics of Poisson stable systems has been an active topic of research in both mathematics and physics. The fundamental importance of this problem is evidenced by the fact that the very idea that a system should return over time to a neighborhood of its initial state is used much more widely than in mathematical theory alone. Thus, in a certain sense, it has become one of the philosophical concepts of modern science.

By now, the rigorous mathematical theory of Poincaré recurrences has been rather fully developed. It describes the statistical regularities of a random sequence of return times both in a neighborhood of a given initial state and in some subset of the system phase space. The first case underlies the so-called *local approach*. The second is used in the *global approach*. The local approach deals with statistical characteristics of recurrence times in some local vicinity of the selected point. In the framework of the global approach, recurrence times are considered in all covering elements of the whole set and their statistics is then studied. In this case, recurrence times depend on a sequence of initial points specified in each covering element of the set, and they are a function of the whole of that set.

Originally, the theorem about recurrences was proved by Poincaré for conservative systems with a given positive measure that is preserved under any transformation. Later, this theorem was generalized to the case of dissipative systems having the property of ergodicity. The fact that none of the results of this theorem depend on the dimension of the system phase space is an important feature of the Poincaré recurrence theory. This circumstance enables one to illustrate the recurrence statistics experimentally using relatively simple low-dimensional dynamical systems as examples.

In this chapter we describe the main mathematical results of the Poincaré recurrence theory, obtained in the frameworks of both the local and global approaches. The regularities of Poincaré recurrence statistics are illustrated for low-dimensional discrete-time systems in the regime of dynamical chaos. The dynamics of a system in the chaotic regime is characterized by Poisson-type stability. Due to the presence of mixing, the chaotic regime ensures that the system is ergodic.

9.2 Local Approach

Poincaré's fundamental theorem proved more than a 100 years ago underlies the recurrence theory. The theorem is formulated as follows:

> For ergodic systems with a given invariant, finite and positive measure, almost any phase trajectory that starts from some initial point of a set in phase space returns to a neighborhood of the initial state infinitely many times.

These trajectories are stable according to Poisson. Almost 30 years later, the Poincaré theorem was generalized by N. Chetaev to a class of periodically driven dynamical systems.

Recurrence times of a phase trajectory in the vicinity of the initial state represent a random sequence. Hence, the characteristics of recurrence times can be described using statistical methods. Figure 9.1 is a schematic representation of a fragment of the phase trajectory as it returns to the vicinity Δ of the initial point during the time $\tau_{ri}(\Delta) = t_{i+1} - t_i$. Since the system is ergodic, the mean recurrence time $\langle \tau_r(\Delta) \rangle$ can be calculated by averaging over time:

$$\langle \tau_r(\Delta) \rangle = \frac{1}{n} \sum_{i=1}^{n} \tau_{ri}(\Delta), \quad n \to \infty. \tag{9.1}$$

9.2.1 Kac's Lemma

One of the important results in Poincaré recurrence theory is the proof that the mean recurrence time in any region Δ of the considered set on which the probability measure is specified is inversely proportional to the probability of the phase trajectory being in this region:

$$\langle \tau_r(\Delta) \rangle \sim \frac{1}{P(\Delta)}. \tag{9.2}$$

9.2 Local Approach

Fig. 9.1 Qualitative illustration of the first return of the phase trajectory Γ to the region Δ belonging to the studied set in $\mathbb{R}^N$

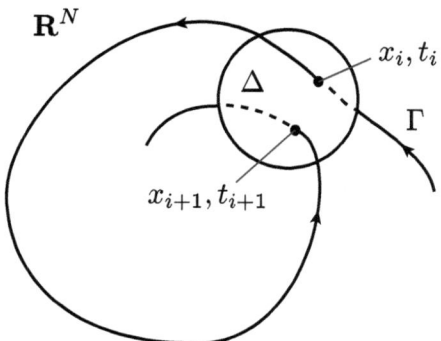

For discrete-time dynamical systems the coefficient of proportionality is strictly equal to unity if the measure satisfies the property of ergodicity. The validity of (9.2) was originally proved under the condition that the system be characterized by an ergodic probability measure and be reversible. More recent studies have shown that the requirement of reversibility is not needed. When the region Δ is small enough, (9.2) can be converted to a form convenient for numerical experiments. We express the probability through the distribution density as follows:

$$P(\mathbf{x}_0, \varepsilon) = \int_\varepsilon p(\mathbf{x}) d\mathbf{x}, \tag{9.3}$$

where $\mathbf{x}_0$ is a given initial point and ε is a small region in the vicinity of the initial point. Assuming that the size of this region is $\varepsilon \ll 1$, one can use the mean value theorem to derive the following expression:

$$P(\mathbf{x}_0, \varepsilon) = p(\mathbf{x})\varepsilon^d. \tag{9.4}$$

When the distribution density is a smooth function, we have $d = N$, where N is the dimension of the system phase space. If the distribution density possesses fractal properties, then $d = d_f$, where d_f is the fractal dimension.

Using (9.3) and (9.4), we can rewrite Kac's lemma in the form

$$\ln \langle \tau_r(\mathbf{x}_0, \varepsilon) \rangle = C - k \ln \varepsilon, \quad k = N \text{ or } d_f, \quad C = \ln[p(\mathbf{x}_0)]^{-1}. \tag{9.5}$$

From (9.5), it follows that the dependence of the mean recurrence time $\langle \tau_r(\mathbf{x}_0, \varepsilon) \rangle$ on the size of the neighborhood ε (on a logarithmic scale) is a straight line with slope k. The shift of this straight line relative to the origin is determined by the coefficient C, which depends on the distribution density. Thus, performing numerical calculations and presenting the results as the dependence (9.5), one can define the dimension d_f, the mean recurrence time for specific values of the neighborhood ε, and the shift coefficient C which contains information on the distribution density $p(\mathbf{x})$.

9.2.2 Exponential Law for Distribution of First Recurrence Times

Finally, let us discuss another fundamental result obtained in the framework of the local approach. For hyperbolic systems possessing the mixing property, it has been proven that the distribution density of the first recurrence times in an ε-neighborhood of the given initial state $\mathbf{x}_0$ satisfies the exponential law

$$p(\tau_r) = \frac{1}{\langle \tau_r \rangle} \exp\left(-\frac{\tau_r}{\langle \tau_r \rangle}\right), \qquad \tau_r \geq \tau^* . \tag{9.6}$$

Equation (9.6) holds in the limit $\varepsilon \to 0$ for all $\tau_r \geq \tau^*$, where τ^* is some value of the recurrence time and $\langle \tau_r \rangle$ is the mean time required for the first recurrence in the ε-neighborhood of the initial state $\mathbf{x}_0$. We note that, compared with Kac's lemma, the exponential law (9.6) is satisfied only if the ergodic system also has the property of mixing.

The law (9.6) enables one to calculate the moments of the random quantity τ_r. In particular, we can derive an expression for the variance σ_r^2, viz.,

$$\sigma_r^2 = \langle (\tau_r - \langle \tau_r \rangle)^2 \rangle = \int_0^\infty p(\tau_r)(\tau_r^2 - \langle \tau_r \rangle^2) d\tau_r = \langle \tau_r \rangle^2 . \tag{9.7}$$

The exponential law (9.6) can be rewritten in the form

$$\ln p(\tau_r) = C - \frac{1}{\langle \tau_r \rangle} \tau_r, \qquad C = \ln \frac{1}{\langle \tau_r \rangle} . \tag{9.8}$$

Equation (9.8) represents a straight line with slope $k = 1/\langle \tau_r \rangle$, shifted by C with respect to the origin.

9.2.3 Numerical Examples

The fundamental results for the local approach to the Poincaré recurrence theory listed above have been repeatedly confirmed by numerical calculations on various examples. Let us consider some of them.

We study the one-dimensional logistic map

$$x_{n+1} = r x_n (1 - x_n) , \tag{9.9}$$

where r is the control parameter. Figure 9.2 shows numerical results for the dynamics of this map. The distribution density $p(x)$ is illustrated in Fig. 9.2a for the parameter value $r = 4$. It corresponds to the theoretical law

9.2 Local Approach

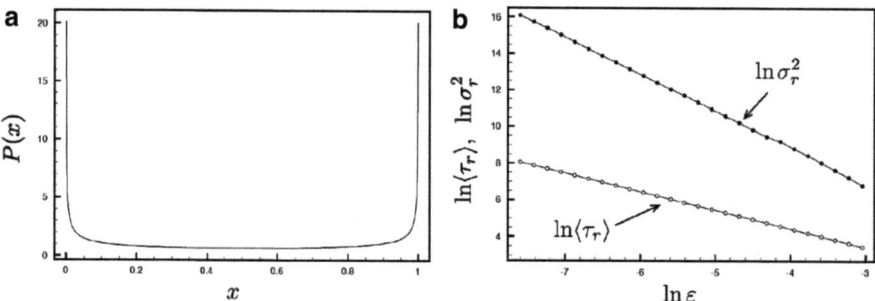

Fig. 9.2 Numerical results for the logistic map (9.9) when $r = 4.0$ and $x_0 = 0.5$. (**a**) Distribution density $p(x)$ on the system attractor. (**b**) Dependences of $\ln\langle\tau_r\rangle$ and $\ln\sigma^2$ on $\ln\varepsilon$. Their slopes are 1.008 and 2.0208, respectively

$$p(x) = \frac{1}{\pi\sqrt{x(1-x)}} \qquad (9.10)$$

and is represented by a continuous smooth function. The dependences of $\ln\langle\tau_r\rangle$ and $\ln\sigma^2$ on $\ln\varepsilon$ are shown in Fig. 9.2b. The plots are straight lines with slopes $k = 1.0$ and $k = 2.0$, respectively (see (9.5) and (9.7)). The difference between the numerical results and the theoretical ones does not exceed 2 %, which is related to the finite calculation accuracy.

The results shown in Fig. 9.2 characterize the dependences of $\langle\tau_r\rangle$ and σ^2 on the ε-neighborhood size of the selected point $x_0 = 0.5$ on the attractor. These data change quantitatively for different points of the attractor, but their linear dependences are preserved. This is due to the fact that, since the distribution density $p(x)$ is non-uniform, the corresponding probabilities $P(x_0, \varepsilon)$ will depend on x_0.

To illustrate this, we consider the one-dimensional cubic map

$$x_{n+1} = (\alpha x_n - x_n^3)\exp\left(-\frac{x_n^2}{B}\right) \qquad (9.11)$$

and calculate $\langle\tau_r\rangle$, which depends on the coordinate of a point on the system attractor, for the parameter values $\alpha = 2.84$, $B = 10$, and fixed $\varepsilon = 0.01$. The numerical results are presented in Fig. 9.3 and clearly confirm Kac's lemma: the mean return time $\langle\tau_r\rangle$ is inversely proportional to the probability $P(x_0, \varepsilon)$.

The exponential law (9.6) of the Poincaré recurrence time distribution is also confirmed numerically. Let us discuss this in more detail. The law (9.6) contains only one parameter $\langle\tau_r\rangle$, and at first sight it appears to be independent of the chosen point x_0 of the considered set and also of the value of ε. In fact it is not. From Kac's lemma (9.2) and (9.5), it follows that the value of $\langle\tau_r\rangle = \langle\tau_r(x_0, \varepsilon)\rangle$ depends on x_0 and ε. As mentioned above, the law (9.6) is valid provided that the system has the mixing property and the limit $\varepsilon \to 0$ is considered. Theoretically, this means that the probability $P(x_0, \varepsilon)$ tends to zero and the mean return time $\langle\tau_r\rangle$ grows to

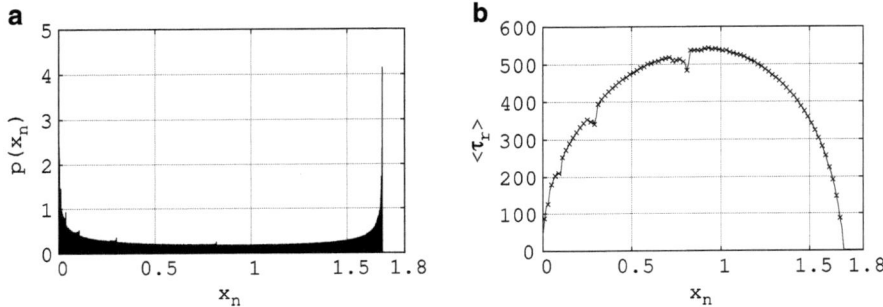

Fig. 9.3 Numerical results for the cubic map (9.11) when $\alpha = 2.84$, $B = 10$, and $\varepsilon = 0.01$. (a) Distribution density $p(x)$. (b) Mean return time $\langle \tau_r(x) \rangle$

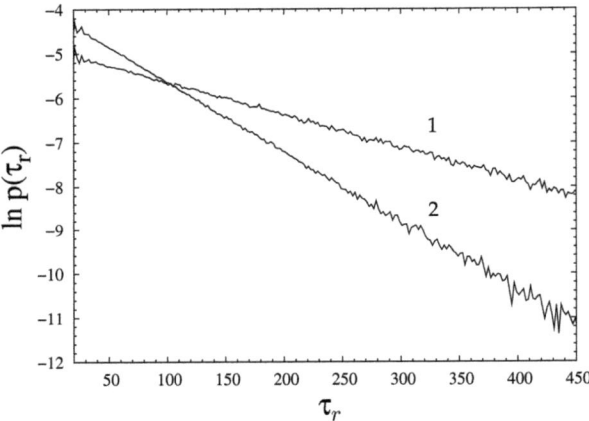

Fig. 9.4 Distribution $p(\tau_r)$ calculated for the system (9.11) at $\alpha = 2.7$ and $B = 10$ for two values in the neighborhood of the point $x_0 = 0.38$: $\varepsilon = 0.01$ (*curve 1*) and $\varepsilon = 0.02$ (*curve 2*)

infinity. From a physical viewpoint, this limit is of no interest. Numerical simulation shows that the exponential law (9.6) also holds to high accuracy in the case of a finite but small $\varepsilon \ll 1$. This conclusion is illustrated in Fig. 9.4, which shows the distribution $p(\tau_r)$ for the system (9.11), calculated for neighborhoods of the attractor point $x_0 = 0.38$ with sizes $\varepsilon = 0.01$ and $\varepsilon = 0.02$.

As can be seen from Fig. 9.4, the validity of the exponential law (9.6) is corroborated by the numerical data. If $\tau_r \geq \tau^* \simeq 40$, the plots $\ln p(\tau_r)$ are straight lines with slope $k = 1/\langle \tau_r \rangle$. For $\varepsilon = 0.01$, we have $\langle \tau_r \rangle = 137.4$. As ε increases to $\varepsilon = 0.02$, the mean return time decreases to the value $\langle \tau_r \rangle = 68.06$. The decrease in $\langle \tau_r \rangle$ leads to a corresponding growth in the slope of the straight line in the plots shown in Fig. 9.4.

As follows from (9.8), if the normalized time $\tau = \tau_r/\langle \tau_r \rangle$ is introduced, the dependence $\ln p(\tau)$ is a straight line of unit slope, shifted by $C = \ln(1/\langle \tau_r \rangle)$ relative to the origin. This is illustrated in Fig. 9.5.

9.3 Global Approach: Afraimovich–Pesin Dimension of Recurrence Times

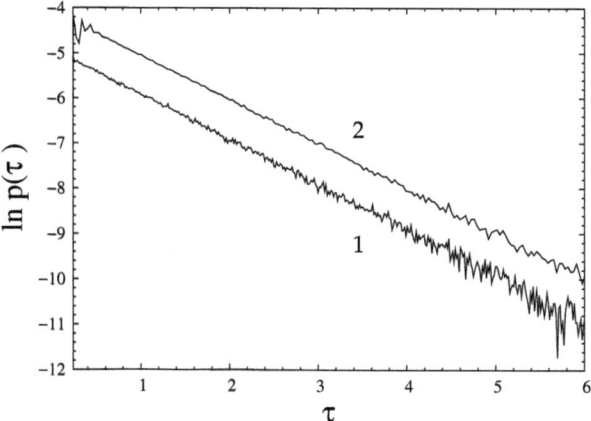

Fig. 9.5 Numerical data of Fig. 9.4 in the normalized time $\tau = \tau_r / \langle \tau_r \rangle$

9.3 Global Approach: Afraimovich–Pesin Dimension of Recurrence Times

There is another approach to analyzing Poincaré recurrences when a set is divided into ε-elements and minimal recurrences are then averaged over these partitions. Since the averaging is performed over the whole set, this approach can be referred to as global. The idea behind the global approach idea is as follows. The considered set of phase trajectories of a dynamical system (for example, the system attractor) is covered with cubes (or balls) of size $\varepsilon \ll 1$. This covering must include the whole of the considered set, as shown in Fig. 9.6.

A minimal time of the first recurrence of a phase trajectory in the ε_i-neighborhood is calculated for each covering element ε_i ($i = 1, 2, \ldots, m$). Then the mean minimal time of the first recurrence is defined over the whole set of covering elements ε_i as follows:

$$\langle \tau_{\inf}(\varepsilon) \rangle = \frac{1}{m} \sum_{i=1}^{m} \tau_{\inf}(\varepsilon_i) . \tag{9.12}$$

Note that the choice of the mean minimal return time is not mandatory. Any characteristic time can be used instead.

A qualitative approximation to (9.12) can be proposed under certain assumptions:

$$\langle \tau_{\inf}(\varepsilon) \rangle \sim \phi^{-1}(\varepsilon^{d/\alpha_c}) , \tag{9.13}$$

where α_c is the dimension of a recurrence sequence, as introduced by Afraimovich and Pesin, and d is the dimension of the considered set. The function ϕ in (9.13) can be given by one of the following forms:

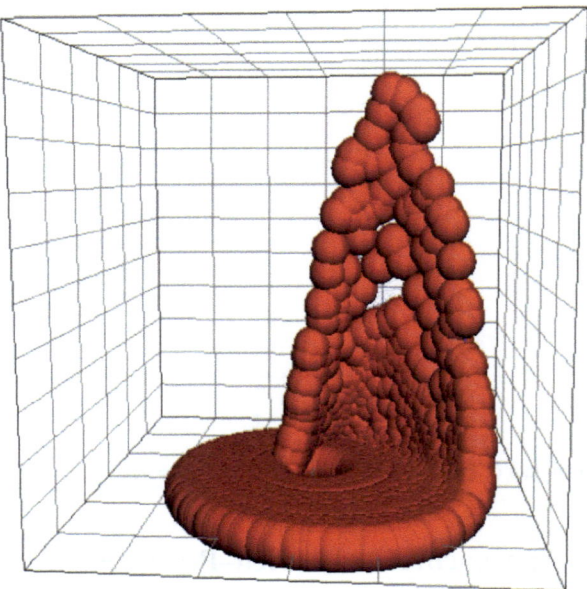

Fig. 9.6 The Rössler attractor covered with balls of diameter ε

$$\phi(t) \sim \frac{1}{t}, \quad \phi(t) \sim \exp(-t), \quad \phi(t) \sim \exp(-t^2), \quad \ldots \quad (9.14)$$

The appropriate choice of $\phi(t)$ depends on the topological entropy[1] h_T of the system, as well as on the multifractality of the considered set, if such a property exists. If $h_T = 0$, then $\phi(t) \sim 1/t$, and from (9.13), we have

$$\langle \tau_{\inf}(\varepsilon) \rangle \sim \varepsilon^{-d/\alpha_c} \quad \text{or} \quad \ln\langle \tau_{\inf}(\varepsilon) \rangle \sim -\frac{d}{\alpha_c} \ln \varepsilon \, . \quad (9.15)$$

If $h_T > 0$, then $\phi(t)$ is typically defined as $\phi(t) \sim \exp(-t)$. It has been shown that in this case the expression (9.13) can be written in the form

$$\langle \tau_{\inf}(\varepsilon) \rangle \sim -\frac{d}{\alpha_c} \ln \varepsilon \, . \quad (9.16)$$

This means that the dependence of $\langle \tau_{\inf}(\varepsilon) \rangle$ for $h_T = 0$ and $h_T > 0$ is described by different formulas. We have (9.15) in the first case and (9.16) in the second.

---

[1] The topological entropy h_T is a non-negative number that serves as a complexity measure of a system and characterizes the exponential rate of growth in time of a number of distinguished orbits. Roughly speaking, positive values of h_T indicate chaotic dynamics in the system.

9.3 Global Approach: Afraimovich–Pesin Dimension of Recurrence Times

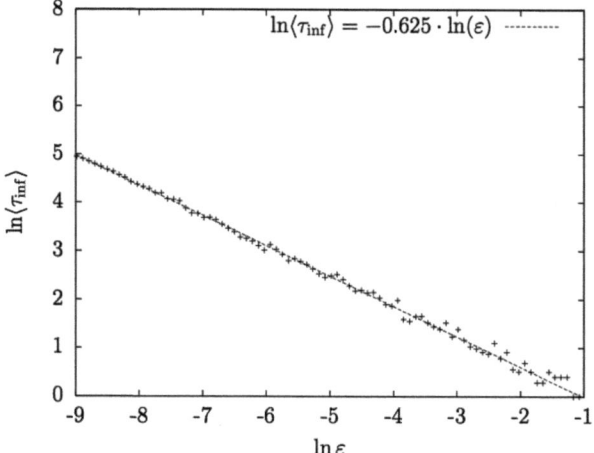

Fig. 9.7 Dependence of $\ln\langle\tau_{\inf}\rangle$ on ε for the map (9.9) at the critical point $r^* = 3.57$

We now consider numerical results that illustrate these features. For this purpose, we turn once again to the map (9.9), in which the transition to chaos is realized through a cascade of period-doubling bifurcations. For parameter values $r < r^* \approx 3.57\ldots$, the system (9.9) possesses stable cycles of period 2^n with $n = 0, 1, 2, \ldots$. The Lyapunov exponent is negative in this region. At the critical point $r = r^*$ of Feigenbaum attractor birth, the Lyapunov exponent vanishes, becoming positive for $r > r^*$. The envelope of the graph $\lambda^+(r)$ for $r > r^*$ obeys the law

$$\lambda^+(r) \sim (r - r^*)^\gamma, \qquad (9.17)$$

where $\gamma = \ln 2/\ln\delta$, $\delta = 4.669\ldots$, and $\gamma \approx 0.45$. As the topological entropy $h_T = 0$ at the critical point $r = r^*$ and $h_T > 0$ for $r > r^*$, both cases described by (9.15) and (9.16) can be realized in the system (9.9).

Numerical results for $\langle\tau_{\inf}\rangle$ versus ε are presented in Fig. 9.7 for the critical point $r = r^*$ where $h_T = 0$. As can be seen from the figure, for small ε ($\ln\varepsilon < -5$), this dependence can be approximated on a logarithmic scale by a straight line with slope $k = 0.625$, which agrees with the theoretical expression (9.15). Calculations of $\langle\tau_{\inf}\rangle$ for $r > r^*$ confirm that (9.16) also holds to a high degree of accuracy. The corresponding calculation results are shown in Fig. 9.8 for $r = 4.0$.

As can be seen from Fig. 9.8, for small ε, the dependence of $\langle\tau_{\inf}\rangle$ on $\ln\varepsilon$ is a straight line with slope $k = d/\alpha_c = 1.431$. Let us discuss this result in more detail. Figure 9.2a shows the distribution density $p(x)$ on the attractor of the system (9.9) for $r = 4.0$. As can be seen, the attractor everywhere densely covers the unit interval $0 \leq x \leq 1$ and so has dimension $d = 1.0$. In this case, knowing the slope and the dimension, one can estimate the Afraimovich–Pesin dimension (AP dimension) α_c, which is equal to $\alpha_c = 1/k \simeq 0.69 = \ln 2$. It is known from the theory that

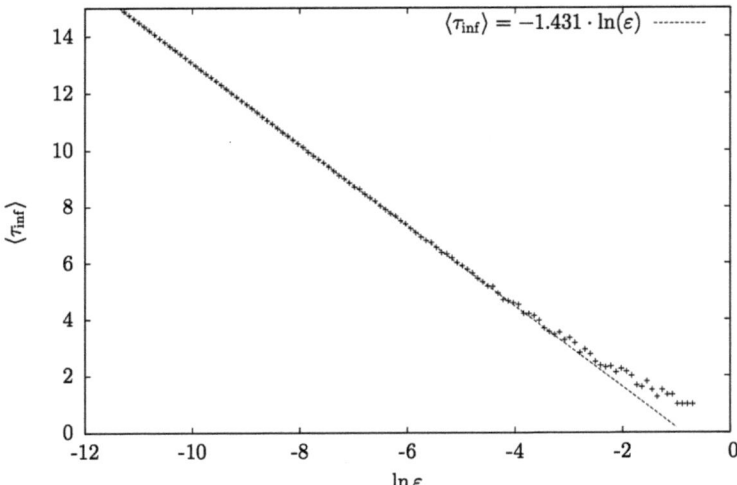

Fig. 9.8 Dependence of $\langle \tau_{\inf} \rangle$ on $\ln \varepsilon$ for the map (9.9) when $r = 4.0$

$\lambda^+ = \ln 2$ for $r = 4.0$. Thus, from the calculation data presented in Fig. 9.8, α_c turns out to be equal to the positive Lyapunov exponent value $\lambda^+ = \ln 2$. This is not accidental and reflects the fundamental interrelation between the AP dimension and the Lyapunov exponents.

9.4 Afraimovich–Pesin Dimension and Lyapunov Exponents

Statistical characteristics of Poincaré recurrences in both the local and the global approach are a consequence of the temporal dynamics of a system and reflect its properties. The mathematical definition of the AP dimension introduced under the global approach must also be related to the characteristics of the temporal evolution of the dynamical system. This interrelation is interesting both from the point of view of nonlinear dynamics as a whole and in connection with a number of applications. It can be rigorously proved that in the general case the AP dimension can be estimated from above by the topological entropy $h_T \geq \alpha_c$. The latter reflects the exponential growth of a number of periodic orbits and in this sense characterizes the complexity of the considered regime of the dynamical system. Experiments testify that, at the 'physical' level of rigor, $h_T \simeq \alpha_c$ for chaotic systems. Therefore, the AP dimension α_c also characterizes the degree of complexity. Moreover, it is known that the topological entropy is estimated by the Kolmogorov–Sinai entropy h_{KS}:

$$h_T = \sup(h_{KS}), \qquad (9.18)$$

which can in turn be evaluated as the sum of positive Lyapunov exponents (Pesin's theorem):

9.4 Afraimovich–Pesin Dimension and Lyapunov Exponents

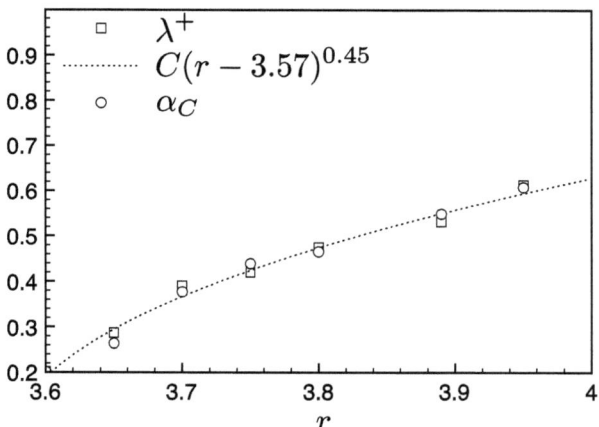

Fig. 9.9 Lyapunov exponent λ^+ and the AP dimension α_c as functions of the parameter r of the map (9.9)

$$h_{KS} = \sum_{i=1}^{n} \lambda_i^+ . \tag{9.19}$$

If this is true, then on this physical level of rigor, one can assume that $\alpha_c \simeq \sum_{i=1}^{n} \lambda_i^+$. This implies that α_c is a quantitative characteristic of the degree of instability of phase trajectories of the system. Numerical experiments confirm this.

We calculate α_c for the one-dimensional map (9.9) in the parameter range $3.6 < r < 4.0$ and compare the numerical data with the Lyapunov exponent value. Calculation results are shown in Fig. 9.9. The figure also presents both calculated data and the theoretical dependence (9.17) for λ^+. The AP dimension α_c is estimated using (9.16). Having defined the slope coefficient $k = d/\alpha_c$, the value of α_c is found by assuming that $d = 1$.

The data shown in Fig. 9.9 confirm that $\alpha_c \simeq \lambda^+$. It is interesting to find out how α_c is related to h_{KS} (9.19). For this purpose, we consider the system of two coupled maps

$$\begin{aligned} x_{n+1} &= r_1 - x_n^2 + \gamma(y_n^2 - x_n^2) , \\ y_{n+1} &= r_2 - y_n^2 + \gamma(x_n^2 - y_n^2) , \end{aligned} \tag{9.20}$$

We fix $r_1 = 1.9$ and $\gamma = 0.002$ and vary r_2 as the control parameter. For $r_2 \leq 1.46$, the system (9.20) has one positive ($\lambda_1 > 0$) and one negative ($\lambda_2 < 0$) Lyapunov exponent. As r_2 increases, the system (9.20) demonstrates the regime of hyperchaos that is characterized by two positive Lyapunov exponents. In this case, h_{KS} is equal to the sum $\lambda_1 + \lambda_2$. If our assumptions are true, then $\alpha_c \simeq \lambda_1 + \lambda_2$. The calculation results presented in Fig. 9.10 corroborate this. α_c is calculated from the experimental data for the slope $k = d/\alpha_c$. The dimension d is defined as the capacity. Thus, from the physical point of view, the AP dimension can be well estimated by h_{KS}.

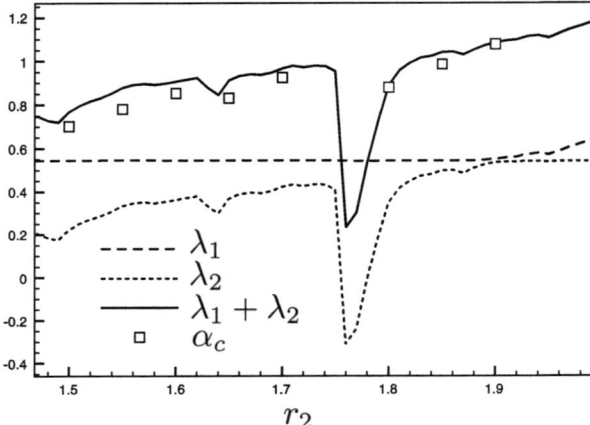

Fig. 9.10 Lyapunov exponents λ_1 and λ_2, their sum $\lambda_1 + \lambda_2$, and the AP dimension α_c for the system (9.20) as functions of the parameter r_2 for $r_1 = 1.9$ and $\gamma = 0.002$

9.5 Summary

In this chapter we have described the underlying results relating to Poincaré recurrences. The fundamentals of the recurrence theory in the local approach are presented along with the key ideas and the results of analysis in the global approach. The local theory that has been developed over the past 100 years is largely complete and has become a classic. The global approach, developed only over the past 15–20 years, remains a subject of current mathematical research and contains a number of uncompleted issues and problems. This chapter does not cover the applied aspects of the recurrence theory. In recent years, this trend in nonlinear dynamics has been the subject of intense interest. It has been shown that the statistical characteristics of a sequence of Poincaré return times can be used to diagnose the effects of synchronization and stochastic resonance, calculate the fractal dimension of chaotic attractors, and carry out the task of controlling chaotic oscillations. The fundamental results described in this chapter underlie the practical use of Poincaré recurrence theory, while the mathematical results of this theory in the local and global approaches are fully described in the monograph [1].

Reference

1. Afraimovich, V., Ugalde, E., Urias, J.: Fractal Dimension for Poincaré Recurrences. Elsevier, Amsterdam/London (2006)

Chapter 10
Fractals in Nonlinear Dynamics

10.1 Introduction

The geometry of research objects is one of their important characteristics. Geometric properties of the object under study occupy a central place when constructing models, regardless of the specific research subject, and are, in a certain sense, an interdisciplinary characteristic. The geometries of particle trajectories, high-rise buildings, natural landscapes, attractors in phase space, crystal structures, etc., are all important when developing and analyzing models in the natural sciences and engineering.

It was natural that, in the context of the most common theoretical concepts, geometry should become an independent part of mathematics. In addition to classical Euclidean geometry, mathematicians have developed the geometry of multi-dimensional spaces with non-Euclidean metrics (the Riemann and Lobachevsky spaces), not to mention the geometry of unusual sets called fractals. As often happens, rigorous mathematical results have not always immediately attracted the attention of practitioners in the natural sciences. In particular, this also concerns the geometry of fractals. This is understandable since nontraditional geometrical concepts were originally abstractions, devoid of specific practical significance.

With time, it became clear that fractal geometry was not just an exotic mathematical oddity, but in fact reflects the properties of real systems and objects. Several types of fractals can be distinguished, namely, constructive (artificially constructed), dynamical (generated by dynamical systems), stochastic (formed according to random laws), and natural (found in nature). By the end of the twentieth century, the theory of fractals was being intensively developed thanks to the achievements of nonlinear dynamics and computer methods, and the discovery of deterministic chaos had a significant effect on the interest in fractal geometry. It was established that strange attractors and nonhyperbolic attractors in the phase space of a dynamical system are objects with a nontrivial and, as a rule, fractal geometry. Boundaries of basins of attraction of attractors and boundaries of regions corresponding to different regimes in the parameter spaces of dynamical systems often have a fractal structure. Fractal sets have even found their application in practical problems, e.g., design of fractal antenna systems, fractal methods of data compression. Computer simulation methods provide a way to reproduce and explore a wide variety of fractals. Furthermore, fractals are very beautiful and are widely used in modern design.

In this chapter we consider those aspects of the theory of fractal sets needed to describe the characteristics of dynamical chaos. We introduce the notion of a fractal and consider examples of fractals of different types. We also discuss in more detail the nature of fractality in dynamical systems. Special attention is paid to the analysis of fractal dimensions of limit sets which are widely used when exploring dynamical chaos.

10.2 Definition of a Fractal: Classic Examples of Fractal Sets

The term 'fractal' was introduced by B. Mandelbrot in the 1970s to designate nontrivial geometric objects. His definition of fractals sets is based on the property of scaling (self-similarity or scale invariance). Hence, a *fractal* is a set having a *self-similar structure*. An element of this set is repeated infinitely on increasingly smaller scales. However, this definition considerably narrows the range of possible fractal sets, excluding from consideration all natural and stochastic fractals.

Another definition that does not require a strictly self-similar structure is based on the notion of *metric dimension*. According to this, a *fractal set* is a compact set in the metric space, for which the metric dimension is strictly greater than the topological dimension. The two dimensions coincide for 'normal' sets, i.e., *manifolds*. A manifold is a generalization of such notions as a point, a line, a surface, etc. Each point of the manifold has a neighborhood that is topologically equivalent to a ball in $\mathbb{R}^n$. The number n is the *topological dimension* of the manifold, i.e., the number of coordinates needed to specify the position of a point on the manifold. The topological dimension is always expressed by a non-negative integer. The metric (fractal) dimension is introduced to distinguish manifolds and fractals and to characterize the properties of the latter. We consider it later on in this chapter. For the moment we note only that, although the fractal dimension is always integer for manifolds and coincides with their metric dimension, it more often turns out to be fractional for fractals (but not always). The word 'fractal', which comes from the Latin *fractus* meaning crushed or fractional, expresses this property.

Constructive fractals have long been known to mathematicians. The simplest and best known example of a constructive fractal is the *Cantor set*, proposed by G. Cantor in 1883. The Cantor set is created as follows: a single line segment is divided into three equal parts and the middle third is deleted. This is the first step of an iterative procedure. These steps are subsequently repeated for the remaining first and third parts of the line segment. This process is continued infinitely many times and the result is a set called a perfect Cantor set (Fig. 10.1).

10.2 Definition of a Fractal: Classic Examples of Fractal Sets

Fig. 10.1 Construction of the perfect Cantor set

At the nth step we get 2^n segments, each of length $(1/3)^n$. However, as $n \to \infty$ the length of the segments tends to zero and their number to infinity. What is the resulting set, consisting as it does of an infinite number of segments of infinitely small length? This is a very interesting question. On the one hand, we have deleted the entire unit interval when constructing the Cantor set. Indeed, a third of the interval is removed at the first step, 2/9 of the interval at the second step, and so on. As a result, the total length removed is

$$L = \frac{1}{3} \sum_{n=1}^{\infty} \left(\frac{1}{3}\right)^{n-1} = 1 \, . \qquad (10.1)$$

This implies that the Cantor set has zero measure (zero length). On the other hand, due to the algorithm for its construction, the Cantor set corresponds in a one-to-one manner to the set of all points of the original unit interval. Thus, the Cantor set has the cardinality of the continuum.

The Cantor set has another important property. At the $(n+1)$th step of the algorithm, we can select an element of the structure which can be reduced to the structure at the nth step by rescaling. Thus, the Cantor set is self-similar in the sense that it reproduces the original structure at finer scales.

If the unit interval is divided into three unequal parts, one can obtain a more complex two-scale Cantor set (multifractal). However, the fundamental properties of the fractal hold in this case too.

The Swedish mathematician Helge von Koch introduced the *Koch curve* in 1904. This is another classic example of a fractal, constructed as follows. As in the case of the Cantor set, a single line segment is divided into three equal segments, but now the middle segment is replaced by the two sides of an equilateral triangle of the same length as the segment being removed (Fig. 10.2). The total length of the constructed curve is 4/3. The same 'rule' is now applied to each of the four resulting segments. As a result, we can obtain a more complex curve consisting of 8 segments with total length 16/9. Repeating this procedure an infinite number of times, we can get a complex curve of infinite length. At the same time, this curve belongs to a bounded part of the plane, which has a finite area. Like the Cantor set, the Koch curve exhibits the property of self-similarity. If this algorithm is applied to an equilateral triangle, we obtain the fractal set called the *Koch snowflake*.

Although the Koch curve has an infinite length and is located in a bounded part of the plane, it does not fill the plane, i.e., it is not everywhere dense in any part

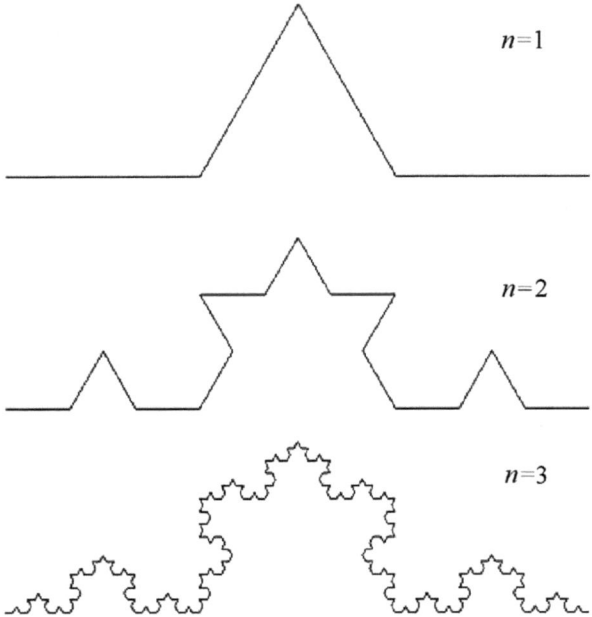

Fig. 10.2 Construction of the Koch curve

of the plane. A curve everywhere densely filling the unit square, i.e., falling in an arbitrarily small neighborhood of any point of the square, is known as the *Peano curve*. This is also a fractal. Although the Peano curve fills the square, its topological dimension is 1, and not 2. A method for constructing the Peano curve, proposed by D. Hilbert, is illustrated in Fig. 10.3.

Mathematicians have created a collection of constructive fractals which are constructed by adding or removing some element of a self-similar structure an infinite number of times. For example, if the middle segment is removed from a square or an equilateral triangle, the resulting set is called the *Sierpinski carpet* (Fig. 10.4). If the middle segment is deleted from a cube, one can get the *Menger sponge*, etc.

In 1961, in the context of nonlinear dynamics, the mathematician S. Smale proposed a construction that leads to a fractal called *Smale's horseshoe*. This procedure is thought of as a two-dimensional map. However, it is not specified explicitly by a function, but rather is described as a certain algorithm. For this reason, it is difficult to decide whether Smale's horseshoe map is a constructive or a dynamical fractal.

Smale's horseshoe is formed as shown in Fig. 10.5a. At each iteration the map f stretches the rectangle S horizontally, squeezes it vertically, and finally bends and folds it into the shape of a horseshoe. The bent part remains outside S, thus ensuring a linear map on $S \cap f^{-1}(S)$. An infinite sequence of iterations results in the limit set

10.2 Definition of a Fractal: Classic Examples of Fractal Sets

Fig. 10.3 Construction of the Peano curve

Fig. 10.4 Sierpinski carpets

$$\Sigma^+ = \bigcap_{k=0}^{\infty} f^k(S) . \tag{10.2}$$

This consists of an infinite number of infinitely thin horizontal strips. It is thus the product of the Cantor set and an interval. If the inverse map f^{-1} is considered, then one can get the analogous set

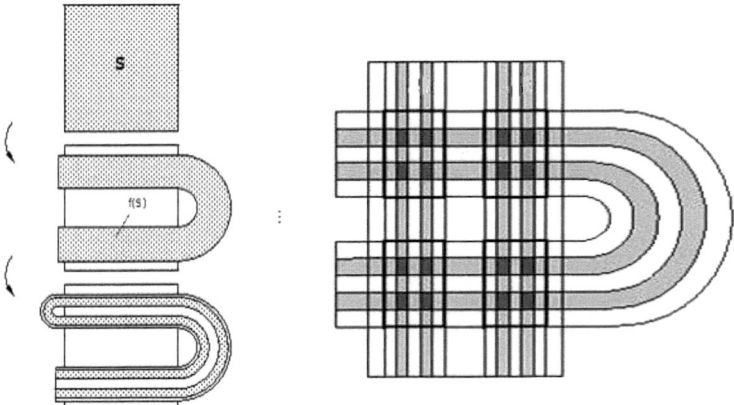

Fig. 10.5 Smale's horseshoe. (**a**) Cantor set Σ^+ construction for direct iterations of the map and (**b**) invariant Cantor set $\Sigma = \bigcap_{k=-\infty}^{\infty} f^k(S)$

$$\Sigma^- = \bigcap_{k=0}^{\infty} f^{-k}(S), \qquad (10.3)$$

consisting of an infinite number of vertical strips. Their intersection $\Sigma = \Sigma^+ \cap \Sigma^-$ is the set of those points that remain in S for all direct and inverse iterations of the map f (Fig. 10.5b). This is the largest invariant set contained in S. It represents the product of two Cantor sets Σ^+ and Σ^-, and is itself a Cantor set. It has been proven that the set Σ includes a countable set of saddle periodic orbits of all possible periods, and that these are everywhere dense in it. Moreover, it contains a lot of non-periodic (chaotic) orbits with the cardinality of the continuum.

Smale's horseshoe has played a very important role in understanding many of the properties of nonlinear systems with complex dynamics. The horseshoe map constructed by Smale was the first example of a reversible two-dimensional map with irregular dynamics. The horseshoe can arise due to transverse intersections of manifolds of saddle periodic orbits. In many cases, horseshoe formation indicates chaotic dynamics.

10.3 The Nature of Fractality in Dynamical Systems

As already mentioned, fractals generated by deterministic dynamics of nonlinear systems are referred to as *dynamical*. Dynamical fractals include attractors and other limit sets in phase spaces of dimension $N > 2$ for flows and $N \geq 2$ for discrete-time systems. When speaking of irregular attractors, the notions 'strange' and 'chaotic' are often treated separately as characteristics of their different properties. Chaoticity means that trajectories on an attractor are exponentially unstable, while

10.3 The Nature of Fractality in Dynamical Systems

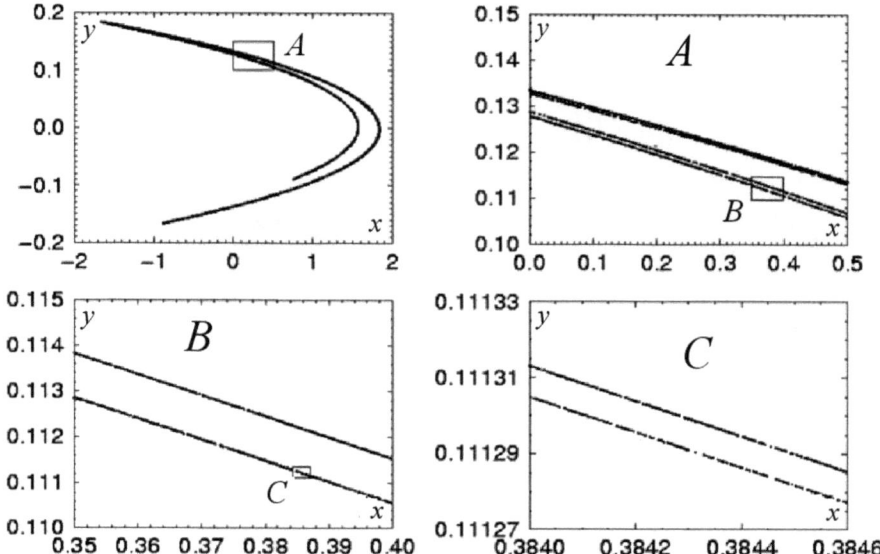

Fig. 10.6 Scaling of the attractor in the map (8.8) for $a = 1.71$ and $b = 0.1$

strangeness refers to its nontrivial (i.e., fractal) geometry. An irregular attractor can be understood as an attractor of a DS which has at least one of these two properties. However, numerous studies have shown that these properties usually come together, even though there are some special cases where one can speak of strange nonchaotic attractors or, on the contrary, chaotic nonstrange attractors. These kinds of attractors have already been discussed in previous chapters.

Consider the chaotic attractor of the two-dimensional Henon map (8.8). We can see the property of scale invariance (scaling) when the same structure is repeated on a finer scale (Fig. 10.6). The Henon attractor is itself similar to a fractal set generated by Smale's horseshoe.

The fractal nature of a chaotic attractor can be explained as follows. Consider a chaotic attractor in a system given by ordinary differential equations and with phase space $\mathbb{R}^3$. Any invertible two-dimensional map (including the Henon map) can be thought of as the Poincaré map arising in a section of a three-dimensional flow. Continuous trajectories diverge on an attractor in $\mathbb{R}^3$. Due to phase volume contraction, the attractor dimension must be less than the dimension of the system phase space, i.e., less than three.

However, trajectories cannot diverge on a two-dimensional manifold (surface) without intersecting. Thus, the chaotic attractor cannot be a manifold, and must therefore be a fractal. This means that regular (simple) attractors are smooth submanifolds of the phase space, and irregular (strange in a generalized sense) attractors are fractals.

Boundaries of the basins of attraction of several coexisting attractors are fractal, and this is also a typical feature of any nonlinear DS. The example of the fractal

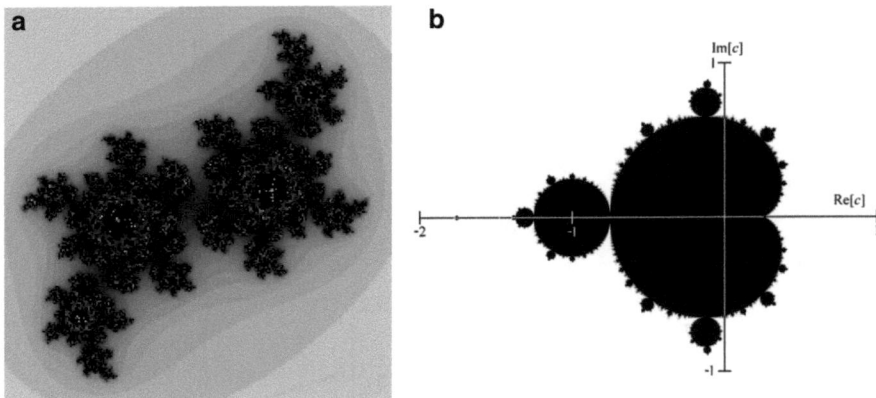

Fig. 10.7 Fractals generated by the complex map (10.4): (**a**) the Julia set and (**b**) the Mandelbrot set

basin boundaries of two chaotic attractors in the Henon map was discussed in the previous chapter (see Fig. 8.11). The fractality of basin boundaries is related to homoclinic intersections of manifolds of saddle cycles and to the Smale's horseshoe generated in their neighborhood. Note that the fractal boundary can separate basins of attraction, not only of chaotic, but also of regular attractors.

Many beautiful dynamical fractals are associated with the following simple Julia map:

$$Z_{n+1} = C - Z_n^2 , \qquad (10.4)$$

where Z is a complex variable and C is a complex parameter. The map (10.4) was studied by G. Julia in the early twentieth century. However, the visualisation of fractals generated by (10.4) became possible only when modern computer equipment came on the scene. The Julia set (Fig. 10.7a) is an example of a fractal boundary between the basins of attraction of the attractor at infinity (light area) and the periodic motion (dark area). The tone is defined by the number of iterations required to reach the attractor. The Mandelbrot set illustrates a fractal structure in the parameter space. The dark region (Fig. 10.7b) is defined as a set of points in the complex plane, corresponding to the values of the parameter C for which the solution of (10.4) is bounded.

Fractal sets are not only generated by deterministic evolution operators. They can also appear in stochastic systems governed by random forces. Such fractals are called *stochastic*. The trajectory of a Brownian particle (as well as an arbitrary trajectory of a diffusive random process) is an example of a stochastic fractal. Many natural fractals appear to belong to this type, e.g., the fragment of coastline shown in Fig. 10.8.

10.4 Fractal Dimensions of Sets

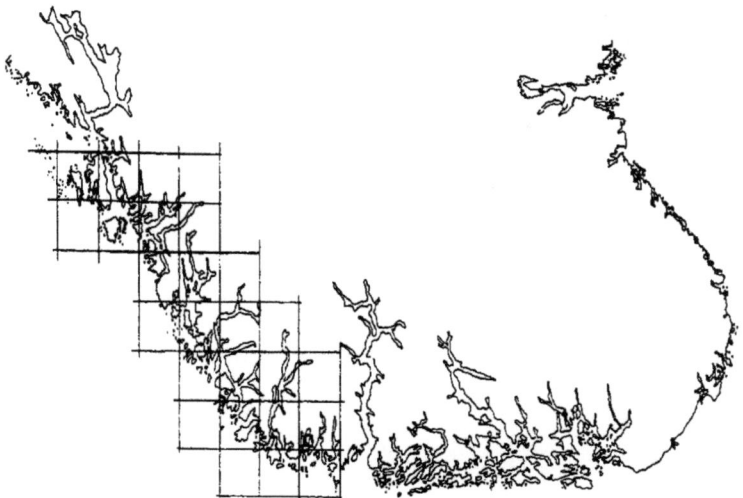

Fig. 10.8 Example of a natural fractal

10.4 Fractal Dimensions of Sets

In the rigorous mathematical theory, the metric or fractal dimension is understood as the *Hausdorff–Besicovitch dimension* which will be defined later. However, this is a rather complex and abstract notion, and the Hausdorff dimension cannot be calculated numerically for an arbitrary dynamical fractal. For this reason, several definitions of the fractal dimension have been introduced that allow numerical calculations. All of them are estimates of the Hausdorff–Besicovitch dimension. Definitions of the fractal dimension can either be based solely on the metric properties of sets, or they can also take into account the probability measure, i.e., the frequency with which a phase trajectory visits various parts of the set. The latter is important for dynamical fractals. Thus, fractal dimensions can be divided into two subgroups: (i) purely metric and (ii) probability–metric. Here we discuss the most frequently used types of fractal dimensions.

10.4.1 The Hausdorff–Besicovitch Dimension

Suppose we have a set of points M in the space $\mathbb{R}^n$. We cover all the points of this set with n-dimensional cubes with sides $\varepsilon_i \leq \varepsilon$ (Fig. 10.9). We now introduce the quantity

Fig. 10.9 Definition of Hausdorff dimension D_H

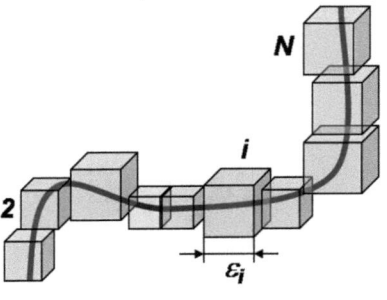

Fig. 10.10 Definition of capacity D_C

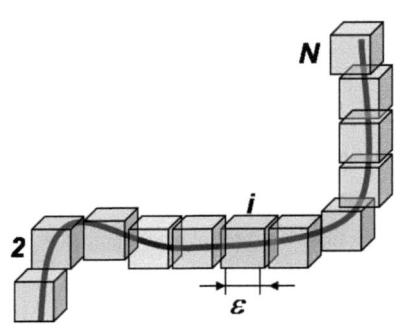

$$l_D = \lim_{\varepsilon \to 0} \inf \sum_{i=1}^{N} \varepsilon_i^D ,$$

where N is the number of covering elements. The infimum implies taking the smallest value over all possible coverings. Hausdorff showed that there is a critical value $D = D_H$ such that $l_D = 0$ for $D > D_H$ and $l_D = \infty$ for $D < D_H$. D_H is called the *Hausdorff–Besicovitch dimension*. For manifolds, its values are integer and equal to the topological dimension, while for fractals, the values of D_H are larger than the topological dimension and often fractional.

10.4.2 Capacity D_C

The capacity is a simplification of the notion of Hausdorff dimension. We cover the set M in $\mathbb{R}^n$ with identical n-dimensional cubes with side ε (Fig. 10.10). A point can be covered with one cube, a line with $N \sim 1/\varepsilon$ cubes, a surface with $N \sim 1/\varepsilon^2$ cubes, etc. If we have a fractal set, then by analogy $N \sim 1/\varepsilon^{D_C}$, where D_C may assume a fractional value.

Let $N \approx a/\varepsilon^{D_C}$, where $a = $ const. Taking the logarithm and passing to the limit $\varepsilon \to 0$, we obtain an expression for D_C that can be taken as the definition of the capacity:

10.4 Fractal Dimensions of Sets

$$D_C = \lim_{\varepsilon \to 0} \frac{\log N(\varepsilon)}{\log \varepsilon^{-1}}. \tag{10.5}$$

$D_C = D_H$ for manifolds and, in many cases, for fractals. In general, the following inequality holds: $D_C \geq D_H$, e.g., the capacity is an upper bound on the Hausdorff dimension. Obviously, the capacity is a purely metric dimension.

The capacity coincides with the Hausdorff dimension and can be found analytically for the simple examples of fractal sets given above. Consider the Cantor set. At each step we choose the value of ε (the length of the covering element) equal to the length of the deleted part of the interval. Then at the kth step, $\varepsilon = 1/3^k$. We get

$$D_C = \lim_{k \to \infty} \frac{\log 2^k}{\log 3^k} = \frac{\log 2}{\log 3} \approx 0.630\,92\ldots.$$

Similar arguments can be applied to the Koch curve. As a result, we have

$$D_C = \frac{\log 4}{\log 3} \approx 1.261\,85\ldots.$$

Finally, consider Smale's horseshoe. Suppose the horizontal stretching coefficient of the unit square is 2 and the vertical contraction coefficient is 2ν, where $\nu > 1$. Then the capacity of the set Σ^+ is

$$D_C(\Sigma^+) = 1 + \frac{\log 2}{\log 2\nu}.$$

10.4.3 Information Dimension D_I

This characteristic is similar to the capacity, but takes into account the probability measure on the studied set. We cover the point set in $\mathbb{R}^n$ with identical n-dimensional cubes of side ε and find the probability of a point falling in each cube. Then we can introduce the distribution entropy

$$H(\varepsilon) = -\sum_{i=1}^{N} P_i \log_2 P_i,$$

where P_i is the probability of finding a point of the set in the ith cube. The information dimension is defined as follows:

$$D_I = \lim_{\varepsilon \to 0} \frac{H(\varepsilon)}{\log \varepsilon^{-1}}. \tag{10.6}$$

Fig. 10.11 Definition of the correlation dimension D_{cor}

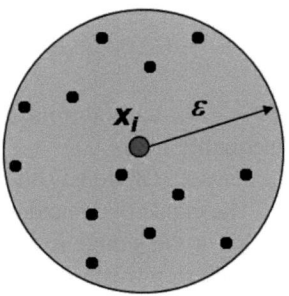

Clearly, $D_I = D_C$ for a uniform probability distribution, i.e., $P_i = 1/N$. For any other distribution, the entropy is less, and hence, $D_I < D_C$.

10.4.4 Correlation Dimension D_{cor}

This characteristic is convenient to use when analyzing time series or any experimental data. Suppose we have a sequence of m points belonging to a set in $\mathbb{R}^n$: $x_1, x_2, \ldots, x_i, \ldots, x_m$. Let $m_\varepsilon(x_i)$ denote the number of points of the sequence falling in a sphere of radius ε centered on some x_i belonging to the same sequence (Fig. 10.11).

For $\varepsilon \to 0$ and $m \to \infty$, the number $m_\varepsilon(x_i)$ is as follows: $\sim m\varepsilon^0$ for a countable set of points, $\sim m\varepsilon^1$ for a line, $\sim m\varepsilon^2$ for a surface, $\sim m\varepsilon^3$ for a three-dimensional manifold, and so on. In the general case, one can write: $m_\varepsilon(x_i) \approx am\varepsilon^{D_i}$, where $a = \text{const}$.

In the general case, the quantity D_i depends on the choice of point x_i. Consider the average value m_ε over all the points, viz.,

$$m_\varepsilon = \frac{1}{m} \sum_{i=1}^{m} m_\varepsilon(x_i) ,$$

and in the limit $\varepsilon \to 0$, $m \to \infty$, write

$$m_\varepsilon = am\varepsilon^{D_{cor}} ,$$

where the exponent D_{cor} can take non-integer values. Then

$$\varepsilon^{D_{cor}} = \frac{1}{am^2} \sum_{i=1}^{m} m_\varepsilon(x_i) .$$

The number of points falling in the sphere of radius ε centered on x_i can be represented by

10.4 Fractal Dimensions of Sets

$$m_\varepsilon(x_i) = \frac{1}{m} \sum_{j=1,\, j\neq i}^{m} \chi(\varepsilon - |x_i - x_j|),$$

where χ is the Heaviside function. We obtain

$$\varepsilon^{D_{\text{cor}}} = \lim_{\varepsilon \to 0} \frac{1}{a} C(\varepsilon),$$

where

$$C(\varepsilon) = \lim_{m \to \infty} \frac{1}{m^2} \sum_{i=1}^{m} \sum_{j=1,\, j\neq i}^{m} \chi(\varepsilon - |x_i - x_j|).$$

The function $C(\varepsilon)$ is called the *correlation integral*. The correlation dimension is defined as

$$D_{\text{cor}} = \lim_{\varepsilon \to 0} \frac{\log C(\varepsilon)}{\log \varepsilon}. \tag{10.7}$$

10.4.5 Generalized Dimension D_q

The dimensions D_C, D_I, and D_{cor} can be generalized by introducing the dimension of order q and using the generalized entropy of order q (the Renyi entropy):

$$H_q(\varepsilon) = \frac{1}{1-q} \log\left(\sum_{i=1}^{N} P_i^q\right),$$

where P_i is the probability of finding a point of the set in the ith covering element. Then the dimension of order q is

$$D_q = \lim_{\varepsilon \to 0} \frac{H_q(\varepsilon)}{\log \varepsilon^{-1}}. \tag{10.8}$$

It can be shown that $D_0 = D_C$, $D_1 = D_I$, and $D_2 = D_{\text{cor}}$.

Besides the dimensions considered above, there are other dimensions such as the pointwise dimension, the Hausdorff dimension of the core, the core capacity, etc. However we shall not discuss these here, but move on to the Lyapunov dimension, which is especially often used in numerical simulation and experimental studies of dynamical systems.

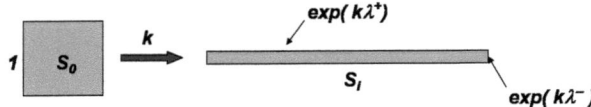

Fig. 10.12 Proof of the equality $D_L = D_C$ for a two-dimensional map with constant stretching and contraction

10.4.6 Lyapunov Dimension D_L

The fractal dimension of an attractor of a DS in the phase space $\mathbb{R}^n$ can be estimated using the spectrum of Lyapunov characteristic exponents (LCE). This estimate is called the *Lyapunov dimension* D_L. First we arrange the LCE spectrum in decreasing order: $\lambda_1 \geq \lambda_2 \geq \ldots \geq \lambda_n$. The Lyapunov dimension is given by the Kaplan–Yorke formula

$$D_L = k + \frac{1}{|\lambda_{k+1}|} \sum_{i=1}^{k} \lambda_i, \qquad (10.9)$$

where the integer part k of the dimension is defined from the conditions

$$\sum_{i=1}^{k} \lambda_i \geq 0, \qquad \sum_{i=1}^{k+1} \lambda_i < 0.$$

Thus, k is the maximum number of exponents in the LCE spectrum whose sum is non-negative.

Applying the Kaplan–Yorke definition for regular attractors, we get the following values for the Lyapunov dimension, which coincide with the topological dimension of the corresponding set:

- $D_L = 0$ for an equilibrium point $(-, -, -, -, \ldots)$,
- $D_L = 1$ for a limit cycle $(0, -, -, -, \ldots)$,
- $D_L = 2$ for a two-dimensional torus $(0, 0, -, -, \ldots)$,
- $D_L = N$ for an N-dimensional torus $(0, 0, 0, \ldots 0, -, \ldots)$,

where the signature of the LCE spectrum is given in the brackets.

The equality $D_L = D_C$ holds for any DS with constant stretching and contraction. For example, consider a two-dimensional map characterized by constant stretching by $\exp(\lambda^+)$ and contraction by $\exp(|\lambda^-|)$ at each iteration, where λ^+ and λ^- are positive and negative Lyapunov exponents. Thus, the unit square is transformed into a long thin strip (Fig. 10.12) for k iterations of the map. In addition to the stretching and contraction, the resulting strip is bent and folded (not shown in the figure).

At the kth iteration, we can cover the resulting strip of area S_k with N squares of side $\varepsilon = \exp(k\lambda^-)$, where

$$N(\varepsilon) = \exp\left[k(\lambda^+ + |\lambda^-|)\right].$$

As $k \to \infty$, we find that $\varepsilon \to 0$, and consequently we can find the capacity of the limit set:

$$D_C = \lim_{i \to \infty} \frac{\log N(\varepsilon)}{\log \varepsilon^{-1}} = \frac{\lambda^+ + |\lambda^-|}{|\lambda^-|} = 1 + \frac{\lambda^+}{|\lambda^-|} = D_L.$$

This therefore coincides with the Lyapunov dimension D_L.

For the Smale map with stretching coefficient $\alpha = 2$ and contraction coefficient $\beta = 2\nu$, $\nu \geq 1$, we similarly obtain

$$D_C(\Sigma^+) = D_L = 1 + \frac{\log 2}{\log 2\nu}.$$

In the general case, the equality $D_L = D_C$ may be violated for systems with varying divergence.

10.5 Relationship Between Different Dimensions

As we have seen, fractal sets and fractal dimensions play a very important role in nonlinear dynamics. However, in most cases none of the dimensions can be calculated exactly. In addition, the question remains unclear: which of the above-mentioned dimensions is closer to the Hausdorff dimension and better reflects the fractal structure of the studied set? Grassberger and Procaccia have shown that, in general, the following relation holds:

$$D_{\text{cor}} \leq D_I \leq D_C. \tag{10.10}$$

As already mentioned, the equality $D_I = D_C$ is valid in the case of a uniform probability distribution over all the covering elements. The correlation dimension D_{cor} takes into account the joint probability of a pair of points falling in the ith element of the partition, so must be less than D_I. For the Lyapunov dimension D_L, the equality $D_L = D_I$ usually holds, although there are exceptions. We note that the capacity D_C is an upper bound for the Hausdorff dimension. We cannot draw conclusions from the inequality (10.10) about the relationship between D_I, D_{cor} and D_H.

Numerical experiments show that, as a rule, values of all the considered dimensions are very close for most chaotic attractors and are fractional for typical chaotic attractors. As an example, we give calculation results for the

Anishchenko–Astakhov oscillator (1.30). In the chaotic attractor mode for $m = 1.5$ and $g = 0.2$, the following values of the dimensions are obtained:

- Capacity $D_C = 2.306 \pm 0.015$,
- Information dimension $D_I = 2.300 \pm 0.013$,
- Correlation dimension $D_{cor} = 2.277 \pm 0.017$,
- Lyapunov dimension $D_L = 2.33 \pm 0.02$.

We thus find the inequality

$$D_L \geq D_C \geq D_I \geq D_{cor}.$$

However, within the possible calculation error, it can be assumed approximately that

$$D_L \approx D_C \approx D_I \approx D_{cor}.$$

10.6 Summary

In this chapter we have discussed the common concepts of fractal sets and illustrated in detail the fractal sets generated by nonlinear dynamical systems. We have introduced and analyzed the most widespread definitions of the fractal dimension of a set and described the relationships between the dimensions corresponding to different definitions. When choosing which dimension should be used, the capabilities of numerical calculations are usually taken into consideration. The Lyapunov dimension or the capacity of the set are most convenient to use when modelling dynamical systems numerically. The correlation dimension is usually used to estimate the fractal dimension of the attractor on experimental data. Presumably, differences between the values of the various dimensions also carry information about the structure of the studied set and can be considered as an important tool for its detailed study. However, the meaning of these fine differences and their relationship with the properties of chaos are insufficiently understood at the present time. The fundamental and applied aspects of the theory of fractals are described in [1–9].

References

1. Crownover, R.M.: Introduction to Fractals and Chaos. Jones and Bartlett, Boston-London (1995)
2. Drazin, P.G.: Nonlinear Systems. Cambridge University Press, Cambridge (1992)
3. Edgar, G.A. (ed.): Classics on Fractals. Addison-Wesley, Reading (1993)
4. Hilborn, R.C.: Chaos and Nonlinear Dynamics: An Introduction for Scientists and Engineers. Oxford University Press, Oxford (2002/2004)
5. Mandelbrot, B.B.: The Fractal Geometry of Nature. Freeman, San Francisco (1982)

6. Moon, F.C.M.: Chaotic and Fractal Dynamics: An Introduction for Applied Scientists and Engineers. Wiley, New York (1992)
7. Moon, F.C.M.: Chaotic Vibration: An Introduction for Applied Scientists and Engineers. Wiley, New York (2004)
8. Pesin, Ya.B.: Dimension Theory in Dynamical Systems: Contemporary Views and Applications. University of Chicago Press, Chicago (1997)
9. Schroeder, M.: Fractals, Chaos, Power Laws. Freeman, New York (1991)

Chapter 11
The Anishchenko–Astakhov Oscillator of Chaotic Self-Sustained Oscillations

11.1 Introduction

In general form, self-sustained oscillatory systems with one degree of freedom are described by the equation

$$\ddot{x} + \Phi(x,\boldsymbol{\alpha})\dot{x} + \Psi(x,\boldsymbol{\alpha}) = 0 , \qquad (11.1)$$

where x is a variable oscillating periodically, $\Phi(x,\boldsymbol{\alpha})$ and $\Psi(x,\boldsymbol{\alpha})$ are nonlinear functions characterizing the action of forces providing periodic self-sustained oscillations, and $\boldsymbol{\alpha}$ is a vector of parameters $(\alpha_1, \alpha_2, \ldots, \alpha_n)$. For example, the van der Pol oscillator is defined by two control parameters $\alpha_1 = \varepsilon$ and $\alpha_2 = \omega_0^2$ and functions $\Phi(x,\boldsymbol{\alpha})$ and $\Psi(x,\boldsymbol{\alpha})$ specified by

$$\Phi(x,\varepsilon) = -(\varepsilon - x^2) , \qquad \Psi(x,\omega_0^2) = \omega_0^2 x . \qquad (11.2)$$

Equation (11.1) can be generalized to a certain class of dynamical systems with 1.5 degrees of freedom as follows:

$$\begin{aligned}\ddot{x} + F_1(x,z,\boldsymbol{\alpha})\dot{x} + F_2(x,z,\boldsymbol{\alpha}) &= 0 , \\ \dot{z} &= F_3(x,\dot{x},z,\boldsymbol{\alpha}) ,\end{aligned} \qquad (11.3)$$

where F_i ($i = 1, 2, 3$) are nonlinear functions in the general case. The phase variable $z(t)$ in (11.3) is related to the variable $x(t)$ by means of the differential operator. This means that z depends on x inertially. If the relation between $z(t)$ and $x(t)$ is inertialess and is described by the algebraic polynomial

$$z = \sum_{n=0}^{K} C_n x^n(t) = \varphi(x) , \qquad (11.4)$$

then Eqs. (11.3) reduce to the oscillator equation in the phase plane (11.1). If z depends on x inertially, then Eqs. (11.3) describe self-sustained oscillations in the three-dimensional phase space and constitute the generalization of (11.1) for this case.

Following K.E. Theodorchik, the generalized Eqs. (11.3) will be referred to as the *dynamical system with inertial nonlinearity*. Indeed, as can be seen from (11.3), they describe systems whose parameters depend inertially on oscillating variables. For example, the nonlinear dissipation function in (11.3) can be written in the form $F_1[x, \alpha(x)]$, where $\alpha(x)$ is the parameter inertially dependent on the variable x. This property was used by K.E. Theodorchik to define self-sustained oscillatory systems with inertial nonlinearity.

Oscillators with 1.5 degrees of freedom can realize regimes of quasiperiodic and chaotic self-sustained oscillations. Thus, oscillators with inertial nonlinearity are the simplest models of quasiperiodic and chaotic self-sustained oscillations. If systems like (11.1) have attractors in the form of a limit cycle, systems like (11.3) can also exhibit self-sustained oscillations whose image is a two-dimensional torus and a chaotic (strange) attractor.

There are several models of oscillators with inertial nonlinearity that can realize regimes of deterministic chaos. Among them the most popular systems are the Lorenz model, the Rössler model, Chua's model, and the Anishchenko–Astakhov oscillator. The equations of the Lorenz model (1963) are as follows:

$$\begin{aligned} \dot{x} &= \sigma(y - x) \,, \\ \dot{y} &= -y - xz + rx \,, \\ \dot{z} &= -bz + xy \,, \end{aligned} \quad (11.5)$$

where r, σ, and b are the control parameters. The following model (1976) was suggested by O.E. Rössler and is now more often used:

$$\begin{aligned} \dot{x} &= -(y + z) \,, \\ \dot{y} &= x + ay \,, \\ \dot{z} &= -cz + b + xz \,, \end{aligned} \quad (11.6)$$

where a, b, and c are the parameters. Chua's system or Chua's circuit (1986) is described by the following three first order differential equations:

$$\begin{aligned} \dot{x} &= \alpha[y - h(x)] \,, \\ \dot{y} &= x - y + z \,, \\ \dot{z} &= -\beta y \,, \end{aligned} \quad (11.7)$$

where α and β are the system parameters and $h(x)$ is the piecewise-linear characteristic

$$h(x) = \begin{cases} bx + a - b, & x \geq 1, \\ ax, & |x| \leq 1, \\ bx - a + b, & x \leq -1. \end{cases}$$

It is easy to see that the Lorenz model, the Rössler system, and Chua's system can be reduced to (11.3) by applying simple transformations. In this sense all these models are systems with inertial nonlinearity. Another common property of the indicated systems is that they all have singular solutions in the form of a saddle or saddle-focus separatrix loop. This important feature is a fundamental reason for the existence of chaotic solutions in these systems, whose image is a strange attractor. In this chapter we discuss in detail the dynamics of the Anishchenko–Astakhov oscillator (1981), which is a typical example of a system with a saddle-focus separatrix loop.

11.2 Theodorchik's Oscillator

The *classical oscillator with inertial nonlinearity* was proposed and described by K.E. Theodorchik. Self-sustained oscillations are provided in the system by introducing into an oscillatory circuit a thermoresistance $R(T)$ with properties that depend nonlinearly and inertially on the current through it. The circuit diagram of Theodorchik's oscillator with inertial nonlinearity is shown in Fig. 11.1. The equation for the current $i(t)$ in the circuit has the form

$$\frac{d^2 i}{dt^2} + \left[\frac{R(T)}{L} - \frac{MS_0}{LC}\right] \frac{di}{dt} + \left[(LC)^{-1} + L^{-1} \frac{\partial R(T)}{\partial T} \frac{dT}{dt}\right] i = 0, \quad (11.8)$$

where S_0 is the slope of the amplifier curve which is assumed to be linear, m is the mutual induction of the feedback circuit, $R(T)$ is the thermoresistance depending on the temperature T, and L and C are the induction and capacity of the oscillatory circuit, respectively.

Assuming the dependence $R(T)$ to be linear

$$R(T) = R_0 + LbT, \qquad b = \text{const.}, \quad (11.9)$$

and setting the heat exchange process to obey Newton's law

$$\rho q \frac{dT}{dt} + kT = R(T)i^2,$$

where q is the specific heat capacity of the thermistor filament, ρ is its mass, and k is the heat transfer coefficient, we obtain a closed system of equations of the form

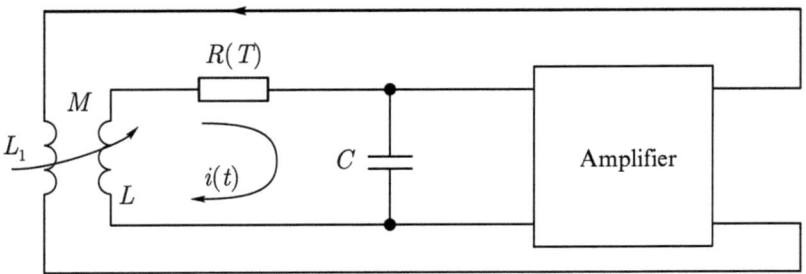

Fig. 11.1 Classical circuit diagram of Theodorchik's oscillator with inertial nonlinearity

$$\frac{d^2 i}{dt^2} + \omega_0^2 i = (\mu - bT)\frac{di}{dt} - bi\frac{dT}{dt},$$

$$\frac{dT}{dt} + \gamma T = \alpha(T) i^2. \tag{11.10}$$

Here, the following notation is used:

$$\mu = \omega_0^2 S_0 M - R_0/L, \quad \omega_0^2 = 1/LC, \quad \gamma = k/\rho q,$$
$$\alpha(T) = \alpha_0 + bLT/\rho q, \quad \alpha_0^2 = R_0/\rho q. \tag{11.11}$$

Using the dimensionless variables

$$x = ai, \quad \dot{y} = -x, \quad z = bT/\omega_0, \quad \tau = \omega_0 t, \quad a = (\alpha b \rho q/\omega_0 k)^{1/2}, \tag{11.12}$$

Eqs. (11.10) take the form

$$\dot{x} = mx + y - xz,$$
$$\dot{y} = -x, \tag{11.13}$$
$$\dot{z} = -gz + gx^2,$$

where $m = \mu/\omega_0 = \omega_0 S_0 M - R_0/\omega_0 L$, $g = \gamma/\omega_0$, and $\dot{x} = dx/d\tau$.

In the three-dimensional two-parameter system (11.13), the parameter m is proportional to the difference between the energies introduced and dissipated in the circuit, and g is a parameter characterizing the relative time of the thermistor relaxation. The parameters m and g will hereafter be called the oscillator excitation and inertia parameters, respectively.

We note the following. As can be seen from the first equation of the system (11.13), since the amplifier characteristic is linear, the growth of the variable x is restricted inertially by the term xz. Hence, stable self-sustained oscillations can appear in Theodorchik's oscillator by means of the inertial nonlinearity. In this case the oscillator amplifier is characterized by a linear dependence of the output signal on the input signal. The slope of the amplifier S_0 is constant:

11.2 Theodorchik's Oscillator

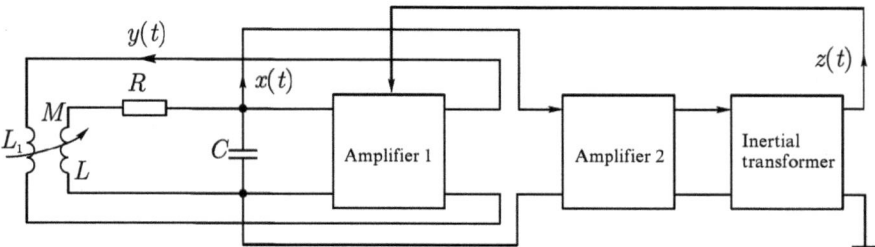

Fig. 11.2 Modified scheme for the oscillator with inertial nonlinearity

$$i_{\text{out}} = S_0 U_{\text{in}}, \qquad \frac{\mathrm{d}i_{\text{out}}}{\mathrm{d}U_{\text{in}}} = S_0, \tag{11.14}$$

where U_{in} is the input voltage and i_{out} is the current at the amplifier output.

The scheme for Theodorchik's oscillator in Fig. 11.1 is not quite convenient for experiments. It includes the thermistor and does not allow one to vary the parameter g in the model (11.13), this being a constant for a given specific thermistor. This inconvenience can be avoided by using a different oscillator circuit.

Consider the scheme shown in Fig. 11.2. The oscillatory circuit presented there has no nonlinear elements, in contrast to the classical case (Fig. 11.1). Amplifier 1 is controlled by the additional feedback circuit including a linear amplifier 2 and an inertial transformer. The differential equations for the oscillator can be written down explicitly by particularizing the slope dependence of amplifier 1, viz., $S(x, z)$. The signal from the inertial transformer output controls the slope of amplifier 1 according to

$$S(x) = S_0 - bz, \tag{11.15}$$

where b is a constant and z is the normalized voltage at the output of the inertial transformer. Let us assume that the inertial transformation can be described by the equation

$$\frac{\mathrm{d}z}{\mathrm{d}t} = -\gamma z + \gamma x^2. \tag{11.16}$$

The circuit realizing this transformation includes a two-half-period square-law detector and an RC-filter.

The voltage x at the input of amplifier 1 is related to the current y through the coil L_1 by the equation

$$\frac{\mathrm{d}^2 x}{\mathrm{d}t^2} + \frac{R}{L}\frac{\mathrm{d}x}{\mathrm{d}t} + \omega_0^2 x = M\omega_0^2 \frac{\mathrm{d}y}{\mathrm{d}t}. \tag{11.17}$$

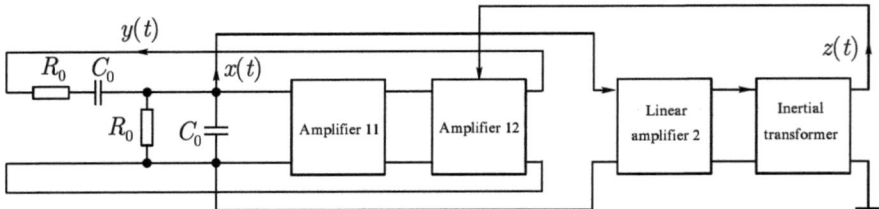

Fig. 11.3 Circuit diagram of an RC-oscillator with inertial nonlinearity

If the current y is linearly dependent on the voltage x, then

$$y = Sx. \tag{11.18}$$

In the case of (11.15), the dependence (11.18) becomes more complicated, viz.,

$$y = S_0 x - bxz. \tag{11.19}$$

After simple transformations, the set of Eqs. (11.16), (11.17), and (11.19) gives a closed system that strictly coincides with Eqs. (11.13) for Theodorchik's oscillator:

$$\begin{aligned} \dot{x} &= mx + y - xz, \\ \dot{y} &= -x, \\ \dot{z} &= -gz + gx^2, \end{aligned} \tag{11.20}$$

where $m = \omega_0 S_0 M - R_0/\omega_0 L$, $g = \gamma/\omega_0$, and $\tau = \omega_0 t$. In (11.20) the parameter g can already be varied in full-scale experiments since it depends on the filter time constant of the detector.

The oscillator scheme of Fig. 11.2 which includes the oscillatory circuit also appears to be inconvenient to use experimentally. As our studies have shown, the scheme containing a Wien bridge as an equivalent of the oscillatory circuit is preferable (Fig. 11.3).

The form of Eqs. (11.20) is maintained if an RC-chain in the form of a Wien bridge is used as a selective element. Generation conditions are provided in this case by applying two amplifier stages, as shown in Fig. 11.3. For a symmetrical Wien bridge, the control parameters m and g in (11.20) are expressed more simply and in an experimentally more convenient way via the circuit parameters

$$m = K_0 - 3, \qquad g = R_0 C_0 / \tau_f, \qquad \tau_f = R_f C_f,$$

where K_0 is the amplification coefficient of a two-stage amplifier, and $R_0 C_0$ and τ_f are the time constants of the Wien bridge and the detector filter, respectively. In full-scale experiments, the parameters m and g are easy to vary and measure by varying the amplification coefficient and the time constant of the filter.

11.2 Theodorchik's Oscillator

Thus, Theodorchik's oscillator can be implemented electronically in several ways which lead to the same mathematical model – the three-dimensional two-parameter dynamical system (11.20). Let us analyze this system. It is easy to see that it is characterized by only one equilibrium state at the origin. Linearizing the system at the equilibrium gives the characteristic polynomial

$$(g + s)(s^2 - ms + 1) = 0 , \qquad (11.21)$$

with eigenvalues (roots)

$$s_{1,2} = \frac{m}{2} \pm i\sqrt{1 - \frac{m^2}{4}} , \qquad s_3 = -g . \qquad (11.22)$$

From the physical point of view, $g > 0$. Thus, if $m > 0$, the equilibrium point is stable. For $0 < m < 2$, it is an unstable focus with two-dimensional unstable and one-dimensional stable manifolds. For $m > 2$, the equilibrium becomes a saddle.

At the bifurcation point $m = 0$, the eigenvalues $s_{1,2}$ become purely imaginary, $s_{1,2} = \pm i$ and

$$\left.\frac{\partial \mathrm{Re}\, s_{1,2}}{\partial m}\right|_{m=0} = \frac{1}{2} ,$$

i.e., the eigenvalues cross the imaginary axis with nonzero velocity. A *classical bifurcation with limit cycle birth*, the Andronov–Hopf bifurcation, is realized. Numerical simulations and experiments indicate that, as a result of this bifurcation, a stable limit cycle is softly born in the system (11.20). Its amplitude grows proportionally to $\sqrt{m}$ and its period is $T_0 \approx 2\pi$. The line $m = 0$ on the parameter plane (m, g) is a bifurcation line of the limit cycle birth.

The numerical analysis of (11.20) has shown that the dynamics of Theodorchik's oscillator is similar in many respects to that of the van der Pol oscillator: a limit cycle is softly born from the stable equilibrium and is a single attractor in the system for $m > 0$ and $g > 0$. However, the DS (11.20) has some important peculiarities. The first is that stable self-sustained oscillations are achieved in the system due to inertial nonlinearity under conditions when the main amplifier of the oscillator has a linear characteristic. The second feature is that the equilibrium at the origin is a saddle focus (or a saddle). As we will see later on, these peculiarities may lead to more complicated dynamics.

Figure 11.4 shows numerical results for the limit cycle evolution in (11.20) when the parameter m increases for fixed $g = 0.3$. As can be seen from the figure, the limit cycle projections on the plane (x, y) resemble a limit cycle in the van der Pol oscillator. The cycle amplitude grows in proportion to $\sqrt{m}$ in the region $m \leq 0.7$. For large values of $m > 1$, the cycle is distorted and the number of harmonics at frequencies $n\omega_0$ ($n = 2, 3, \ldots$) increases in the oscillation spectrum, while their intensities also grow. This is also typical for the van der Pol oscillator.

We note that Eqs. (11.13) and (11.20) of Theodorchik's oscillator can be simply and easy reduced to the general form

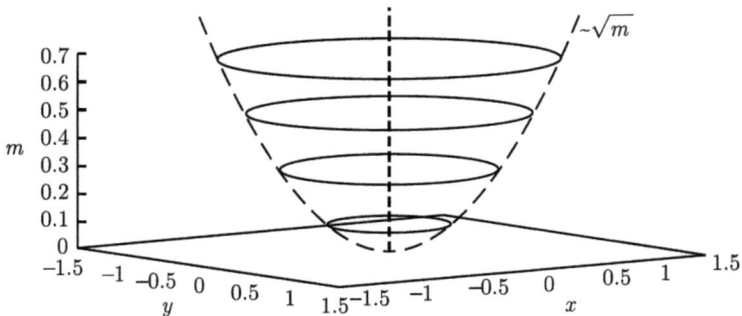

Fig. 11.4 Evolution of the limit cycle in the system (11.20) when m is increased for fixed $g = 0.3$

$$\ddot{x} - (m-z)\dot{x} + (1-gz)x + gx^3 = 0 ,$$
$$\dot{z} = -gz + gx^2 ,$$
(11.23)

where $F_1 = z - m$, $F_2 = (1 - gz + gx^2)x$, and $F_3 = g(x^2 - z)$. The first equation corresponds in its form to the equation of a Duffing-type nonlinear oscillator, in which the dissipation and the frequency depend on the variable x inertially. The form of this dependence is specified by the second equation of (11.23).

11.3 Modification of the Oscillator with Inertial Nonlinearity: The Anishchenko–Astakhov Oscillator

Let us return to the scheme of Theodorchik's oscillator shown in Fig. 11.2. We assume a more general case when the main amplifier 1 of the oscillator is nonlinear. Let the slope of its characteristic (excluding the additional inertial feedback) be

$$S^1(x) = S_0 - S_1 x^2 .$$
(11.24)

Such an oscillator is identical to the van der Pol oscillator (4.8).

Let the inertial transformation be performed according to the equation

$$\frac{dz}{dt} = -\gamma z + \varphi(x) ,$$
(11.25)

where $\varphi(x)$ is a nonlinear function. Then, instead of the system (11.20), we can obtain a closed system which has the following form in the dimensionless variables:

$$\dot{x} = mx + y - xz - dx^3 ,$$
$$\dot{y} = -x ,$$
$$\dot{z} = -gz + g\Phi(x) ,$$
(11.26)

11.3 Modification of the Oscillator with Inertial Nonlinearity: The...

where

$$m = \omega_0 S_0 M - \frac{R_0}{\omega_0 L}, \quad g = \frac{\gamma}{\omega_0}, \quad d = \omega_0 S_1 M, \quad \Phi(x) = \frac{\varphi(x)}{\omega_0}, \quad \tau = \omega_0 t.$$

In (11.26), $d = d(S_1)$ is a parameter corresponding to the degree of influence of the inertialess nonlinearity of the main amplifier characteristic, and $\Phi(x)$ is a nonlinear function of the inertial transformer, defined by the form of the function $\varphi(x)$ in (11.25).

By excluding the variable y, the equations of the system (11.26) can be written in the form (11.3), viz.,

$$\begin{aligned}\ddot{x} + (m - z - 3dx^2)\dot{x} + \left[1 - gz + g\Phi(x)\right]x &= 0, \\ \dot{z} &= -gz + g\Phi(x).\end{aligned} \quad (11.27)$$

Both the dissipation and the frequency of the nonlinear oscillator (11.27) are inertially dependent on the variable x. In the case of a strong system inertia ($\tau_f \gg T_0$), as $g \to 0$, the system (11.27) reduces to the van der Pol oscillator equation (4.8):

$$\ddot{x} - \alpha(1 - \beta x^2)\dot{x} + x = 0, \quad (11.28)$$

where $\alpha = m$ and $\beta = 3d/m$. In addition, the system (11.27) coincides with the van der Pol equation for any type of function $\Phi(x)$.

As follows from the initial system of equations (11.26), it is completely identical to Eqs. (11.20) of Theodorchik's oscillator for $d = 0$ and $\Phi(x) = x^2$. Another limiting case is an inertialess oscillator corresponding to the condition $g \to \infty$. As can be seen from (11.26), $z = -\Phi(x)$ under this condition and we obtain

$$\ddot{x} - \left[m - \Phi(x) - 3dx^2\right]\dot{x} + x = 0. \quad (11.29)$$

This equation coincides with the van der Pol equations only when

$$\Phi(x) = x^2. \quad (11.30)$$

Studies have shown that the Anishchenko–Astakhov oscillator (11.26) can realize, not only periodic, but also chaotic self-sustained oscillations. This depends on the form of the function $\Phi(x)$, which must be different from (11.30), and on the degree of inertia (g must take values in a certain interval $g_1 \leq g \leq g_2$).

We shall now analyze the oscillatory regimes of the oscillator (11.26) in more detail.

11.3.1 Periodic Regimes of Self-Sustained Oscillations and Their Bifurcations

Studies have shown that the DS (11.26) can generate different types of both periodic and chaotic oscillations when the nonlinearity function $\Phi(x)$ is specified. The oscillatory regimes can undergo various bifurcations when the system parameters are varied. Thus, it is absolutely necessary to conduct the bifurcation analysis of the system (11.26).

Here we outline the purpose of bifurcation analysis and describe algorithms for carrying out such analysis. We shall attempt to describe characteristic oscillatory regimes (from a set of possibilities) and their modifications as the parameters are varied. To do this, we define the *partition of the parameter plane* into regions of qualitatively different types of motion, indicating their *phase portraits* and particularizing the *types of regime bifurcations* at the region boundaries. For two-parameter systems, the common algorithm for constructing bifurcation diagrams comprises the following steps:

1. Find singular points of the system, study their stability, and reveal characteristic bifurcations of stability loss, and in particular, bifurcations with periodic motion (cycle) birth.
2. Examine the character of the cycle birth bifurcation that defines its stability.
3. Carry out a one-parameter study of the evolution of cycles along different parameters and find the points of characteristic bifurcations.
4. Perform the two-parameter study of cycles, including the construction of bifurcation lines corresponding to different types of codimension-one bifurcation. The points of codimension-two bifurcations are to be found on these lines.

In systems with three parameters, the two-parameter analysis is repeated for selected values of the third parameter and bifurcation situations of higher codimension are thereby explored.

The mathematical model of the modified oscillator with inertial nonlinearity (11.26) is a nonlinear three-dimensional dissipative system with three independent parameters, which specifies the flow in $\mathbb{R}^3$:

$$-\infty < x < \infty, \quad -\infty < y < \infty, \quad 0 \leq z < \infty,$$

where the variable z is defined on the positive half-axis since, from the physical point of view, it represents a detected voltage $x(t)$ at the filter output. The divergence of the velocity field of the flow (11.26) depends on the parameters and phase coordinates as follows:

$$\text{div}\,\mathbf{F} = m - g - 3dx^2 - z. \tag{11.31}$$

11.3 Modification of the Oscillator with Inertial Nonlinearity: The...

Studies using a quasi-linear approximation $m < g \ll 1$ show that the system is globally dissipative and that, for any initial condition from the region of definition of the phase variables, the following is always valid:

$$\text{div}\,\mathbf{F} < 0 .$$

If the function $\Phi(x)$ has no terms linear in the variable x, then the system of equations (11.26) linearized near the singular point will completely coincide with the linearization of the system (11.20). Consequently, we can obtain a characteristic equation similar to (11.21):

$$(g + s)(s^2 - ms + 1) = 0 . \tag{11.32}$$

Its eigenvalues are therefore

$$s_{1,2} = \frac{m}{2} \pm \frac{i}{2}(4 - m^2)^{1/2} , \qquad s_3 = -g . \tag{11.33}$$

Given the importance of analyzing the characteristics and bifurcations of the equilibrium state, we repeat that here for the equations of the oscillator (11.26). The real parts of all the eigenvalues are negative and the equilibrium point is stable in the region of the parameter plane $g > 0, -2 < m < 0$. From the physical point of view, the parameter g is always positive, being the ratio of the characteristic times of the system, i.e., the ratio of the oscillation period to the relaxation time of the filter. The parameter m may be either smaller (if the oscillator is not excited) or greater than zero in the generation regimes which are of particular interest. In the interval $0 < m < 2$, the singular point is a *saddle focus* with two-dimensional unstable and one-dimensional stable manifolds. The line $m = 2$ corresponds to the replacement of the saddle focus by a saddle node.

As can be seen from (11.33), the eigenvalues $s_{1,2}$ cross the imaginary axis at the bifurcation point $m = 0$ with the nonzero velocity

$$\left.\frac{\partial \text{Re}\, s_{1,2}(m)}{\partial m}\right|_{m=0} = 1/2 .$$

The third eigenvalue $s_3 = -g$ is separated from the imaginary axis. A classical Andronov–Hopf bifurcation is realized, i.e., the bifurcation of a cycle birth from the saddle focus. The linear analysis of the cycle birth bifurcation is insensitive to either the presence of dissipative nonlinearity (the eigenvalues are independent of the coefficient d) or the form of function $\Phi(x)$, provided that it does not involve a term linear in x. Hence, the line $g > 0, m = 0$ is a bifurcation line of the cycle birth in the region of the control parameters of the system which is physically realizable and which represents a positive quadrant $m \geq 0, g > 0$ on the plane.

We start by studying the system (11.26) for the case when $d = 0$:

$$\dot{x} = mx + y - xz,$$
$$\dot{y} = -x, \qquad (11.34)$$
$$\dot{z} = -gz + gI(x)x^2, \qquad I(x) = \begin{cases} 1, & x > 0, \\ 0, & x \leq 0, \end{cases}$$

restricting ourselves to the two-parameter analysis.

The calculation of the fixed point stability within the linear approximation is practically the only task which can be solved analytically for the system under study. Further research studies will be carried out numerically and experimentally (using the electronic model of the oscillator).

To solve the problem of the stability of a cycle that is being born, we must analyze the type of Andronov–Hopf bifurcation. Numerical calculations have shown that the *first Lyapunov quantity* $L_1(g)$ is *negative* at the singular point, everywhere along the line of cycle birth. The limit cycle being born is thus stable (*supercritical bifurcation*).

The first Lyapunov quantity can be approximately calculated analytically using Bautin's algorithm.[1] Approximate analytical and numerical results agree qualitatively. Thus, in the systems (11.26) and (11.34), a *stable limit cycle is softly born* on the line $m = 0$, $g > 0$, with a radius that grows in proportion to $\sqrt{m}$ and with period, according to the theorem, equal to

$$T_0 \approx \frac{2\pi}{|s_{1,2}(0)|} = 2\pi.$$

It has been found by numerically integrating the system (11.34) that, for values $0 < m < 2$ and $0 < g < 2$, the stable periodic oscillations with amplitude $\sim \sqrt{m}$ and period $T_0 \approx 2\pi$ solve the Cauchy problem with initial conditions close to the singular point at zero. For values $m < 0.5$, agreement with the theorem of cycle birth is virtually perfect to within the accuracy of calculation. When $m > 0.5$, small deviations from theoretical predictions appear in the dependences of the amplitude and the period of the cycle $\Gamma_0(m)$.

Let us do a one-parameter study of the evolution of the cycle family $\Gamma_0(m, g)$ to find the points of typical bifurcations. We thus explore different sections of the parameter plane spanned by m and g by fixing m and calculating the cycle and its multipliers when g is varied. Calculations show that the cycle family $\Gamma_0(m, g)$ has

---

[1] Problems arise when using the above algorithm due to a discontinuity of the second derivative $\Phi(x) = I(x)x^2$ in (11.26). If one approximates $\Phi(x)$ by the exponential function $\exp(x) - 1$ and restricts to the first three terms of its Taylor expansion, then the calculation can be carried through and it can be shown that $L_1(g) < 0$ for any $g > 0$.

11.3 Modification of the Oscillator with Inertial Nonlinearity: The...

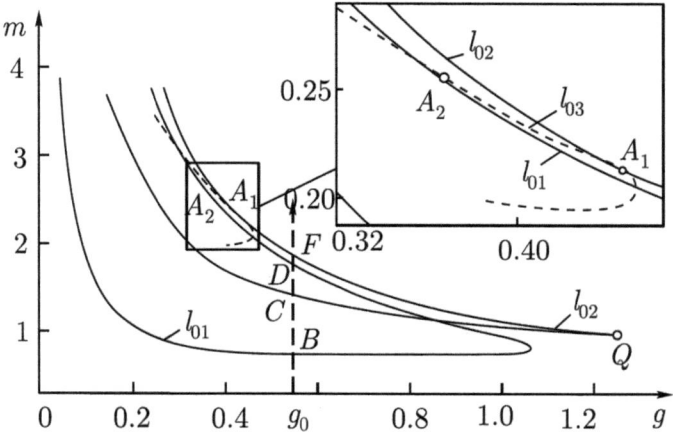

Fig. 11.5 Bifurcation diagram for the family of period-one cycles of the system (11.34). l_{01} is the period-doubling bifurcation line, l_{02} is the multiplicity line, l_{03} is the neutrality line, and A_1, A_2 and Q are the codimension-two bifurcation points

the following bifurcations:

- The multiplier μ_1 with the greatest modulus becomes -1 at the bifurcation point, which corresponds to the *oscillation-period-doubling bifurcation*.
- The multiplier μ_1 of the cycle Γ_0 takes the value $+1$, which corresponds to the merging and vanishing (or the birth) of stable and unstable cycles.
- There are situations where the product of the cycle multipliers satisfies the condition $|\mu_1\mu_2| = 1$ when the parameter is varied.

The latter is called the *neutrality condition* since it corresponds to the case when the sum of the Lyapunov exponents of the cycle Γ_0 becomes zero: $\lambda_1 + \lambda_2 = 0$. If the multipliers are complex conjugate here, then the *bifurcation of a two-dimensional torus birth* is realized. For the real multipliers μ_1 and μ_2, there is no bifurcation situation available and the cycle is a saddle. For the cycle family $\Gamma_0(m, g)$, both cases are observed, i.e., the bifurcation of torus birth occurs in the system (11.34)!

Having determined the parameter values corresponding to the bifurcations of cycles $\Gamma_0(m, g)$, we proceed to a two-parameter analysis – the construction of bifurcation lines in the parameter space. The *bifurcation diagram* is shown in Fig. 11.5 for the family of period-one cycles Γ_0 born as a result of Andronov–Hopf bifurcation. One of the cycle multipliers becomes -1 on the line l_{01} (the *period-doubling bifurcation line*). The cycle Γ_0 is saddle inside the region bounded by the line l_{01} and is stable outside this region, since both of its multipliers belong to the inside of the unit circle. On the line l_{02}, $\mu_1 = +1$. The merging and subsequent vanishing of stable and saddle cycles take place here. Alternatively, if one moves along parameters in the opposite direction, a pair of cycles is born from the closeness of trajectories. The line l_{02} will be referred to hereafter as the line of multiple cycles, or the *multiplicity line*. Inside the region bounded by the multiplicity line, three period-one cycles always exist, the saddle one Γ_0'' and two cycles Γ_0 and Γ_0' which can be stable or saddle.

Fig. 11.6 Qualitative view of the dependence $\mu_1(m)$ for the cycle Γ_0 in the section $g = g_0$. Critical points B, C, D, and F correspond to those indicated in Fig. 11.5

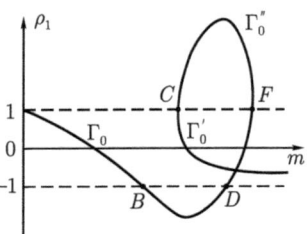

The situation is made clear in Fig. 11.6, where the calculated dependence is shown qualitatively for the largest multiplier μ_1 of the cycle Γ_0 when moving along the parameter m for fixed g_0, as indicated in Fig. 11.5. A pair of cycles Γ_0' and Γ_0'' is born at the point C, the above cycles merge and vanish at the point F, and the cycle Γ_0 undergoes the period-doubling bifurcation at the point B, becoming a saddle, but finding its stability again at the point D.[2] The situation is described when the parameter m increases. The typical points of bifurcations B, C, D, and F are marked in Fig. 11.5.

The multiplicity line l_{02} forms a characteristic angle with the vertex at the point Q where all three of the cycles Γ_0, Γ_0', and Γ_0'' merge into one cycle. Point Q is bifurcational and has codimension 2. In catastrophe theory, this point is called the *cusp point*. The presence of a cusp corresponds to the simplest and most frequently encountered catastrophe in multi-dimensional systems, and the only possible catastrophe in general two-dimensional systems. The interrelation between the cusp catastrophe and the dynamics of the system under study will be discussed below.

Figure 11.5 shows a segment of the bifurcation line l_{03} where the neutrality condition of the cycle Γ_0 is fulfilled, i.e., $|\mu_1\mu_2| = 1$. The line l_{03} is strictly bifurcational in the segment from point A_1 to A_2 where the cycle multipliers are complex conjugate and emerge from the unit circle. At points A_1 and A_2, both multipliers are either -1 (point A_2) or $+1$ (point A_1) which correspond to the resonances 1 : 1 (A_1) and 1 : 2 (A_2). Like point Q, the bifurcation points A_1 and A_2 have codimension 2. Besides the emergence of a pair of multipliers from the unit circle, the resonance conditions are satisfied at these points. As calculations have shown, the bifurcation of torus birth from the cycle Γ_0 leads to a regime of unstable beats. The stable two-frequency oscillations have not been found in the autonomous system (11.34). The two-parameter analysis of the stability of the period-one cycles of the system may be completed, since typical bifurcations have been defined and the corresponding bifurcation diagram has been constructed in the parameter plane (m, g).

---

[2] Such calculations requiring transitions to unstable cycles at points C and F are non-trivial, but possible with a suitable modification of calculation algorithms for the cycle multipliers. Usual integration leads to the loss of cycle and to the abrupt change of regimes here.

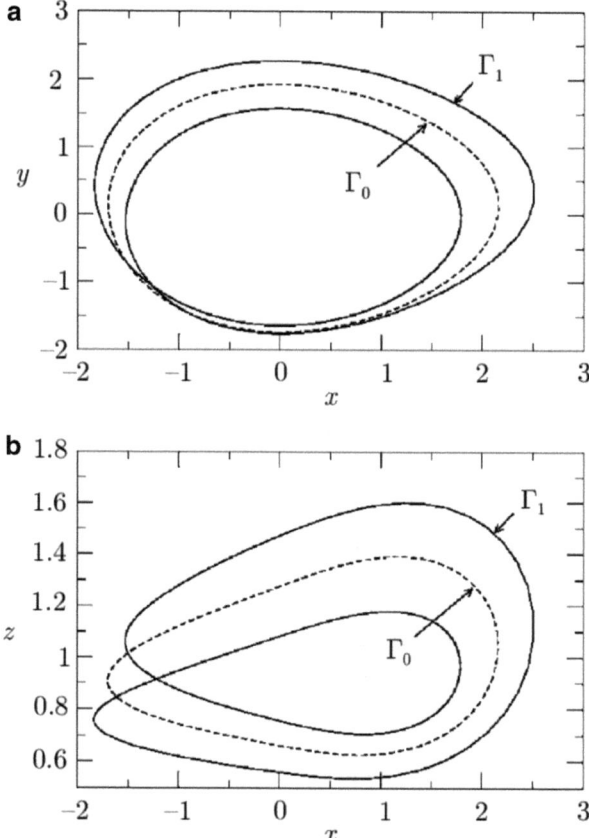

Fig. 11.7 Period-doubling bifurcation of the period-one cycle. Projections of phase trajectories on the plane (x, y) (**a**) and (x, z) (**b**). Γ_0 is the period-one cycle and Γ_1 is the period-two cycle

The cycle Γ_0 is asymptotically stable (the first Lyapunov quantity is strictly negative at the singular point) when approaching the line of doubling l_{01} from below. This means that crossing the line l_{01} can lead to the soft birth of a stable cycle Γ_1 with twice the period ($T_1 \simeq 2T_0$), within the linear approximation. Taking a point on the cycle Γ_0 close to the point of doubling bifurcation as an initial approximation, we find a cycle Γ_1 that is defined by a period-two fixed point in the Poincaré map. Having shifted along the parameter behind the bifurcation line l_{01}, we can determine the cycle Γ_1 numerically. It is really stable, has period close to the doubled period, and goes twice round the period-one cycle, which has lost its stability, in the phase space. The situation described above is illustrated in Fig. 11.7 for $g = 0.2$. Below the bifurcation point $m^* = 0.966...$, the cycle Γ_0 is stable in the system. For higher parameter values ($m > m^*$), the cycle Γ_1 with doubled period is stable and the cycle Γ_0 becomes a saddle.

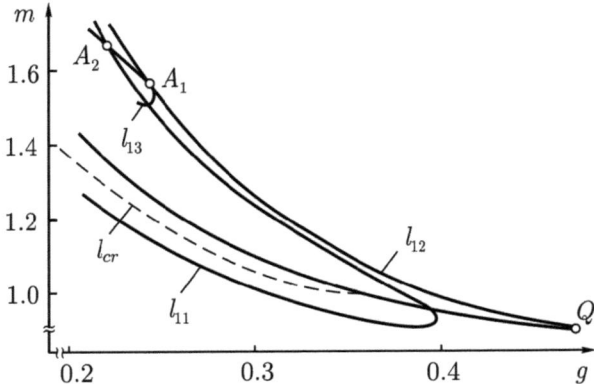

Fig. 11.8 Bifurcation diagram for the family of period-two cycles $\Gamma_1^1(m, g)$, with l_{cr} the line of critical parameter values

Let us explore in a similar manner a family of *period-two* cycles $\Gamma_1(m, g)$, first in the one-parameter manner and then in the two-parameter manner. The bifurcation analysis shows that the character of the bifurcation and the structure of the relative location of the corresponding bifurcation lines in the parameter plane for cycles Γ_1 with doubled period replicate the picture for period-one cycles. The only difference is that two independent families of period-two cycles Γ_1^1 and Γ_1^2 are detected.

The bifurcation diagram of one of the families of cycles $\Gamma_1^1(m, g)$ is shown in Fig. 11.8 and confirms the above. The presence of two families of period-two cycles is confirmed by the dependence of the multipliers on the parameter m in the form of two loops, in contrast to one loop for the cycles with period T_0. Calculations confirm that there are two independent families of period-four cycles inside each region bounded by the bifurcation lines of period-doubling of cycles $\Gamma_1^1(m, g)$ and $\Gamma_1^2(m, g)$. The hierarchy of multiplication of cycle families may be assumed to extend to infinity. Their bifurcation diagrams constitute a system of topologically equivalent *embedded structures* that correspond to universal properties which generalize the regularities of Feigenbaum-type similarity for the case of two parameters. This has been corroborated experimentally and confirmed qualitatively for cycles Γ_k with period $T_k \simeq 2^k T_0$, $k = 0, 1, 2$ and partially for $k = 3$. Quantitative regularities are difficult to ascertain since the cycles must be analyzed numerically with sufficiently long periods and to a high degree of accuracy.

This problem is more conveniently considered as applied to two-parameter two-dimensional model maps. The complexity of the phase space partition into different trajectory types is vizualized in the relatively simple system (11.34) by representing the results geometrically. We may consider a combined three-dimensional space where the dependence $\xi = \xi(m, g)$ is represented graphically with ξ taken as one of the fixed point coordinates in the Poincaré section. For example, if the secant plane $x = 0$ is introduced in the phase space $\mathbb{R}^3$ of the system (11.34), then ξ can be chosen as the coordinate z of a two-dimensional map in the secant plane.

11.3 Modification of the Oscillator with Inertial Nonlinearity: The...

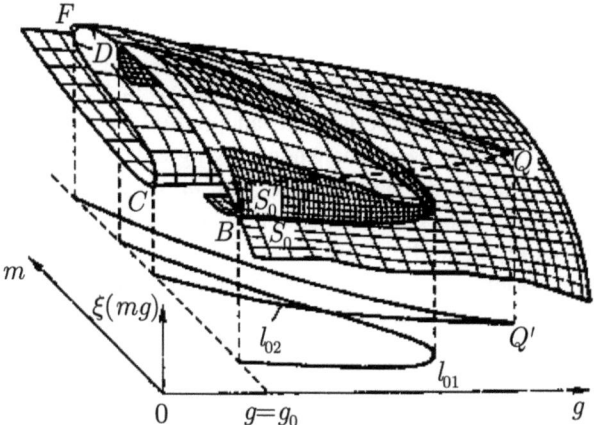

Fig. 11.9 Geometric representation of critical phenomena in the system (11.34). The peculiar points are the same as in Figs. 11.5 and 11.6

Calculations show that a complicated two-dimensional surface S^0 in the above space shown in Fig. 11.9 is a locus of points corresponding to stable and unstable cycles $\Gamma_0(m, g)$ of the system. The surface S^0 emerges from the birth line of the cycle Γ_0 and has two folds and a cusp when $\xi \geq 0$. For clarity, the section of the surface S^0 by the plane $g = g_0$ is shown in Fig. 11.9. Three sheets of the surface, corresponding to the cycles Γ_0 (upper sheet), Γ_0' (lower sheet), and Γ_0'' (internal sheet), are available inside the region between points C and F. Due to the presence of folds, the projection of the complicated surface onto the plane of parameters m and g provides a peculiar line l_{02} consisting of upper and lower branches which intersect at the point Q. A *hysteresis* phenomenon, caused by 'jumps' at point F (when decreasing m) and C (when increasing m) is inevitably connected with the variation of the parameter m ($g = g_0$). These *abrupt switchings* are well known in catastrophe theory and as a matter of fact lead to the cusp catastrophe itself. Besides the fold and cusp, no other peculiarities are observed in two-parameter families when the surface is projected onto the plane.

Thus, the bifurcation line of multiplicity l_{02} in the diagram shown in Fig. 11.5 originates due to the presence of folds and a cusp at the surface S^0. How does the bifurcation line of doubling l_{01} originate? If only period-one cycles Γ_0 are analyzed, this line corresponds to the projection of the corresponding line onto the surface S^0, in accordance with the critical values of the cycle amplitudes, when a multiplier takes the value -1. Taking into account the fact that the cycle Γ_1 with doubled period is softly born in this case, a surface S_0^1 is formed in Fig. 11.9. It crosses the surface S^0 along the line of doubling. The projection of the line of intersection of surfaces S_0^1 and S^0 onto the plane of parameters m and g will produce a bifurcation line l_{01}. No catastrophe takes place in the vicinity of this line, since the transition from a stable regime to another stable one *proceeds softly*.

11.3.2 Period-Doubling Bifurcations: Feigenbaum Universality

As can be seen from the bifurcation diagrams shown in Figs. 11.5 and 11.8, the period-doubling bifurcation of period-one and period-two cycles is realized in the Anishchenko–Astakhov oscillator when moving either along the parameter m with $g = $ const. or along the parameter g with $m = $ const. The regularities of period-doubling bifurcations in the system (11.34) have been studied in detail, both numerically and experimentally.

Let us consider briefly the results of the above experiments. Figure 11.10 shows calculation data for the evolution of limit cycles in the system (11.34), obtained when moving along the parameter m at fixed $g = 0.3$. This figure presents projections of phase trajectories onto the plane of variables (x, y), parts of the time series $x(t)$, and the corresponding power spectra $S_x(f)$ calculated for the times series $x(t)$. The data shown in Fig. 11.10 illustrate the presence of a cascade of period-doubling bifurcations which culminate in the transition to the regime of a chaotic attractor. We calculate the bifurcational values of parameters m_k and g_k for which the multipliers μ_k of the corresponding cycles become -1. They are indicated in Table 11.1. The universal Feigenbaum constant δ is evaluated using the expression

$$\delta = \lim_{k \to \infty} \delta_k = \frac{\Delta_k}{\Delta_{k+1}} = \frac{\alpha_{k+1} - \alpha_k}{\alpha_{k+2} - \alpha_{k+1}}, \qquad (11.35)$$

or $\alpha_k = m_k$ ($g = $ const.) or $\alpha_k = g_k$ ($m = $ const.).

The calculations are performed for $k = 1, 2, 3, 4$, e.g., up to the bifurcation point of doubling of a period-16 cycle. As experiments have shown, the universal Feigenbaum constant δ is already reasonably accurately estimated in this case, although the ratio (11.35) is only valid in the limit $k \to \infty$.

Critical values of parameters m^* (or g^*) corresponding to the bifurcation point of a chaotic attractor birth are estimated using the formula

$$\alpha^* = \frac{\alpha_k \delta - \alpha_{k-1}}{\delta - 1}, \qquad (11.36)$$

where k corresponds to the value $k = 4$.

The calculation data were compared with the experimental results obtained for the electronic model of the oscillator. Figure 11.11 shows projections of the phase portraits of cycles and the corresponding power spectra, illustrating the evolution of oscillation modes when the parameter m is varied ($g = 0.3$). In the experiment, all the parameters and variables are reduced to their dimensionless form according to the mathematical model (11.34). It is easily seen that the experimental data presented in Fig. 11.11 correspond to the calculation results (Fig. 11.10). More detailed information can be obtained by comparing calculated and experimental data for the bifurcational values of the parameter m_k, as given in Table 11.2.

11.3 Modification of the Oscillator with Inertial Nonlinearity: The... 193

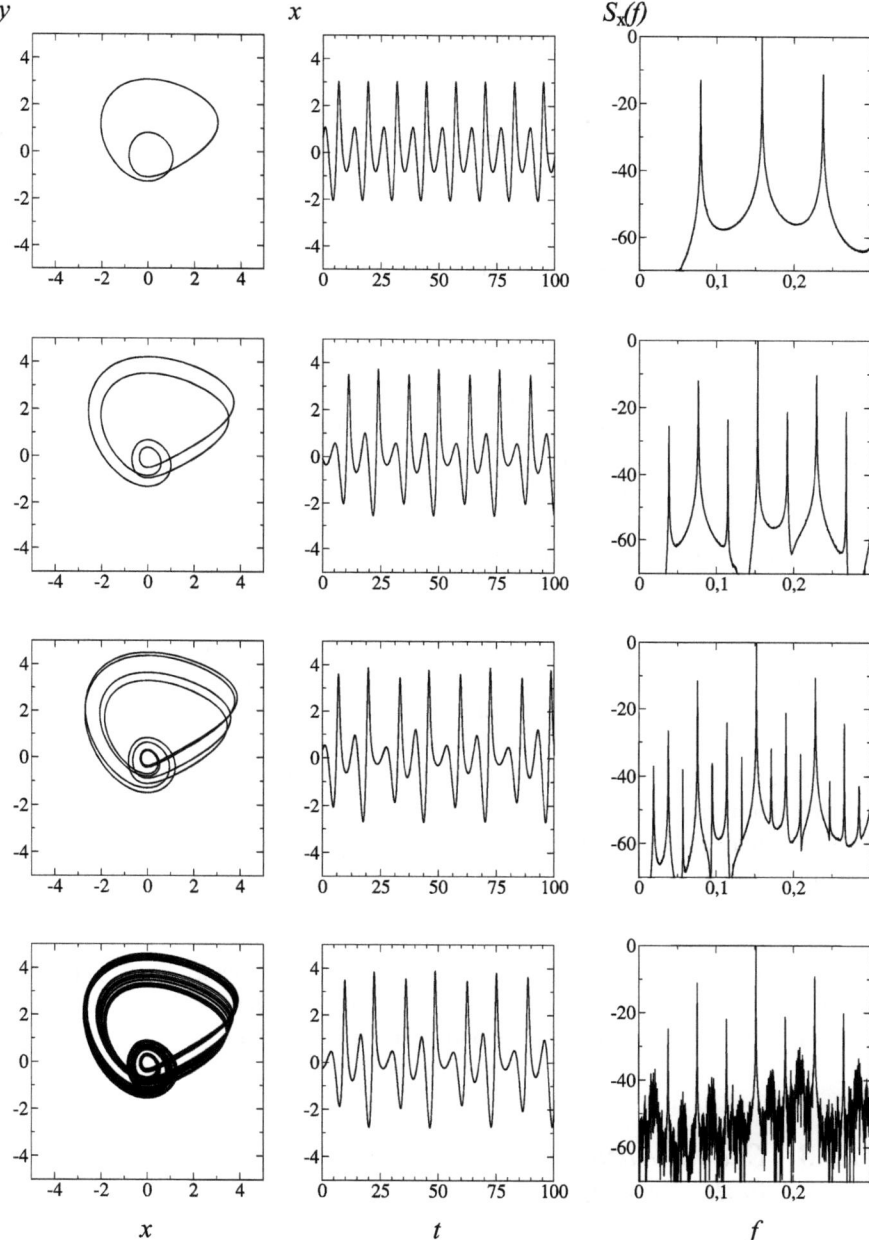

Fig. 11.10 Projections of phase trajectories onto the xy-plane, parts of the time series $x(t)$, and the corresponding normalized power spectra $S_x(f)$ in the system (11.34) for $g = 0.3$ and increasing (*top* to *bottom*) values of the parameter m in the interval $0.8 < m < 1.09$. $S_x(f)$ is measured in dB and x, y, t, f are the dimensionless quantities

Table 11.1 Bifurcational values of the parameters m_k and g_k, the Feigenbaum constant δ_k, and critical points under period doublings in the system (11.34) (numerical calculation)

	$g = 0.3$			$m = 1.45$		
k	m_k	δ_k	m^*	g_k	δ_k	g^*
0	0.7700	–	–	0.1200	–	–
1	1.0200	–	1.0880	0.16898	–	0.18233
2	1.0713	4.873	1.0853	0.18162	3.876	0.18506
3	1.08216	4.724	1.08511	0.18438	4.582	0.18513
4	1.08449	4.66896	1.08512	0.18497	4.66836	0.18513

The bifurcational values of m_k can be reliably measured experimentally up to $k = 3$, corresponding to the doubling bifurcation of a period-8 cycle. As can be seen from Table 11.2, this information is insufficient for defining the Feigenbaum constant either experimentally ($\delta_2 = 5.0 \pm 0.12$) or numerically ($\delta_2 = 4.873$). At the same time, the relation between the calculated and experimental values of m_k up to $k = 3$ indicates that the transition to chaos via a cascade of period-doubling bifurcations occurs in the DS (11.34) according to the Fengenbaum universality.

It is known that the universality of the transition to chaos via a cascade of period-doubling bifurcations was proven by Feigenbaum for a class of smooth one-dimensional maps with a quadratic extremum. Hence, the following question naturally arises: Why are these regularities fulfilled, to a high degree of accuracy, for the three-dimensional dynamical system (11.34)? The answer to this question certainly exists and lies in the fact that the dynamics of the system (11.34) can be characterized to a high degree of accuracy by a one-dimensional map of the Feigenbaum class.

Consider the regime of a chaotic attractor in the system (11.34) for the parameters $m = 1.5$ and $g = 0.3$. We introduce a secant plane in the phase space by the condition $x = 0$ and construct a two-dimensional map $(y_{n+1}, x_{n+1}) = F(y_n, z_n)$ on the secant plane. As calculations have shown, the resulting map is close to one-dimensional. To confirm this, a one-dimensional map $y_{n+1} = f(y_n)$ is constructed numerically using the calculation results of the map in the secant $x = 0$. The results are shown in Fig. 11.12. It is evident from the figure that the map is indeed close to a one-dimensional map and has a smooth quadratic maximum.

11.3.3 Chaotic Attractor and Homoclinic Trajectories in the Oscillator

As a result of the robust intersection of stable and unstable manifolds of saddle cycles (stable and unstable separatrices of saddle fixed points), homoclinic trajectories (points) serve as 'a signal of disaster' or the forerunner of a complicated aperiodic motion of the system. This phenomenon was discovered and studied by A. Poincaré, G. Birkhoff, S. Smale, S.E. Newhouse, and L.P. Shilnikov. The presence of a denumerable set of stable and unstable periodic trajectories of different

11.3 Modification of the Oscillator with Inertial Nonlinearity: The... 195

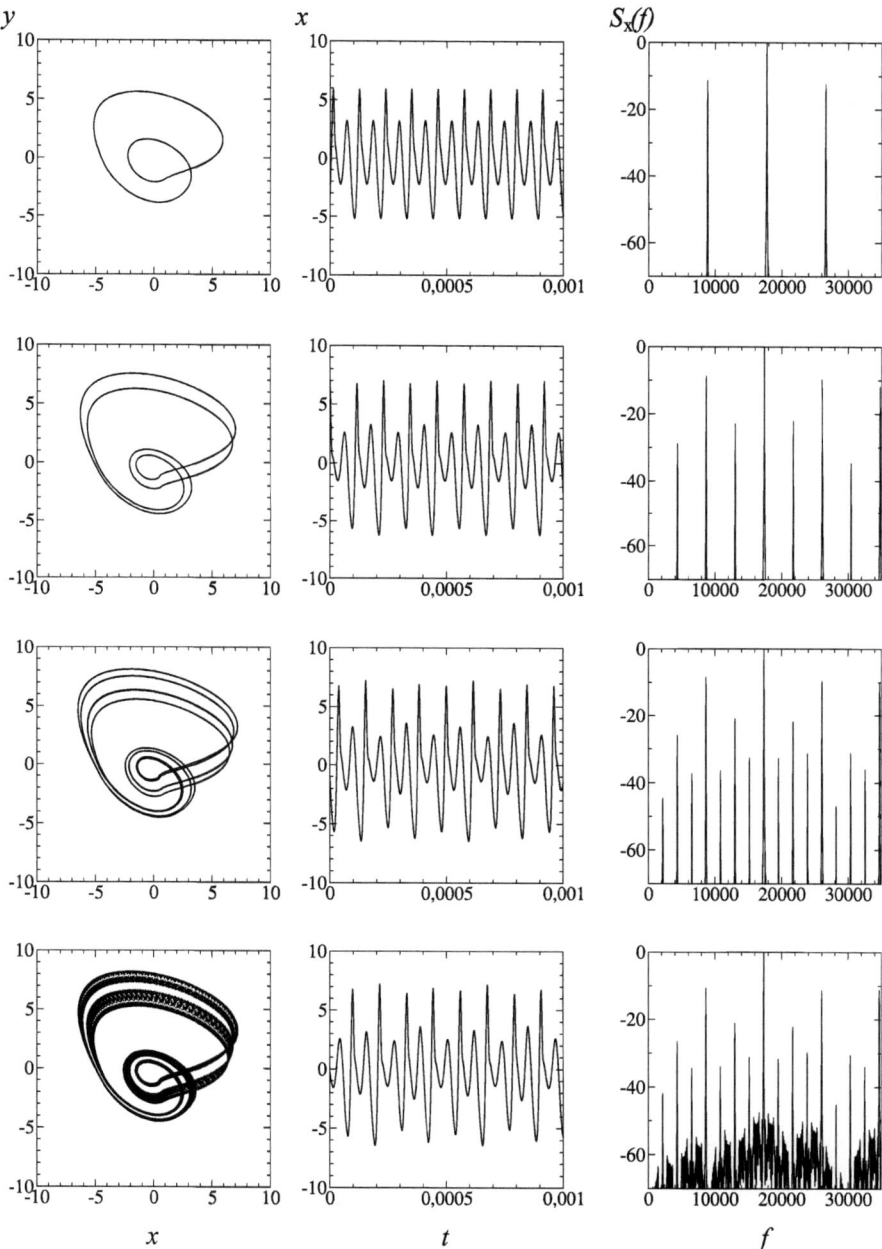

Fig. 11.11 Experimental data obtained for the electronic model of the oscillator (11.34) and corresponding to the data in Fig. 11.10. These results were obtained by converting the analog signal $x(t)$ into a digital one, followed by signal processing on a computer. x, y are the normalized variables. The time t is measured in seconds and f in Hz

Table 11.2 Comparison of calculated and experimental bifurcational values of the parameter m for $g = 0.3$

k	m_k (calculation)	m_k (experiment)	δ_2
0	0.770 ± 10^{-6}	0.77 ± 0.01	
1	1.020 ± 10^{-6}	1.02 ± 0.01	
2	1.0713 ± 10^{-6}	1.07 ± 0.01	$\delta_2 = 4.873$ (calculation)
3	1.08216 ± 10^{-6}	1.08 ± 0.01	
m^*	1.08516 ± 10^{-6}	1.09 ± 0.01	$\delta_2 = 5.0 \pm 0.12$ (experiment)

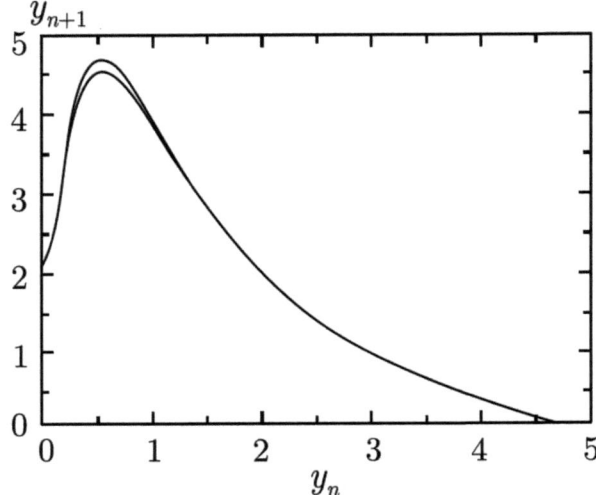

Fig. 11.12 One-dimensional map $y_{n+1} = f(y_n)$, constructed numerically for $m = 1.5$ and $g = 0.3$

periods, including a continuum of Poisson-stable trajectories, in the vicinity of the homoclinic trajectories follows from their availability, with some additional assumptions.

Shilnikov's theorem on a saddle focus is one of the fundamental results in the theory of dynamical chaos. The essence of the theorem is that, if a double-asymptotic trajectory in the form of a saddle-focus separatrix loop exists in a dynamical system, then a non-trivial hyperbolic subset of trajectories appears in its vicinity. This subset may be attracting, so a chaotic attractor, the image of deterministic chaos, will be observed in experiment.

A comprehensive experimental analysis of the onset mechanisms and the topological structure of chaotic attracting sets in the modified oscillator with inertial nonlinearity suggests that a homoclinic trajectory of the *saddle-focus separatrix loop* type can exist in an autonomous dynamical system.

11.3 Modification of the Oscillator with Inertial Nonlinearity: The...

The first attempts to find a separatrix loop in the oscillator equations (11.26) were unsuccessful. Moreover, it became clear that these equations have no such solution. Let us show that this is so. A singular point of the system is characterized by two-dimensional unstable and one-dimensional stable manifolds. Let us reverse the time in (11.26) and specify the initial conditions $x(0) = y(0) = 0$, $z(0) > 0$ for the one-dimensional unstable manifold. System integration confirms that the trajectory goes to infinity along the z-axis. It follows from (11.26) that $z(\tau) = z(0) \exp(g\tau)$. As $\tau \to \infty$, the trajectory does not return to the singular point!

This gives rise to a hypothesis that has been quite successful. The saddle-focus separatrix loop exists in a certain perturbed system. Without perturbation the loop itself vanishes, but the structure of the phase space partition into trajectories is maintained. To confirm these considerations, one has to define the form of the slightly perturbed system, prove the presence of the saddle-focus separatrix loop in it, and ascertain the structure of the attractors and study their evolution when the perturbation is removed. The solution of the above task is ambiguous, but the particular form of the small perturbation turns out to be of little importance due to the robustness.

We add a constant positive term γ to the second equation of the original system (11.26) and consider the following perturbed system:

$$\dot{x} = mx + y - xz, \qquad \dot{y} = -x + \gamma, \qquad \dot{z} = -gz + gI(x)x^2. \qquad (11.37)$$

The singular point of the flow (11.37) is the only one, as before. It is slightly shifted relative to the origin and represents a saddle focus. Its coordinates are $x^0 = \gamma$, $y^0 = \gamma(\gamma^2 - m)$, $z^0 = \gamma^2$. The equilibrium state in the perturbed system (11.37), for $m > 0$, is characterized by two-dimensional unstable and one-dimensional stable manifolds. To find the loop Γ_0^1, we reverse the time and solve the Cauchy problem repeatedly with initial conditions on the unstable one-dimensional manifold for fixed $g = 0.3$ and various m and γ. Having selected a small value of $\gamma = 0.1$, we find the bifurcation point $m^* = 1.176\ldots$ at which the one-rotational saddle-focus loop Γ_0^1 is realized. The three-dimensional image of a double-asymptotic trajectory Γ_0^1 is shown in Fig. 11.13. The loop is naturally destroyed when any of the control parameters of the system (11.37) deviate from their bifurcational values. Detailed calculations of the bifurcation diagrams for the system (11.26) and for the perturbed system (11.37) have confirmed their qualitative equivalence. Hence, it can be argued that the structure and properties of chaos in the system (11.26) are completely determined by the existence of the saddle-focus separatrix loop in the system (11.37).

Experimental and numerical studies have convincingly demonstrated that chaotic self-sustained oscillations of different structure can be generated and that the effect of deterministic chaos is closely related to the presence of the saddle-focus separatrix loop in the system (11.26). As an example, Fig. 11.14 illustrates projections of chaotic trajectories onto the plane (x, y) which correspond to a so-called *spiral-type* attractor. A *funnel* attractor is exemplified in Fig. 11.15, together with a saddle-focus separatrix loop Γ_0. The above results have been obtained numerically. However, they

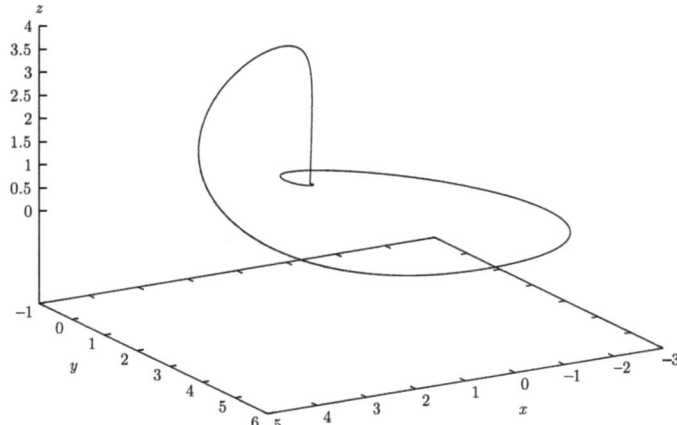

Fig. 11.13 Saddle-focus separatrix loop in the perturbed system (11.37) for $m = 1.176$ and $g = 0.3$

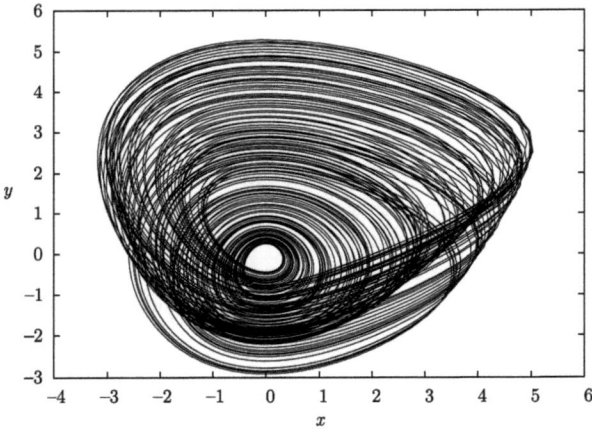

Fig. 11.14 Projection of the spiral-type chaotic attractor onto the plane (x, y) in the system (11.26) for $d = 0$, $m = 1.563$, and $g = 0.17$

can be reproduced in a full-scale experiment. A surprisingly good correspondence is observed between the experimental and numerical results. As an example, Fig. 11.16 shows photographs of the attractor obtained experimentally.

The following important conclusion can be drawn. The simplest type of oscillator with self-sustained chaotic oscillations can be realized if and only if:

- We create an amplifying cascade with a resonant circuit at the input which provides the inverted parabola type characteristic that has a controlled slope with derivative greater than 1,
- We introduce a positive feedback satisfying all the conditions of self-sustained oscillation excitation.

11.3 Modification of the Oscillator with Inertial Nonlinearity: The... 199

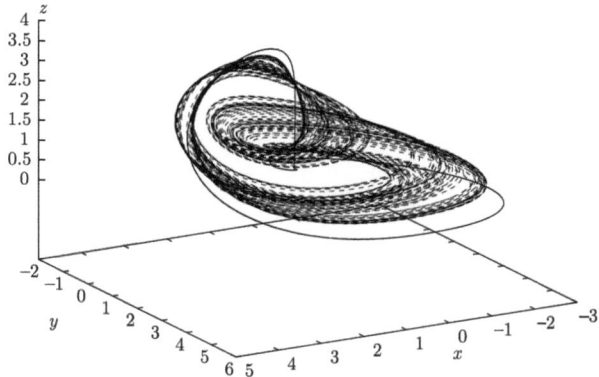

Fig. 11.15 A funnel attractor and a saddle-focus separatrix loop in the systems (11.26) and (11.37), respectively

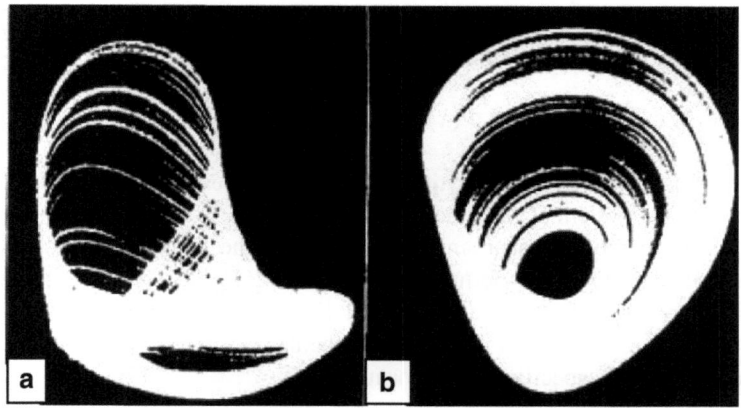

Fig. 11.16 Projections of the spiral attractor onto the plane (x, z) (**a**) and (x, y) (**b**). Full-scale experiment for $m = 1.5$ and $g = 0.2$

The necessary characteristic of the amplifier is realized in the Anishchenko–Astakhov oscillator via the inertial feedback, where a one-half-period detector is used to introduce nonlinearity. We are convinced that this is not the only practical way to achieve the result. There will also be other ways.

To conclude this section, we note the following important fact. The majority of the results were obtained using the oscillator described by the model (11.34). From the standpoint of full-scale and numerical experiments, this is justified. However, there is a certain mathematical peculiarity. The function $\Phi(x) = I(x)x^2$ has a discontinuity in the derivative at the origin, i.e., it is nonsmooth. This circumstance leads to a number of mathematical complications. It has been established that the basic properties of the oscillator are weakly dependent on these mathematical details. In particular, the oscillator exhibits the full range of properties if $\Phi(x)$ is

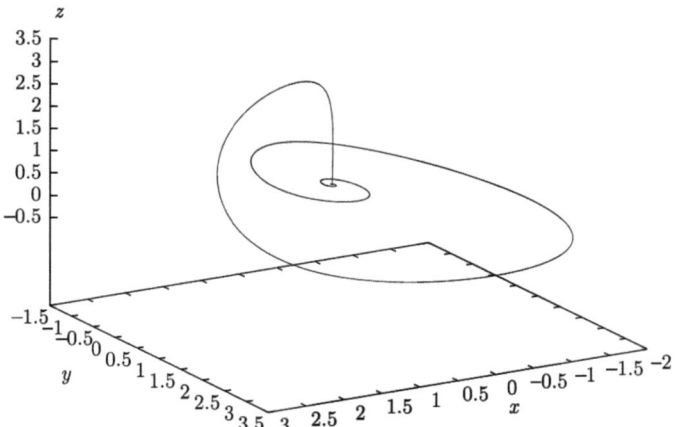

Fig. 11.17 Saddle-focus separatrix loop in the perturbed system (11.37) with $\Phi(x) = \exp(x) - 1$

represented by the exponential function

$$\Phi(x) = \exp(x) - 1 . \tag{11.38}$$

This function is smooth and analytical and can be successfully used to analyze the properties of the system (11.26) theoretically.

Studies have shown that a singular solution in the form of a separatrix loop can also be realized in the perturbed system (11.26) with $\Phi(x) = \exp(x) - 1$. Calculation results are presented in Fig. 11.17. Thus, the regime of deterministic chaos can also be observed in the Anishchenko–Astakhov oscillator when the nonlinear detector is specified in the form (11.38).

11.4 Summary

In this chapter, we present the basic elements required to construct models of dynamical systems with 1.5 degrees of freedom, which, from the point of view of the theory of oscillations, can be attributed to models of oscillators with inertial nonlinearity. We describe an electronic model of the Anishchenko–Astakhov oscillator and describe how it is created, formulating the corresponding equations of the system. A general algorithm is provided for numerical study of the dynamics of an oscillator with chaotic self-sustained oscillations. The mechanism for the appearance of chaotic self-sustained oscillation through a cascade of period-doubling bifurcations is analyzed in detail. The fundamental role of the presence of a saddle-focus separatrix loop in the system is illustrated, and it is explained how this is the primary reason for the occurrence of both the cascade of period-doubling bifurcations and the chaotic attractor, according to Shilnikov's theorem. The Anishchenko–Astakhov oscillator and its dynamics are described in detail in the book [1]. The basic properties of the oscillator are also presented in [2].

References

1. Anishchenko, V.S.: Dynamical Chaos – Models and Experiments. World Scientific, Singapore (1995)
2. Anishchenko, V.S., Astakhov, V.V., Neiman, A.B., Vadivasova, T.E., Schimansky-Geier, L.: Nonlinear Dynamics of Chaotic and Stochastic Systems. Springer, Berlin (2002)

Chapter 12
Quasiperiodic Oscillator with Two Independent Frequencies

12.1 Introduction

Quasiperiodic oscillations are widespread in nature and are an important subject of research in the natural sciences. Their peculiarity lies in the fact that they include two or more independent frequencies in the oscillation spectrum:

$$x(t) = x[\phi_1(t), \phi_2(t), \ldots, \phi_p(t)], \tag{12.1}$$

where $\phi_i(t) = \omega_i t$, $i = 1, 2, \ldots, p$. As a result, $x(t)$ in (12.1) is 2π-periodic in each argument $\phi_i(t)$, but the quasiperiodic process itself is, in the general case, non-periodic, i.e., $x(t) \neq x(t + T_0)$. Generally, the power spectrum of $x(t)$ oscillations includes all independent frequencies ω_i, their harmonics $n\omega_i$ ($n = 2, 3, \ldots$), and combinations of frequencies of type $n\omega_i \pm m\omega_k$, where n and m are positive integers, and $i, k = 1, 2, \ldots, p$.

A simple quasiperiodic oscillation can be exemplified by the motion of any fixed point on the Earth's surface. This point performs a periodic motion that is related to the daily rotation of the Earth around its axis. At the same time, the point is involved in the periodic motion of the Earth around the Sun with a period of 1 year. Quasiperiodic (multifrequency) oscillations arise when a carrier electromagnetic signal is modulated by an information signal (radio engineering, communication technology). They accompany the transition to a turbulent fluid flow in hydrodynamics and they describe complex oscillatory processes in living organisms (biophysics, electrophysiology), etc.

Quasiperiodic oscillations with n independent frequencies are associated with an n-dimensional torus in the phase space of a dynamical system. The analysis of stability, bifurcations, and transitions to chaos as applied to quasiperiodic oscillations is related to the study of bifurcations of n-dimensional tori. Over many years, these problems remained a subject of research for specialists in nonlinear dynamics and turbulence. But after the publication of papers by Ruelle and Takens, interest in the study of quasiperiodic oscillations with ω_i, $i = 1, 2, 3, \ldots, n$, and $n \geq 4$ began to decrease. The reason was that the instability of motions on three- and four-dimensional tori had been demonstrated. At the same time, a number of works revealed that there were stable oscillations with four and six independent frequencies. This fact does not allow one to exclude from consideration the Landau hypothesis on the transition to chaos through high-dimensional tori. Moreover, a number of unsolved problems still remain regarding applications to a two-dimensional torus. We note, for example, the bifurcation of two-dimensional torus doubling that was observed first by the authors of this book and later by many other researchers. Even now, the bifurcational mechanism for doubling of one of the periods of a two-dimensional torus remains unclear. Furthermore, a number of problems remain open regarding the effect of synchronization of two-frequency oscillations.

The problems indicated above and other potential issues can be solved by developing a simple basic model of an oscillator with autonomous two-frequency oscillations. According to the theory of oscillations, one must consider a dynamical system which has a solution in the form of stable two-frequency oscillations. This model is also necessary as, for example, the equations of the van der Pol oscillator to study limit cycles. In this chapter we give a physical justification and introduce a basic model for a quasiperiodic oscillator with two independent frequencies.

12.2 Methods for Realizing Two-Frequency Oscillations and Their Properties

The regime of quasiperiodic oscillations can be obtained by several different methods. The simplest consists in modulating oscillations of an autonomous oscillator by an external periodic signal. Let an autonomous oscillator be described in the form

$$\ddot{x} + F(x, \dot{x}, \alpha)\dot{x} + \omega_0^2 x = 0 , \qquad (12.2)$$

where $F(x, \dot{x}, \alpha)$ is a function providing self-sustained oscillations. The system (12.2) has a solution in the form of a stable limit cycle with period $T = 2\pi/\omega_0$, $\omega_0 = 2\pi f_0$. We now apply an external harmonic (or in the general case any periodic) signal with a relatively small amplitude A_0 to the oscillator (12.2):

$$\ddot{x} + F(x, \dot{x}, \alpha)\dot{x} + \omega_0^2 x = A_0 \sin(\omega_1 t + \phi) . \qquad (12.3)$$

This implies that the autonomous oscillations with frequency f_0 will be modulated by the external signal with frequency f_1. We thus obtain stable two-frequency oscillations. This method is widely used in many different tasks of nonlinear dynamics. However, it has one major disadvantage, which is as follows. Although the system (12.3) demonstrates two-frequency oscillations, one of the frequencies, namely, the external signal frequency, is not independent and retains its value when

12.2 Methods for Realizing Two-Frequency Oscillations and Their Properties

we vary the control parameter α of the autonomous system (12.2). This poses difficulties when solving a certain class of problems where frequencies must be independent and the process must represent self-sustained oscillations.

This inconvenience can be avoided by producing quasiperiodic oscillations in a different way. This method uses a system of two coupled oscillators, written in the form

$$\ddot{x}_1 + F(x_1, \dot{x}_1, \alpha_1)\dot{x}_1 + \omega_{01}^2 x_1 = \gamma \Phi_1(x_1, x_2),$$
$$\ddot{x}_2 + F(x_2, \dot{x}_2, \alpha_2)\dot{x}_2 + \omega_{02}^2 x_2 = \gamma \Phi_2(x_1, x_2),$$
(12.4)

where $\Phi_{1,2}$ are coupling functions and γ is the coupling strength. If the frequencies of each of the two autonomous oscillators are different, $f_{01} \neq f_{02}$, then autonomous two-frequency self-sustained oscillations can be observed in the system (12.4) in the presence of coupling ($\gamma \neq 0$). In this case, the natural frequencies of quasiperiodic self-sustained oscillations will indeed be independent. Note that using a system of coupled oscillators with two, three, or more partial subsystems enables one to realize quasiperiodic self-sustained oscillations with two, three, or more independent frequencies.

Let us discuss an example in which autonomous two-frequency oscillations are obtained by applying the above method. Consider two coupled oscillators with periodic oscillations. A partial system is given by the model of the van der Pol oscillator in the regime of a limit cycle as the image of stable and nearly harmonic oscillations:

$$\dot{x}_1 = y_1,$$
$$\dot{y}_1 = (\varepsilon - x_1^2)y_1 - \omega_1^2 x_1.$$
(12.5)

Here, ε is the excitation parameter, $\omega_1^2 = (2\pi f_1)^2$, and $f_1 = 1/T_0$, where f_1 is the frequency and T_0 is the oscillation period. As is well known, self-sustained oscillations arise in the system (12.5) from the Andronov–Hopf bifurcation at the point $\varepsilon^* = 0$. Their amplitude increases proportionally to $\sqrt{\varepsilon}$ for $\varepsilon > \varepsilon^*$.

The second partial system is described by the same van der Pol oscillator, but the frequencies are different ($f_2 \neq f_1$). We study a self-sustained oscillatory mode in the case of symmetrical coupling between the oscillators:

$$\dot{x}_1 = y_1,$$
$$\dot{y}_1 = (\varepsilon - x_1^2)y_1 - \omega_1^2 x_1 + \gamma(x_2 - x_1),$$
$$\dot{x}_2 = y_2,$$
$$\dot{y}_2 = (\varepsilon - x_2^2)y_2 - \omega_2^2 x_2 + \gamma(x_1 - x_2).$$
(12.6)

The parameter γ is responsible for the degree of internal coupling between the oscillators, the parameter ε is the same for both oscillators, and frequencies ω_1 and ω_2 are chosen to be different, but close enough to each other.

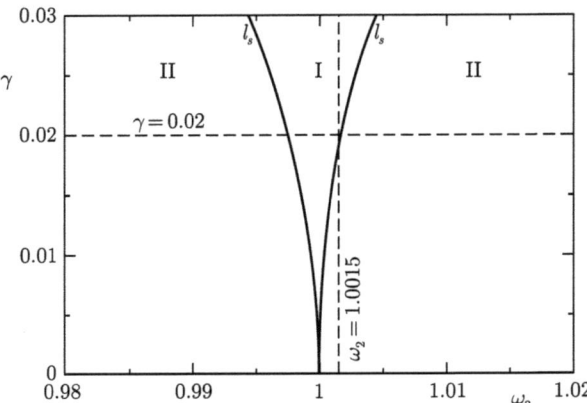

Fig. 12.1 Region of mutual synchronization of oscillators (12.6) for $\varepsilon = 0.1$ and $\omega_1 = 1$. Region I is where the resonance limit cycle exists with winding number $\Theta = 1 : 1$, and region II is the region of quasiperiodic oscillations

Consider regimes of self-sustained oscillations in the system (12.6) for the parameter values $\varepsilon = 0.1$, $\omega_1 = 1$, and $\gamma = 0.02$. The parameter ω_2 is varied in the interval $0.98 < \omega_2 < 1.02$ and one can then explore how partial frequency detuning can influence the system dynamics.

Calculation results for the dynamical regimes are presented in Fig. 12.1 in the parameter plane (γ, ω_2) for fixed values of the other control parameters. The calculations are performed for the case when the frequencies f_1 and f_2 of the partial oscillators are slightly different. Two regions, I and II, are marked in Fig. 12.1. Quasiperiodic oscillations are realized in region II, while region I corresponds to the resonance when $f_1 = f_2$. Let us discuss this fundamental effect.

The so-called *Poincaré winding number* Θ is an important characteristic of two-frequency oscillations. From the experimental point of view, Θ can be characterized as the ratio of the frequencies: $\Theta = \omega_2 : \omega_1 = f_2 : f_1$. When the second frequency in the system (12.6) is varied, the winding number changes and takes both irrational and rational values. If Θ is rational, then $f_2 : f_1 = m : n$, where m and n are integers. In this case we have multiple frequencies $f_2 = f_1 m/n$, i.e., only one frequency is formally independent in the system, viz., f_1, while f_2 is either its harmonic or subharmonic. According to Arnold's theory, we are dealing with the effect of resonance on the torus. When Θ is rational, the resonance regions appear to be some finite regions in the parameter space and characterize the effect of synchronization, which will be discussed in detail in a separate chapter. If Θ is irrational, ergodic quasiperiodic oscillations with two irrationally related frequencies are observed in the system. For rational values of Θ, the effect of synchronization takes place in the form of frequency locking. The motion becomes periodic with one independent frequency.

Figure 12.1 shows the resonance region $\Theta = 1 : 1$, which corresponds to the effect of frequency locking at the basic tone. The first oscillator locks the

12.2 Methods for Realizing Two-Frequency Oscillations and Their Properties

frequency of the second oscillator, and as a result, the frequencies of the interacting oscillators are equal, i.e., $f_1 = f_2$, in the synchronization region (region I in Fig. 12.1). Synchronization region I in the 'coupling (γ)–detuning (ω_2)' parameter plane represents what is known as *Arnold's beak* or *Arnold's tongue* with Poincaré winding number $\Theta = 1 : 1$, which corresponds to synchronization at the basic tone.

Regimes of two-frequency oscillations or beats are observed outside the synchronization region (these are regions II in Fig. 12.1), and the frequencies of the partial oscillators do not coincide ($f_1 \neq f_2$). Two-frequency quasiperiodic oscillations in region II are represented by a non-resonant (in the general case) ergodic two-dimensional torus which is the image of a self-sustained oscillatory mode.

When entering region I from region II (by crossing the bifurcation lines l_s in Fig. 12.1), a structure in the form of stable and saddle limit cycles is born on the two-dimensional torus. Both cycles lie on the torus surface. The stable cycle corresponds to mutual synchronization of the two oscillators and characterizes the stable periodic motion with frequency $f_1 = f_2$ in the frequency-locking regime.

The above is illustrated in Fig. 12.2 which shows projections of phase portraits of the two-dimensional torus T^2 (Fig. 12.2a) and of stable L_0 and saddle L_0^* resonant limit cycles on it (Fig. 12.2b). A highly important fact must be mentioned: the torus T^2 exists both in region II and in region I! Only the stable limit cycle L_0 can be observed in region I in a full-scale experiment. But this cycle lies on the two-dimensional torus surface.

Consider the Poincaré section by the plane $x_1 = 0$ for the regimes shown in Fig. 12.2a, b. Calculated results are presented in Fig. 12.3. The closed invariant curve l which is the image of T^2 in the Poincaré section is pictured in Fig. 12.3a. Figure 12.3b needs to be described in more detail. Inside synchronization region I (Fig. 12.1), two limit cycles, stable and saddle (Fig. 12.2b), exist on the torus. They correspond to the stable fixed point P and the saddle Q in the Poincaré section (Fig. 12.3b). The unstable separatrices of the saddle Q are closed on the stable node P and form a closed invariant curve. This curve is the image of a two-dimensional resonant torus in the synchronization region. If we cross the parameter plane of region I in the direction of the bifurcation lines l_s (Fig. 12.1), the following situation is observed: the saddle Q and the node P approach each other and merge on the bifurcation lines l_s. Upon entering region II, both fixed points disappear through a saddle-node bifurcation. Hence, the synchronization regime is observed in the region where stable and saddle limit cycles exist on the two-dimensional torus, while it is destroyed when these cycles undergo a saddle-node bifurcation.

The results described above regarding analysis of the dynamics of the system (12.6) in the vicinity of the resonance $\Theta = 1 : 1$ will be used to study the effect of synchronization of quasiperiodic oscillations in Chap. 14.

We now describe an autonomous model for the oscillator with quasiperiodic oscillations which will enable us to realize, not only two-frequency oscillations, but also doubling bifurcations of two-dimensional tori. This model illustrates the third method for generating quasiperiodic oscillations, in which the regime of self-modulation is used in the system itself.

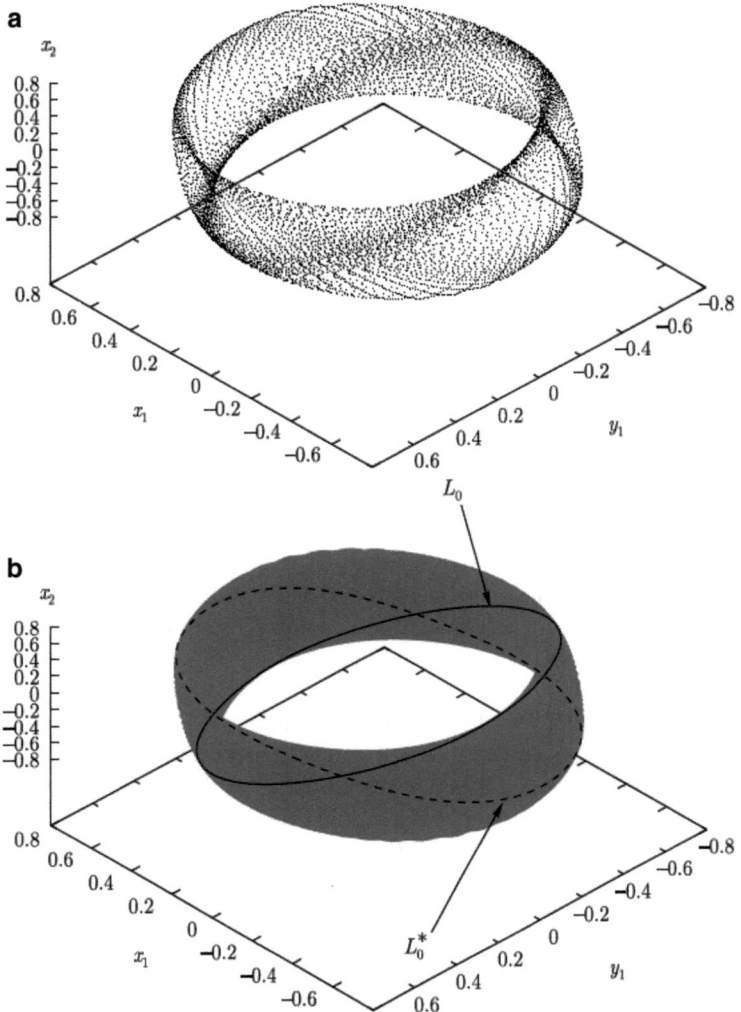

Fig. 12.2 Projections of phase trajectories on the plane of variables (x_1, y_1, x_2) of the system (12.6) for $\varepsilon = 0.1$, $\gamma = 0.02$, and $\omega_1 = 1$: (**a**) for region II ($\omega_2 = 1.003$), and (**b**) for region I ($\omega_2 = 1.0015$). The two-dimensional torus projection outside the resonance region is displayed in *gray*, stable L_0 and saddle L_0^* cycles on the torus in the resonance region are shown by *bold lines*

12.3 Statement of Oscillator Equations

Consider the model for the Anishchenko–Astakhov oscillator whose scheme is presented in Fig. 11.2. The oscillator includes the classical van der Pol oscillator with an additional inertial feedback. The oscillator equations form a three-dimensional system with three parameters, which can be written in the form

12.3 Statement of Oscillator Equations

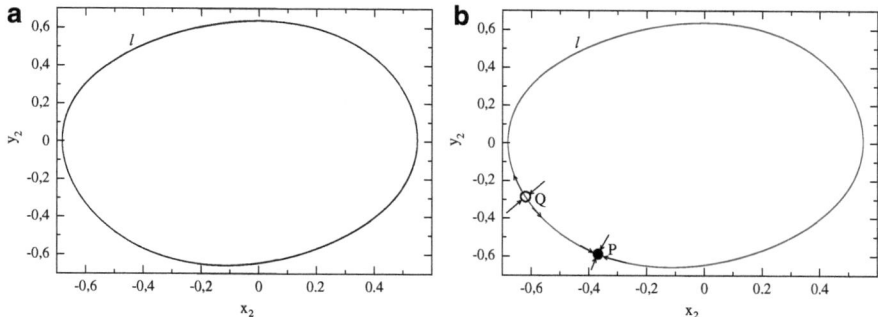

Fig. 12.3 Poincaré sections of the ergodic torus (**a**) and of resonant cycles on it (**b**) for the cases (**a**) and (**b**) of Fig. 12.2, respectively

$$\dot{x} = mx + y - xz - dx^3 , \qquad \dot{y} = -x , \qquad \dot{z} = -gz + g\Phi(x) . \qquad (12.7)$$

The first two equations of the system (12.7) describe the van der Pol oscillator. This can be easily checked by putting $\dot{z} = 0$ and using $\Phi(x) = x^2$. As has been shown, according to Shilnikov's theorem, the transition to chaos can be realized in the system (12.7) provided that the nonlinear function $\Phi(x)$ is asymmetric with respect to the variable x and has the form

$$\Phi(x) = I(x)x^2 , \qquad I(x) = \begin{cases} 1, & x > 0 , \\ 0, & x \leq 0 , \end{cases} \qquad (12.8)$$

or $\Phi(x) = \exp(x) - 1$.

The asymmetry conditions (12.8) ensure the existence of a singular solution in the form of a saddle-focus separatrix loop in the system (12.7), and as a consequence, the realization of a spiral chaos mode. From a physical point of view, this is achieved by the self-consistent influence of feedback on the main amplifier. The feedback signal is given by the third equation in (12.7). This influence is insignificant for small amplitudes of the signal $x(t)$, and (12.7) generates a limit cycle. As the excitation parameter m grows, the intensity of $x(t)$ oscillations increases, the amplitude of the feedback signal $z(t)$ also grows, and its influence on the amplification factor of the main amplifier becomes more significant. The system exhibits a sequence of period-doubling bifurcations of cycles and transition to chaos.

Undamped two-frequency oscillations can be obtained if the system (12.7) involves an element with a natural frequency that differs from the resonance frequency of the oscillator circuit. One of the possible ways is to introduce an oscillatory circuit into the additional feedback. One must make sure that the feedback signal $z(t)$ represents oscillations with an independent frequency which modulate the amplification factor and provide quasiperiodic oscillations.

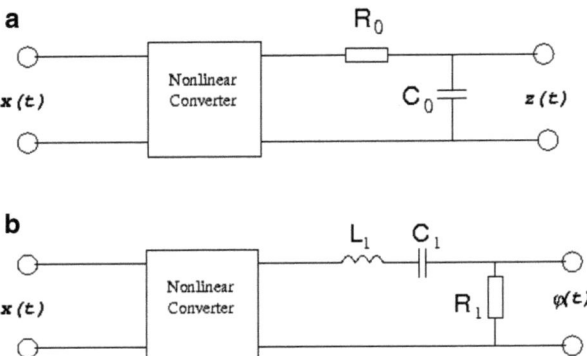

Fig. 12.4 Schematic diagram of the inertial stage of additional feedback (**a**) in the Anishchenko–Astakhov oscillator and (**b**) in the oscillator with quasiperiodic oscillations

Consider the circuit diagrams shown in Fig. 12.4. A schematic diagram is drawn in Fig. 12.4a for an inertial stage of additional feedback in the Anishchenko–Astakhov oscillator. This stage is represented by an *RC*-chain described by a one-dimensional differential equation, viz., the third equation in (12.7). Figure 12.4b shows the diagram for the new inertial stage, which includes an oscillatory circuit with some resonance frequency. The circuit in Fig. 12.4b is described by the equations

$$\dot{z} = \varphi, \qquad \dot{\varphi} = -\gamma\varphi + \gamma\Phi(x) - gz, \tag{12.9}$$

where γ is the attenuation parameter and g is the parameter defining the normalized resonance frequency of the new filter. It is easy to see that Eqs. (12.9) describe a dissipative oscillatory circuit in the regime of forced oscillations:

$$\ddot{z} + \gamma\dot{z} + gz = \gamma\Phi(x). \tag{12.10}$$

We now introduce a new feedback in the oscillator (Fig. 11.2) so that the control signal of the feedback is $\dot{z}(t) = \varphi(t)$. The new oscillator is described by the equations

$$\begin{aligned} \dot{x} &= mx + y - xz - dx^3, \\ \dot{y} &= -x, \\ \dot{z} &= \varphi, \\ \dot{\varphi} &= -\gamma\varphi + \gamma\Phi(x) - gz. \end{aligned} \tag{12.11}$$

The system (12.11) is a nonlinear dissipative dynamical system of dimension $N = 4$, governed by four control parameters: m is the excitation parameter, d is the nonlinear dissipation parameter, γ is the attenuation parameter, and g is the filter

12.4 Bifurcation Diagram of the Quasiperiodic Oscillator

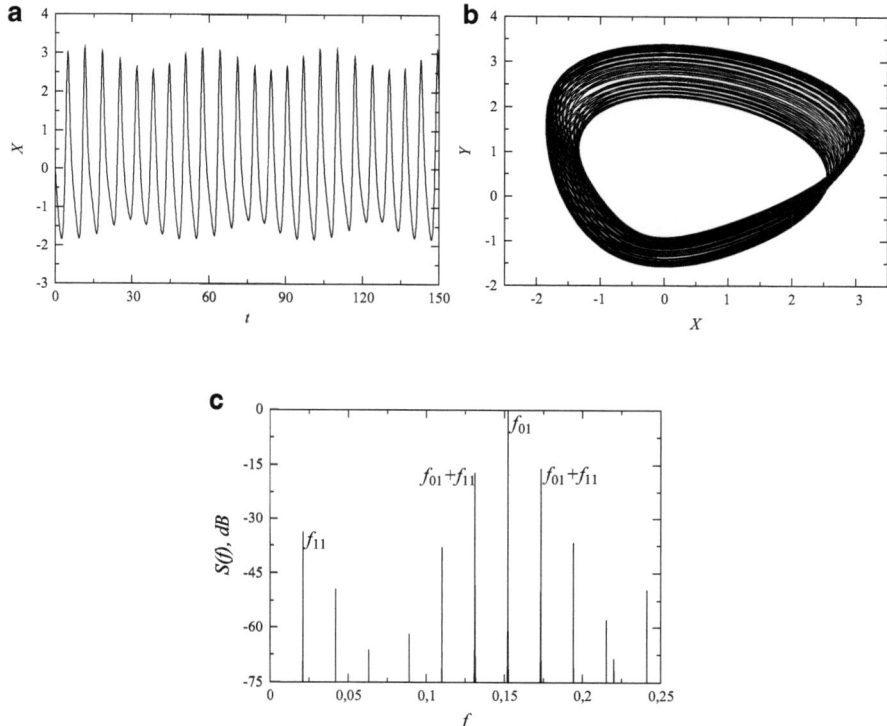

Fig. 12.5 Regime of quasiperiodic two-frequency oscillations: (**a**) time series, (**b**) projection of the phase portrait, and (**c**) power spectrum. The parameter values are $m = 0.06$, $g = 0.5$, $\gamma = 0.2$, and $d = 0.001$

inertia parameter. Two of these parameters are significant in the system (12.11), namely, the oscillator excitation parameter m and the inertia parameter g characterizing the resonance frequency of the filter.

When $\Phi(x)$ is specified by (12.8), the system (12.11) has a solution in the form of stable two-frequency oscillations. The indicated regime is exemplified in Fig. 12.5.

12.4 Bifurcation Diagram of the Quasiperiodic Oscillator

The bifurcation diagram of the system (12.11) is presented in Fig. 12.6 on the plane of two control parameters m and g for fixed $\gamma = 0.2$ and $d = 0.001$. The function $\Phi(x)$ is given in (12.11) by $I(x)x^2$ in (12.8).

A stable limit cycle T_0 is born on the line $m = 0$ through a soft Andronov–Hopf bifurcation. When it crosses the bifurcation line l_1, the cycle undergoes the period-doubling bifurcation. The line l_2 corresponds to the period-doubling bifurcation of

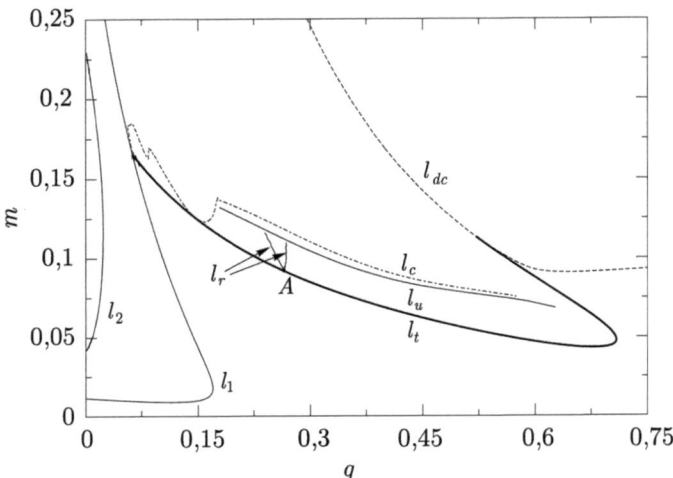

Fig. 12.6 Bifurcation diagram of the oscillator (12.11) for $\gamma = 0.2$ and $d = 0.001$. $l_{1,2}$ are period-doubling bifurcation lines, l_t is the torus birth line, l_u is the torus destruction line, l_c is the chaotic attractor destruction line, l_r are lines bounding the resonance region 1 : 4 on the torus, l_{dc} are lines of multiple cycles, and A is the codimension 2 point corresponding to the condition $\psi = 1 : 4$

the cycle born on the line l_1 (Fig. 12.6). A pair of complex-conjugate multipliers $\mu_{1,2} = \exp(\pm i\theta)$ of cycle T_0 emerges on the unit circle on the bifurcation line l_t, and a two-dimensional torus is softly born in the system (Neimark's bifurcation). Obviously, as one moves along the line l_t, the angle θ takes a set of rational values corresponding to the resonances on the torus. Figure 12.6 exemplifies the resonance region $\theta = 1 : 4$, which is bounded by lines l_r of saddle-node bifurcations of the resonance cycle on the torus and rests on the codimension-two point A. The line l_u is drawn beyond the torus birth line l_t. As one crosses it from the bottom up, the transition to chaos is observed through the destruction of quasiperiodic oscillations. The line l_c corresponds to a crisis (destruction) of the chaotic attractor that emerges on the line l_u. A pair of saddle cycles merges and then disappears on the line l_{dc}.

12.5 Two-Dimensional Torus-Doubling Bifurcation

We fix the parameters $g = 0.5$, $d = 0.001$, and $\gamma = 0.2$ and consider how the torus evolves in the range of the parameter m between lines l_t and l_u. Figure 12.7a–d shows projections of attractors on the plane when passing the bifurcation points of two-dimensional torus doubling. The torus doubling is clearly indicated by the structure of the Poincaré section, as well as by the form of the time series and their power spectra. Here, the torus period-doubling bifurcation corresponds to the

12.5 Two-Dimensional Torus-Doubling Bifurcation

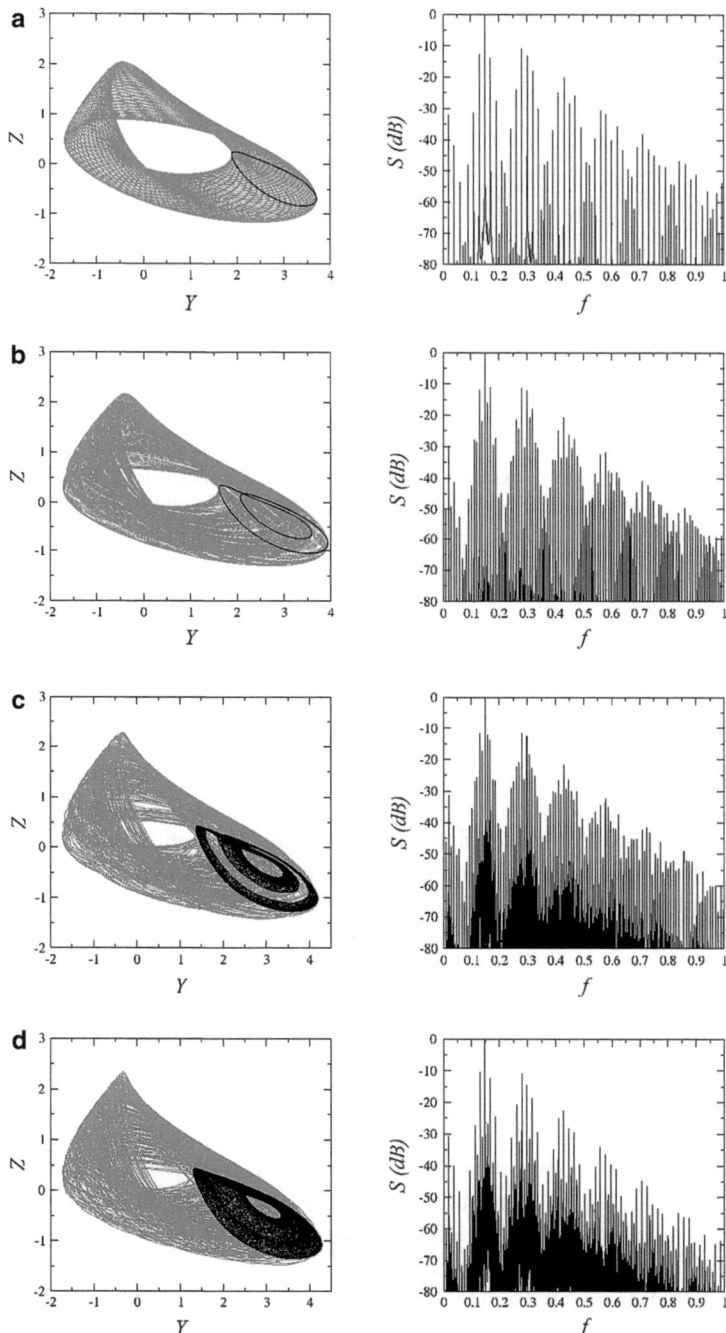

Fig. 12.7 Projections of attractors of the system (12.11) and of their Poincaré sections on the plane, and the corresponding power spectra for different values of the parameter m and for fixed $d = 0.001$, $\gamma = 0.2$, and $g = 0.5$

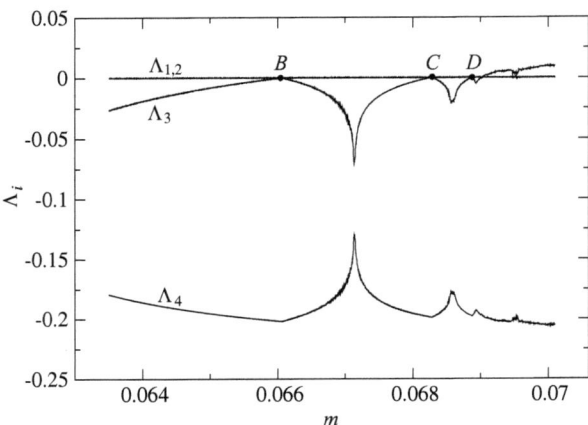

Fig. 12.8 Lyapunov exponent spectrum as a function of the parameter m ($d = 0.001$, $\gamma = 0.2$, $g = 0.5$) in the parameter range between lines l_t and l_u. Points B, C, and D are the torus-doubling bifurcation points

modulation period-doubling bifurcation (or to the period-doubling bifurcation of a cycle in the Poincaré map).

From the standpoint of bifurcation theory, it is important to decide whether the ergodic torus is doubled or whether the resonance on the torus takes place near the bifurcation point, then the resonance cycle is doubled, and finally a doubled torus is formed from this cycle. To answer to this question, the full spectrum of Lyapunov exponents is calculated when passing the torus-doubling bifurcation points.

As can be seen from Fig. 12.8, the three maximal Lyapunov exponents immediately become zero ($\Lambda_1 = \Lambda_2 = \Lambda_3 = 0$) at the bifurcation points B, C, and D. The bifurcational transition is characterized by the following changes in the LCE spectrum signature:

$$0, 0, -, - \quad \Longrightarrow \quad 0, 0, 0, - \quad \Longrightarrow \quad 0, 0, -, -$$
$$\text{torus}_2 \qquad\qquad \text{bifurcation} \qquad\qquad \text{torus}_2$$
$$\text{point}$$

In numerical experiments, the parameter m is varied with the very small step $\Delta m = 3 \times 10^{-6}$. Our calculations indicate that the limit cycle birth (the Lyapunov exponent spectrum is $0, -, -, -$) is not observed when passing the bifurcation point. The ergodic torus undergoes the period-doubling bifurcation and no resonance cycles are detected in the numerical experiment.

The calculation data presented above convincingly prove that the system (12.11) in the autonomous regime can generate stable two-frequency oscillations depending on values of the control parameters. These oscillations may have different values of the winding number and they can exhibit a cascade of two-dimensional torus-doubling bifurcations and the transition to chaos through the torus destruction.

12.6 Summary

In this chapter we have introduced a model of an oscillator with two-frequency quasiperiodic self-sustained oscillations which can be regarded as one of the fundamental models of the theory of nonlinear oscillations. The model is an autonomous system of four ordinary differential equations with four parameters. The system can realize stable quasiperiodic oscillations which show torus-doubling bifurcations and transitions to chaos through the two-dimensional torus destruction when the system parameter is varied. The proposed model can be used to study in detail nonlinear properties of two-frequency self-sustained oscillations, such as synchronization.

Chapter 13
Synchronization of Periodic Self-Sustained Oscillations

13.1 Introduction

Synchronization is one of the fundamental properties of nonlinear systems. It is understood as an adjustment of certain relations between characteristic times, frequencies, or phases of oscillations of interacting systems. The effect of synchronization was discovered by Huygens in the seventeenth century and plays a huge role in nature and technology. The development of electronic communication devices in the first half of the twentieth century did much to spur the development of the theory of synchronization. Later the application to periodic self-sustained oscillations, which has become the classic application of this theory, was developed in detail, dealing also with the presence of noise.

The classical theory of synchronization distinguishes between *forced synchronization* by an external signal and *mutual synchronization* between two interacting self-sustained oscillatory systems. The same effects manifest themselves in both cases. They are related to two classic synchronization mechanisms: *natural (basic) frequency locking* (and hence, phase locking), or *frequency entrainment*, and *suppression of one of two independent frequencies*.

Let $\Phi_1(t)$ and ω_1 be the phase and frequency of a quasiharmonic oscillator, and $\Phi_2(t)$ and ω_2 the phase and frequency of another self-sustained oscillator coupled to the first. Conditions of synchronization can be formulated as follows:

$$m\Phi_1(t) - n\Phi_2(t) = \text{const.} \tag{13.1}$$

or

$$m\omega_1 = n\omega_2, \tag{13.2}$$

where m and n are positive integers. These conditions define the effect of phase and frequency locking and must hold in a certain (finite) region of the control parameter values. This region is called the *synchronization region*. The simplest case, namely,

1 : 1 ($m = n = 1$), corresponds to the *basic region of synchronization* or the *synchronization region at the basic tone*.

The phenomenon of synchronization of self-sustained oscillations has attracted much attention from the research community for a number of years now. This is partly due to the importance of this effect from the applied point of view. This is well exemplified by the synchronization of an electronic clock by an external signal from a highly stable oscillator, which has contributed to the high level of timing accuracy in transport systems. Synchronization of powerful periodic oscillators by a weak signal from an external high-quality reference generator can significantly improve their characteristics, including frequency stability, amplitude and phase fluctuations, and others.

Recently, the interest in synchronization has increased in biology, chemistry, and even in the social and economic sciences. It has been shown that synchronous behavior is found in interacting cells of living tissue, ensembles of neurons, biological populations, etc. However, it is very important to note that the oscillatory processes being analyzed in this case are not always strictly periodic. Naturally, there are many questions concerning the applicability of the classical theory of synchronization to such oscillatory processes. The answers to some of them will be given in the following chapters.

In this chapter we describe in detail the classical theory of synchronization of periodic self-sustained oscillations due to an external harmonic force in a nonautonomous van der Pol oscillator. This effect manifests itself through the fact that the oscillator frequency is adjusted to the external force frequency. The external amplitude may in this case be significantly less than the amplitude of oscillations of the autonomous oscillator. In the simplest case the oscillator oscillates at the external force frequency when it varies within a finite range of the natural frequency of the autonomous oscillator, within the so-called basic region of synchronization.

We restrict ourselves to describing the basic region of synchronization of the nonautonomous van der Pol oscillator using the phase equation, truncated equations for the amplitude and phase, and the full nonautonomous van der Pol equation.

13.2 Forced Synchronization of the van der Pol Oscillator: Truncated Equations for the Amplitude and Phase

Consider the effect of an external harmonic force on the van der Pol oscillator. The forced system is described by the equation

$$\ddot{x} - \left(\varepsilon - x^2\right)\dot{x} + x = b\sin\omega t, \qquad (13.3)$$

where ε is the control parameter of the autonomous oscillator, and b and ω are the amplitude and frequency of the additive external force.

13.2 Forced Synchronization of the van der Pol Oscillator: Truncated...

We rewrite (13.3) as follows:

$$\ddot{x} + x = \left(\varepsilon - x^2\right)\dot{x} + b\sin\omega t\;,$$
$$\ddot{x} + \omega^2 x = \left(\omega^2 - 1\right)x + \left(\varepsilon - x^2\right)\dot{x} + b\sin\omega t\;. \tag{13.4}$$

If the frequency of the external force is close to the natural (basic) frequency of the autonomous oscillator ($\omega \sim 1$), the excitation parameter ε is small and positive, and the external force is weak (its amplitude b is small), then the right-hand side of (13.4) represents a weak perturbation of the harmonic oscillator with frequency ω, and the nonautonomous oscillator is a quasilinear or weakly linear system. In this case a solution of (13.4) can be sought in the form[1]:

$$x(t) = \mathrm{Re}\left[a(t)\exp(i\omega t)\right] = \frac{1}{2}\left[a\exp(i\omega t) + a^*\exp(-i\omega t)\right]\;, \tag{13.5}$$

with the additional condition

$$\dot{a}\exp(i\omega t) + \dot{a}^*\exp(-i\omega t) = 0\;. \tag{13.6}$$

The function $a(t)$ is assumed to be slowly varying compared with the period $T = 2\pi/\omega$. Taking into account the additional condition, we find the first and the second derivatives:

$$\dot{x} = \frac{1}{2}\left[\dot{a}\exp(i\omega t) + \dot{a}^*\exp(-i\omega t) + i\omega a\exp(i\omega t) - i\omega a^*\exp(-i\omega t)\right]$$
$$= \frac{1}{2}\left[i\omega a\exp(i\omega t) - i\omega a^*\exp(-i\omega t)\right]\;,$$

$$\ddot{x} = \frac{1}{2}\left[i\omega\dot{a}\exp(i\omega t) - i\omega\dot{a}^*\exp(-i\omega t) - \omega^2 a\exp(i\omega t) - \omega^2 a^*\exp(-i\omega t)\right]$$
$$= i\omega\dot{a}\exp(i\omega t) - \frac{\omega^2}{2}\left[a\exp(i\omega t) + a^*\exp(-i\omega t)\right]\;.$$

Substituting $\ddot{x}$, $\dot{x}$, and x into (13.4) and expressing $\sin(\omega t)$ in terms of exponential functions, we obtain

---

[1]This representation is similar to the change of variables

$$x = \rho(t)\cos\left[\omega t + \varphi(t)\right]\;, \qquad \dot{x} = -\omega\rho(t)\sin\left[\omega t + \varphi(t)\right]\;,$$

where $\rho(t) = |a(t)|$, $\varphi(t) = \arg a(t)$, which is also widely used in the van der Pol averaging method.

$$i\omega \dot{a} \exp(i\omega t) = \frac{\omega^2 - 1}{2}\Big[a \exp(i\omega t) + a^* \exp(-i\omega t)\Big] + \frac{b}{2i}\Big[\exp(i\omega t) - \exp(-i\omega t)\Big]$$

$$+ \frac{i\omega}{2}\bigg\{\varepsilon - \frac{1}{4}\Big[a^2 \exp(2i\omega t) + 2|a|^2 + (a^*)^2 \exp(-2i\omega t)\Big]\bigg\}$$

$$\times \Big[a \exp(i\omega t) - a^* \exp(-i\omega t)\Big],$$

$$\dot{a} = \frac{\omega^2 - 1}{2i\omega}\Big[a + a^* \exp(-2i\omega t)\Big] - \frac{b}{2\omega}\Big[1 - \exp(-2i\omega t)\Big]$$

$$+ \bigg\{\frac{\varepsilon}{2} - \frac{1}{8}\Big[a^2 \exp(2i\omega t) + 2|a|^2 + (a^*)^2 \exp(-2i\omega t)\Big]\bigg\}\Big[a - a^* \exp(-2i\omega t)\Big].$$

Removing the parentheses and averaging over the period $T = 2\pi/\omega$ on the right- and left-hand sides of the equation, then taking into account the fact that $a(t)$ is a slowly varying function, we obtain the truncated equation for the complex amplitude in the form

$$\dot{a} = -i\frac{\omega^2 - 1}{2\omega}a + \frac{\varepsilon}{2}a - \frac{1}{8}|a|^2 a - \frac{b}{2\omega}. \tag{13.7}$$

Representing the complex quantity $a(t)$ in polar coordinates, viz.,

$$a(t) = \rho(t) \exp[i\varphi(t)], \tag{13.8}$$

we derive the truncated equations

$$\dot{\rho} = \frac{\varepsilon}{2}\rho - \frac{1}{8}\rho^3 - \beta \cos \varphi, \tag{13.9}$$

$$\dot{\varphi} = -\Delta + \frac{\beta}{\rho} \sin \varphi, \tag{13.10}$$

where $\Delta = (\omega^2 - 1)/2\omega$ is the detuning between the frequency of the external force and the natural frequency of the oscillator, and $\beta = b/2\omega$ is the intensity (or amplitude) of the forcing. The quantity φ, which will hereafter be referred to as the *phase*, is the difference between the phases of the oscillator and the force.

Synchronization, e.g., an adjustment of the oscillator frequency to the frequency of the external force, can be observed for a small periodic force. In this case the force significantly influences the phase, but has only a small effect on the amplitude. Thus, the synchronization process can be described in the phase approximation.

13.2.1 Analysis of Synchronization in the Phase Approximation

When the external forcing is weak, one can assume that the amplitude $\rho(t)$ corresponds to the radius of the limit cycle of the autonomous oscillator $\rho(t) = 2\sqrt{\varepsilon}$. Then the effect of synchronization can be analyzed in the phase approximation:

$$\frac{d\varphi}{dt} = -\Delta + \frac{\beta}{2\sqrt{\varepsilon}} \sin\varphi . \qquad (13.11)$$

Equation (13.11) describes one dynamical variable φ, so the phase space dimension is 1. The system dynamics can be considered either on the real axis from $-\infty$ to $+\infty$, or, taking into account the periodicity of the function $\sin\varphi$, on the circle with radius $\rho = 2\sqrt{\varepsilon}$. The system behavior depends on parameters Δ, β, and ε, where Δ is the frequency detuning, or mismatch, i.e., the difference between the natural frequency and the forcing frequency, β is the amplitude of the external force, and ε is the excitation parameter of the autonomous oscillator.

Let us consider the phase dynamics in relation to the detuning Δ and the external amplitude β for a fixed value of the parameter ε corresponding to quasiharmonic oscillations of the autonomous oscillator. The phase dynamics φ is pictured in Fig. 13.1 for $\varepsilon = 0.1$, $\beta = 0.01$ and for different values of the frequency detuning. For a small detuning, the phase is constant in time (lines 5 and 6 for $\Delta = \mp 0.03$). As the frequency detuning grows and when $|\Delta|$ exceeds a certain critical value $|\Delta_c|$, the phase behavior $\varphi(t)$ changes qualitatively. In fact, its value starts varying in time.

According to the sign of the parameter Δ (if the frequency of the forcing is greater or less than the natural frequency of the oscillator), the phase φ either decreases or increases in time. For a small subcriticality $|\Delta - \Delta_c|$, the time series $\varphi(t)$ shows long intervals during which the phase is nearly constant. These long epochs intermingle with relatively short time intervals where the phase changes by 2π. As the supercriticality grows, the intervals of constant phase decrease, and the mean rate of phase change increases.

Thus, there is an interval of values of the detuning $|\Delta| < \Delta_c$ where the phase is constant, $\varphi(t) = \mathrm{const.}$, and its derivative (the rate of phase change) is zero. This means that the oscillator oscillates periodically at the frequency of the external force, and synchronization is therefore observed. Outside the synchronization region the phase varies in time, and the oscillations become quasiperiodic. The mean rate of phase change $\langle\dot\varphi(t)\rangle$ defines the second independent frequency, the so-called *beat frequency*. As seen from Fig. 13.1, when the supercriticality is small, the phase varies very slowly, which corresponds to a very low beat frequency. As the supercriticality grows, the beat frequency increases.

In the phase approximation, synchronous motions correspond to equilibrium states or fixed points of the dynamical system (13.11). Regimes of synchronization must be related to stable equilibrium states. We shall now find the equilibrium states

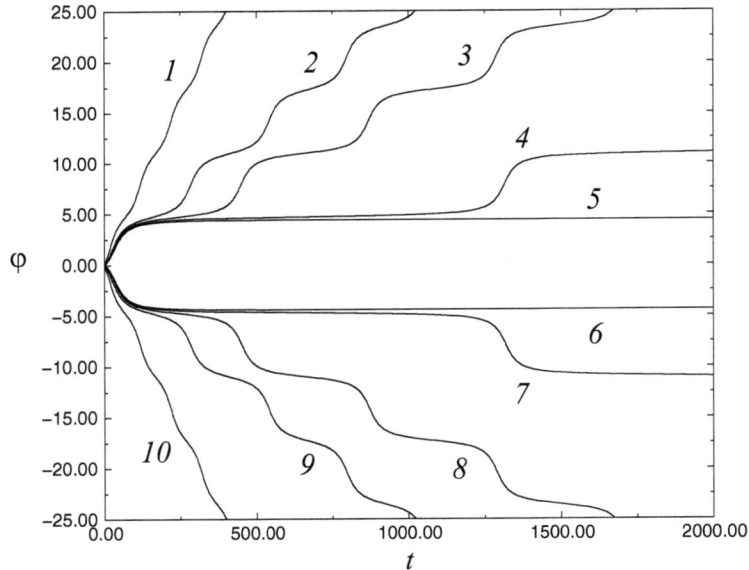

Fig. 13.1 Temporal phase dynamics $\varphi(t)$ for $\varepsilon = 0.1$, $\beta = 0.01$ and for different values of the frequency detuning: $\Delta = \mp 0.07$ for lines 1 and 10, $\Delta = \mp 0.04$ for lines 2 and 9, $\Delta = \mp 0.035$ for lines 3 and 8, $\Delta = \mp 0.032$ for lines 4 and 7, and $\Delta = \mp 0.03$ for lines 5 and 6

of the system (13.11), define their region of existence on the parameter space, and explore their stability and bifurcations when the system parameters are varied.

Fixed points of the system (13.11) will be considered in a one-dimensional phase space on the circle with radius $\rho = 2\sqrt{\varepsilon}$, which is constructed on the plane with coordinates $\rho \sin \varphi$, $\rho \cos \varphi$ (see Fig. 13.2). Equilibrium states can be found by setting the right-hand side of (13.11) equal to zero:

$$-\Delta + \frac{\beta}{2\sqrt{\lambda}} \sin \varphi = 0 . \tag{13.12}$$

From (13.12), we find two fixed points with coordinates

$$\varphi_1 = \arcsin \frac{2\Delta\sqrt{\varepsilon}}{\beta} , \tag{13.13}$$

$$\varphi_2 = \pi - \arcsin \frac{2\Delta\sqrt{\varepsilon}}{\beta} . \tag{13.14}$$

They exist if $|2\Delta\sqrt{\varepsilon}/\beta| \leq 1$ or $|\Delta| \leq \beta/2\sqrt{\varepsilon}$. Consequently, for a fixed ε, the region of existence of the fixed points in the parameter plane (β, Δ) is bounded by the line given by the equation

13.2 Forced Synchronization of the van der Pol Oscillator: Truncated...

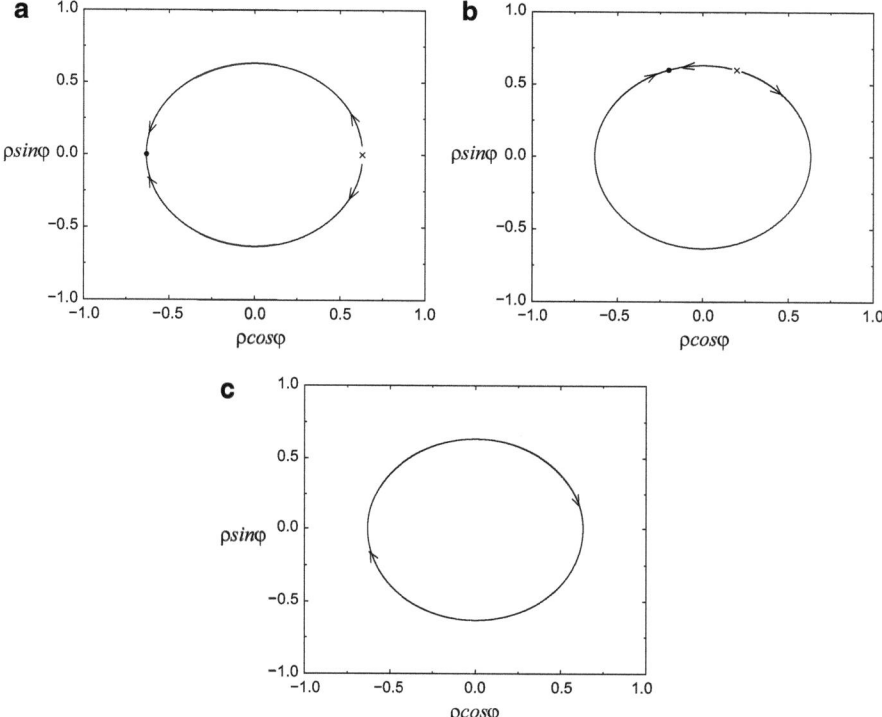

Fig. 13.2 Phase portraits of the system (13.11) for $\varepsilon = 0.1$, $\beta = 0.01$ and for different values of the detuning: (**a**) $\Delta = 0$, (**b**) $\Delta = 0.03$, (**c**) $\Delta = 0.032$. Stable and unstable fixed points are marked by symbols • and ×, respectively

$$\beta = 2\sqrt{\varepsilon}|\Delta| \ . \tag{13.15}$$

The synchronization region, whose borders are determined by (13.15), is constructed in Fig. 13.3.

We now analyze the stability of fixed points. The behavior of the system (13.11) is considered in the neighborhood of fixed points φ_i ($i = 1, 2$) in a linear approximation. We represent the dynamical variable $\varphi(t)$ in the form $\varphi(t) = \varphi_i + \tilde{\varphi}(t)$, where $\tilde{\varphi}(t)$ is a small deviation from a fixed point φ_i. We rewrite (13.11) as follows:

$$\frac{d}{dt}(\varphi_i + \tilde{\varphi}) = -\Delta + \frac{\beta}{2\sqrt{\varepsilon}}\sin(\varphi_i + \tilde{\varphi}) \ ,$$

$$\frac{d\tilde{\varphi}}{dt} = -\Delta + \frac{\beta}{2\sqrt{\varepsilon}}(\sin\varphi_i \cos\tilde{\varphi} + \sin\tilde{\varphi}\cos\varphi_i) \ .$$

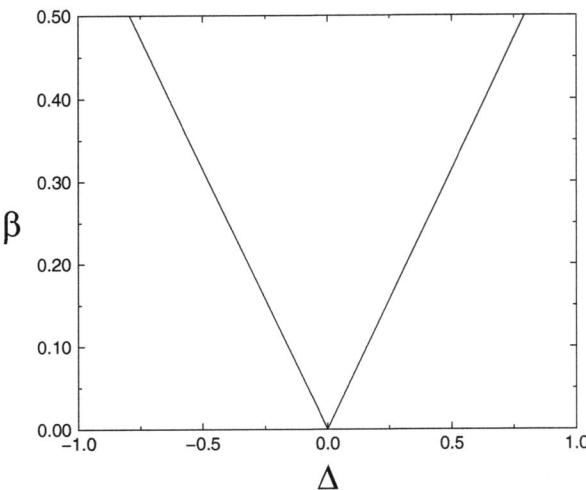

Fig. 13.3 Synchronization region in the plane of the control parameters Δ, β for $\varepsilon = 0.1$

Expanding $\cos \tilde{\varphi}$ and $\sin \tilde{\varphi}$ in a Taylor series and taking the first-order terms in $\tilde{\varphi}$, we obtain the linearized equation

$$\frac{d\tilde{\varphi}}{dt} = -\Delta + \frac{\beta}{2\sqrt{\varepsilon}} (\sin \varphi_i + \tilde{\varphi} \cos \varphi_i) \ .$$

From the definition of the fixed point, it follows that

$$-\Delta + \frac{\beta}{2\sqrt{\varepsilon}} \sin \varphi_i = 0 \ .$$

As a result, we get the equation

$$\frac{d\tilde{\varphi}}{dt} = \left(\frac{\beta}{2\sqrt{\varepsilon}} \cos \varphi_i \right) \tilde{\varphi} \ , \quad (13.16)$$

which has solution

$$\tilde{\varphi}(t) \sim \exp \left[\left(\frac{\beta}{2\sqrt{\varepsilon}} \cos \varphi_i \right) t \right] \ . \quad (13.17)$$

Thus, the stability of equilibrium states depends on the sign of $\cos \varphi_i$. If $\cos \varphi_i > 0$, then the small deviation $\tilde{\varphi}(t)$ grows in time and the equilibrium φ_i is unstable. If $\cos \varphi_i < 0$, the small deviation decays in time and the equilibrium is stable. From the expressions (13.13) and (13.14), it follows that $\cos \varphi_1 > 0$ and the fixed point φ_1 is unstable. The point φ_2 is stable as $\cos \varphi_2 < 0$.

Stable and unstable fixed points are shown in Fig. 13.2a, b in the system phase space for different values of the parameter Δ. As can be seen from (13.13) and

13.2 Forced Synchronization of the van der Pol Oscillator: Truncated...

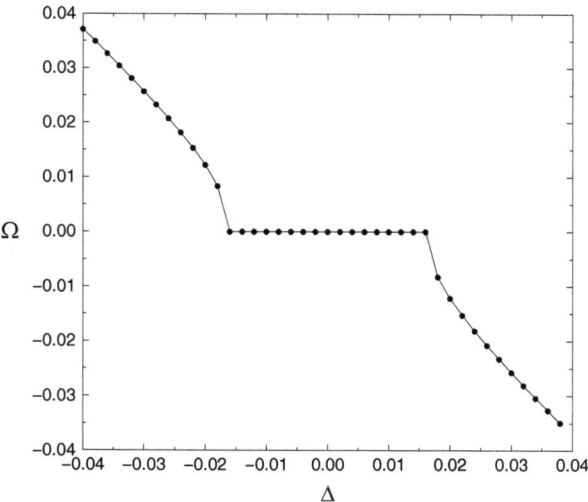

Fig. 13.4 Beat frequency $\Omega = \langle \dot{\varphi}(t) \rangle$ as a function of the frequency detuning Δ of the system (13.11) for $\beta = 0.01$ and $\varepsilon = 0.1$

(13.14), for $\Delta = 0$, the coordinate of the unstable point is $\varphi_1 = 0$, while that of the stable point is $\varphi_2 = \pi$ (Fig. 13.2a). As the detuning increases, when $2\Delta\sqrt{\varepsilon}/\beta$ tends to unity, the unstable and stable fixed points move towards each other along the circle: the unstable point rotates counter-clockwise and the stable one clockwise. For $\varphi = \pi/2$, they collide and disappear when $2\Delta\sqrt{\varepsilon}/\beta$ exceeds unity (Fig. 13.2c). The fixed points disappear when we exit the synchronization region. The representative point rotates along the circle with average velocity $\langle \dot{\varphi}(t) \rangle$, which defines the beat frequency.

The dependence of $\Omega = \langle \dot{\varphi}(t) \rangle$ on the frequency detuning Δ is plotted in Fig. 13.4 for the system (13.11) when $\beta = 0.01$ and $\varepsilon = 0.1$. The frequency $\Omega(t)$ is the difference between the mean frequency of self-sustained oscillations and the frequency of the external force. It is also called the *beat frequency*. The interval of Δ values where $\Omega = 0$ corresponds to the synchronization region. Outside this region, there are beats with frequency Ω.

13.2.2 Bifurcational Analysis of the System of Truncated Equations

The synchronization can only be described in the phase approximation using (13.11) for a very weak external force, when its influence on the amplitude of oscillations of the autonomous oscillator can be neglected. The appearance or disappearance of synchronization is related to a bifurcation giving rise to the birth or vanishing of stable and unstable fixed points in a one-dimensional phase space. In the plane of the control parameters (β, Δ), there is a synchronization tongue bounded by the

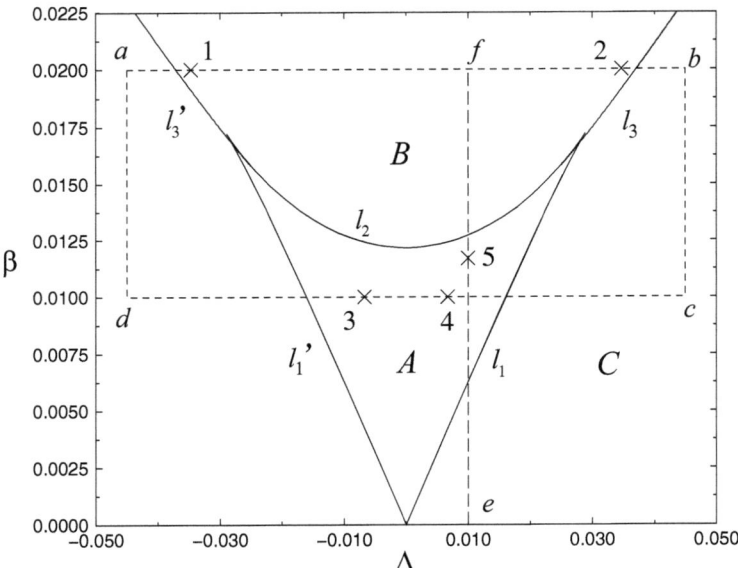

Fig. 13.5 Bifurcation lines for stable and unstable fixed points of the system of truncated equations (13.9)–(13.10) in the plane of the control parameters β, Δ when $\varepsilon = 0.1$

straight lines $\beta = 2\sqrt{\varepsilon}|\Delta|$, on which the indicated bifurcation takes place for two fixed points.

If the amplitude equation is taken into consideration, i.e., if the system of truncated equations for the amplitude and phase (13.9)–(13.10) is considered, then much more complex bifurcational phenomena can be observed on the plane of the control parameters β, Δ.

For the system of truncated equations (13.9)–(13.10), the synchronization regimes are also associated with the equilibrium states $\varphi(t) = $ const. and $\rho(t) = $ const. in the phase plane. Synchronization regions and bifurcation lines for synchronous motions are shown in Fig. 13.5 on the plane of the control parameters β, Δ for fixed $\varepsilon = 0.1$.

Synchronization is observed for parameter values from regions A and B which are bounded by the region C of quasiperiodic motions. In region A there are three fixed points: a stable node P_N (which corresponds to the synchronization regime), a saddle point P_S, and an unstable point P_R, which, depending on the values of β and Δ, may be either an unstable node or an unstable focus. Figure 13.6a is a typical phase portrait for the synchronization region A. The regime of synchronization in region B also corresponds to the stable fixed point P_N, but the structure of the phase space has become fundamentally different (Fig. 13.6b). There are no unstable points P_R and P_S here. Quasiperiodic oscillations are observed in region C. They correspond to a stable limit cycle C (Fig. 13.6c) in the system of truncated equations

13.2 Forced Synchronization of the van der Pol Oscillator: Truncated...

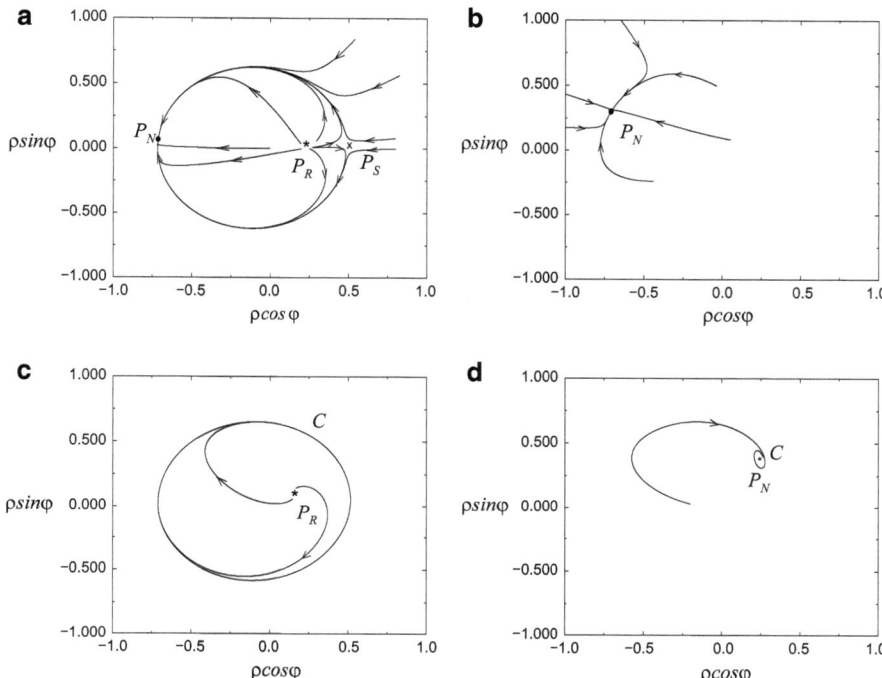

Fig. 13.6 Typical phase portraits of the system (13.9)–(13.10) for $\varepsilon = 0.1$ and for different values of β and Δ from regions A, B, and C in Fig. 13.5: (**a**) $\beta = 0.01, \Delta = 0.001$ (region A). (**b**) $\beta = 0.02, \Delta = 0.01$ (region B). (**c**) $\beta = 0.01, \Delta = 0.03$ (region C). (**d**) $\beta = 0.02, \Delta = 0.037$ (region C)

(13.9)–(13.10). Besides the cycle C, the phase portrait of the system includes the unstable focus P_R.

We now consider bifurcations of typical regimes of the system when we move through the plane of parameters β, Δ in Fig. 13.5. Figure 13.7 shows the phase-parameter diagram plotted versus the detuning parameter Δ for fixed $\beta = 0.01$. This picture corresponds to movement along the line dc in the plane of the parameters β, Δ. For a small detuning Δ (within the interval between points 3 and 4 marked by × symbols in Figs. 13.5 and 13.7), the fixed point P_R is an unstable node. The corresponding phase portrait is shown in Fig. 13.6a. As Δ increases or decreases when we pass through points 3 and 4, the unstable node is transformed into an unstable focus. As Δ is varied further, including exit from region A when we cross the lines l_1 and l'_1, the unstable focus P_R undergoes no bifurcations. The violation of the synchronization regime is defined by the behavior of the fixed points P_N and P_S. On the phase portrait, the unstable manifolds of the saddle P_S close on the stable node P_N. When Δ increases, the saddle and the stable node approach each other and merge into a saddle node on the border l_1. This situation is followed by the bifurcation giving rise to birth of a stable limit cycle, which is accompanied

Fig. 13.7 Phase-parameter diagram for fixed points P_N, P_S, and P_R of the system (13.9)–(13.10) when the detuning parameter Δ is changed and $\beta = 0.01$, $\varepsilon = 0.1$

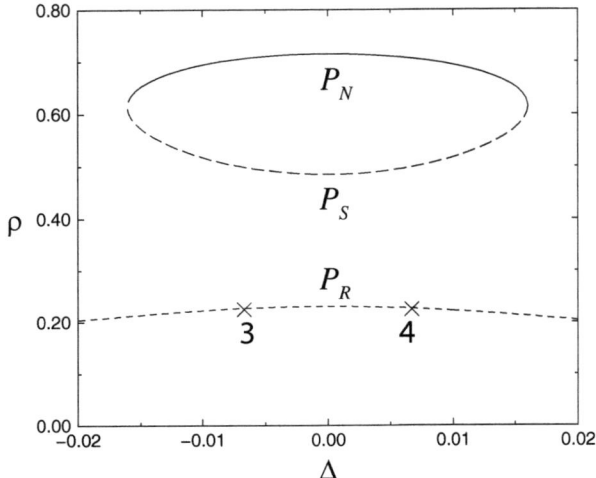

Fig. 13.8 Phase-parameter diagram for fixed points P_N, P_S, and P_R of the system (13.9)–(13.10) when the parameter β is changed and $\Delta = 0.01$, $\varepsilon = 0.1$

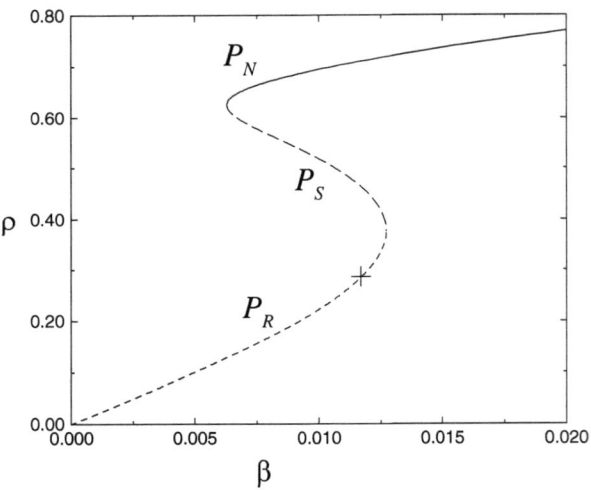

by the disappearance of a complex singular point of saddle-node type. The limit cycle is hardly born and has a finite radius immediately after the bifurcation point (Fig. 13.6c). The corresponding finite amplitude oscillations have a very long period near the bifurcation point. This case corresponds to the hard emergence of slow beats of the variable $x(t)$ in the initial system. The bifurcations at the boundaries of regions A and C considered above illustrate the mechanism of synchronization through locking. Region A of the synchronization tongue is also called the *locking region*.

The phase-parameter diagram for the fixed points of the system (13.9)–(13.10) is plotted in Fig. 13.8 when the parameter β is varied for fixed $\Delta = 0.01$. In the plane of the parameters β, Δ (Fig. 13.5), this corresponds to moving along the line ef. In

13.2 Forced Synchronization of the van der Pol Oscillator: Truncated...

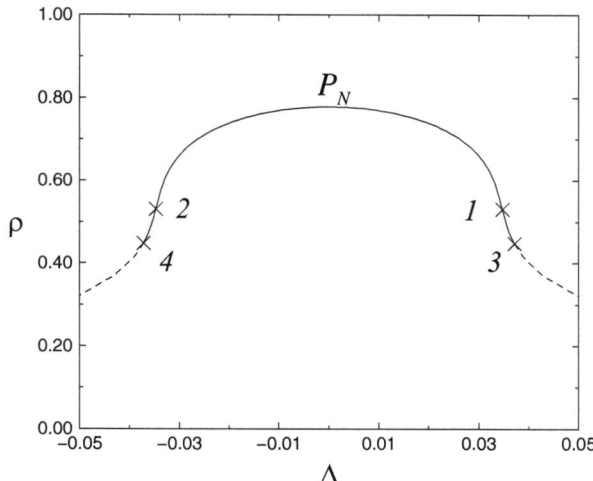

Fig. 13.9 Phase-parameter diagram of the fixed point P_N of the system (13.9)–(13.10) when the detuning Δ is varied for $\beta = 0.02$ and $\varepsilon = 0.1$

region C, below the line l_1 there is a stable limit cycle C and an unstable focus P_R situated inside it. As the amplitude of the external force β increases, the unstable focus P_R moves along the phase plane. When we cross the line l_1, no bifurcations occur with it. The limit cycle C disappears on the boundary of the synchronization region, and a complex singular point of saddle-node type is born. Above the line l_1, the stable node P_N and the saddle P_S diverge in the phase plane. When β increases beyond the point marked by the symbol ×, the unstable focus P_R becomes an unstable node and approaches the saddle P_S. When we cross the line l_2 in Fig. 13.5, a saddle-node bifurcation occurs, causing the unstable node P_R and the saddle P_S to merge and disappear. Only one singular point, the stable node P_N, remains in the phase plane. This point in region B of Fig. 13.5 continues to correspond to the regime of forced synchronization of the van der Pol oscillator. Region B of the synchronization tongue is called the *suppression region*. The mechanism of violation (or occurrence) of synchronization is different here than on the border of regions A and C.

Figure 13.9 is the phase-parameter diagram of the fixed point P_N when Δ is varied for fixed $\beta = 0.02$. On the plane of the parameters β, Δ (Fig. 13.5), this corresponds to moving along the line ab. When Δ lies within the interval bounded by points 1 and 2, the equilibrium P_N is a stable node. Beyond point 2 and before crossing the line l_3, the stable node becomes a stable focus. When we cross the line l_3, a supercritical Andronov–Hopf bifurcation takes place and the point P_N loses its stability. A stable limit cycle is born from the unstable focus in its neighborhood. The radius of the cycle grows proportionally to the square root of the supercriticality. The corresponding phase portrait is shown in Fig. 13.6d. The time series $x(t)$ of the driven van der Pol oscillator exhibits a soft occurrence of beats. The second independent mode of quasiperiodic motions grows smoothly from zero. When we move in the reverse direction along the parameter and when we enter region B, the

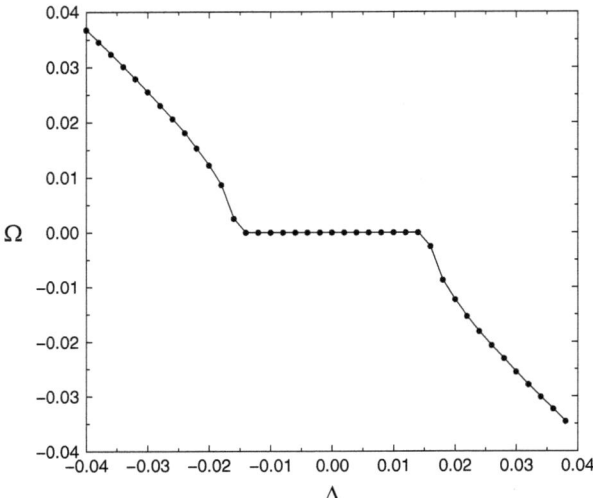

Fig. 13.10 Dependence of $\Omega = \langle \dot{\varphi}(t) \rangle$ on the frequency detuning Δ of the system (13.9)–(13.10) for $\varepsilon = 0.1$ and $\beta = 0.01$

self-sustained oscillatory mode is completely suppressed and the system oscillates at the external frequency. This can happen only at large amplitudes of the external force.

The dependence of $\Omega = \langle \dot{\varphi}(t) \rangle$ on the frequency detuning Δ is plotted in Fig. 13.10 for the system of truncated equations (13.9)–(13.10) when $\varepsilon = 0.1$ and $\beta = 0.01$. The horizontal portion corresponds to the synchronization region when $\Omega = 0$ and the $x(t)$ oscillations occur at the frequency ω of the external force. The width of the plateau in the plot depends on the parameter β. As it increases, the synchronization region expands. The selected values of $\beta = 0.01$ and $\varepsilon = 0.1$ correspond to the locking region (region A in Fig. 13.5).

13.2.3 Bifurcational Analysis of the Nonautonomous van der Pol Oscillator

Let us return to the full equation (13.3) of the van der Pol oscillator subject to the external harmonic force, viz.,

$$\ddot{x} - \left(\varepsilon - x^2\right)\dot{x} + x = b \sin \omega t ,$$

and examine the way the behavior of this system depends on the amplitude and frequency of the external harmonic action for a fixed value of the control parameter ε of the autonomous oscillator.

The nonautonomous van der Pol oscillator has a three-dimensional phase space. Thus, in addition to fixed points and limit cycles, a two-dimensional torus can also

13.2 Forced Synchronization of the van der Pol Oscillator: Truncated...

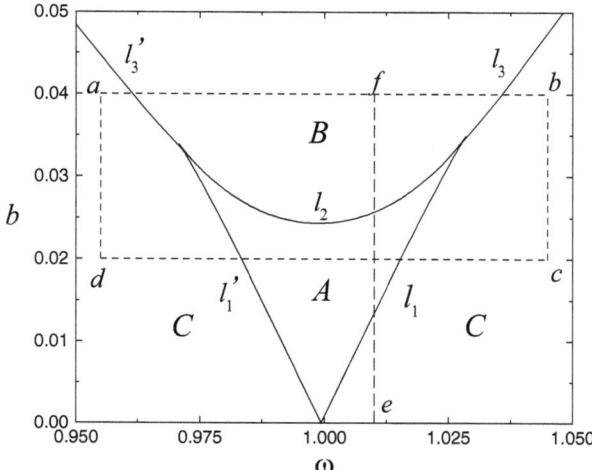

Fig. 13.11 Basic synchronization tongue on the amplitude b–frequency ω parameter plane of the external force of the system (13.18) for $\varepsilon = 0.1$

exist here. The equation of the nonautonomous oscillator (13.3) can be written in the form of a system of three first-order differential equations:

$$\dot{x} = y, \qquad \dot{y} = (\varepsilon - x^2) y - x + b \sin(z), \qquad \dot{z} = \omega. \qquad (13.18)$$

Here, x, y, z are dynamical variables which determine the dimension of the phase space. The dynamics of this system is quite complicated and varies considerably over a wide range of parameter values. We restrict ourselves to the study of synchronization modes and their bifurcations in the neighborhood of the so-called basic synchronization tongue, when the frequency of synchronous oscillations coincides with the frequency of the external action. We consider only quasiharmonic regimes of the autonomous oscillator when the parameter ε takes moderate values.

The synchronization mode corresponds to a stable limit cycle C_N. The latter and a saddle limit cycle C_S lie on the surface of a two-dimensional torus which is formed by the closing of the unstable manifolds of the saddle cycle C_S on the stable cycle C_N. We shall consider bifurcations of these limit cycles when we move across the plane of control parameters b, ω for fixed $\varepsilon = 0.1$.

Bifurcation lines of synchronous motions are plotted in Fig. 13.11 on the external amplitude–external frequency plane for a fixed value of the parameter ε. The system exhibits synchronization modes in regions A and B and quasiperiodic oscillations in region C. In contrast to the truncated equations for the amplitude and phase, it should be noted that, in the system (13.18), stable and unstable synchronous motions correspond to stable and unstable limit cycles, and quasiperiodic oscillations to a two-dimensional ergodic torus in the three-dimensional phase space of the system. For parameter values in region A, three limit cycles exist in the system phase space. These are the stable cycle C_N, the saddle cycle C_S, and an absolutely unstable cycle C_R. Projections of the limit cycles on the plane (x, y) are shown in Fig. 13.12.

Fig. 13.12 Projections of limit cycles on the plane (x, y) of the system (13.18) for $b = 0.02$, $\omega = 1.01$, and $\varepsilon = 0.1$

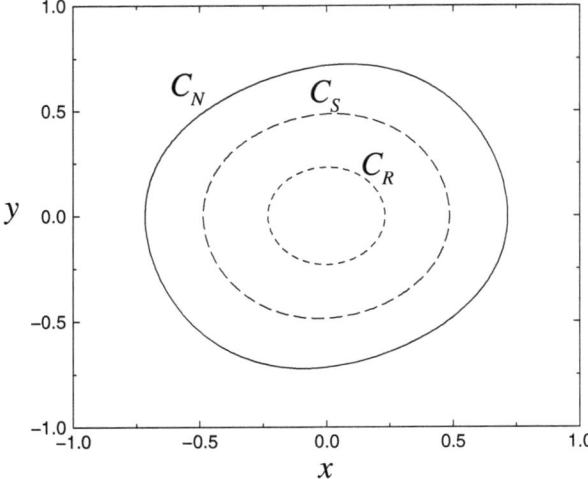

Fig. 13.13 Phase-parameter diagram for limit cycles C_N, C_S, C_R of the system (13.18) when the external frequency ω is varied for $b = 0.02$ and $\varepsilon = 0.1$

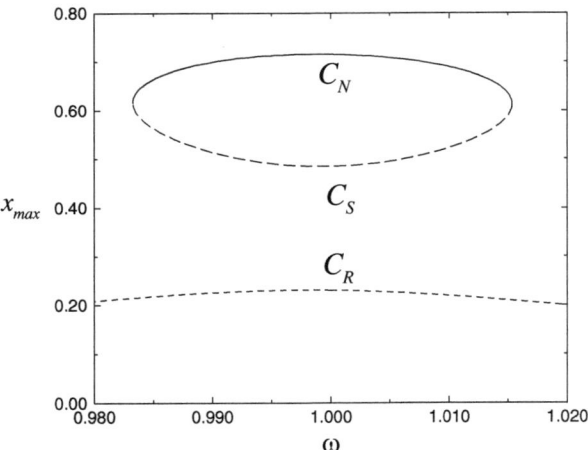

The phase-parameter diagram is plotted in Fig. 13.13, showing the dependence on the external force frequency ω when the amplitude is fixed at $b = 0.02$, which corresponds to moving along the line dc in Fig. 13.11. The ordinate axis in the bifurcation diagram of Fig. 13.13 represents maximal values of the dynamical variable $x(t)$ of the corresponding limit cycles C_N, C_S, and C_R. As the external frequency ω changes upon crossing the boundaries l_1 and l'_1 of the synchronization region, the unstable limit cycle C_R undergoes no bifurcations. The transition from synchronous to quasiperiodic oscillations is related to bifurcations of the other two limit cycles. The diagram shows that, when ω increases or decreases upon approaching the bifurcation lines l_1 or l'_1, C_N approaches C_S. They merge on the boundary into a singular saddle-node cycle (one of the cycle multipliers takes the value $+1$) and disappear when we pass through the bifurcation line. This saddle-

13.2 Forced Synchronization of the van der Pol Oscillator: Truncated...

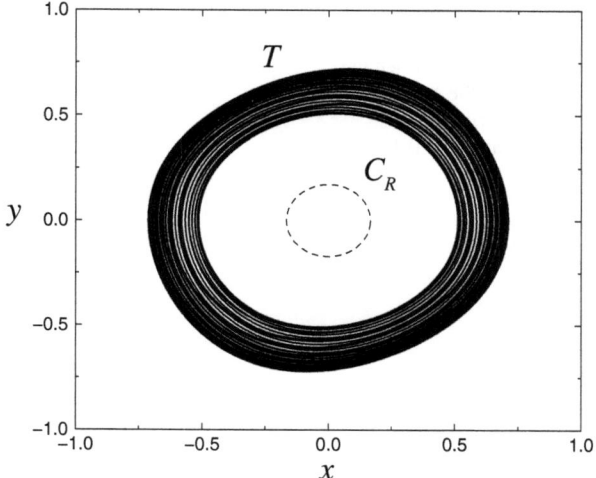

Fig. 13.14 Phase portrait (attracting torus T and unstable limit cycle C_R) of the system (13.18) after a saddle-node bifurcation of the cycles C_N and C_S for the parameters $\varepsilon = 0.1$, $b = 0.02$, and $\omega = 1.017$

node bifurcation of the merging and disappearance of limit cycles C_N and C_S occurs on the two-dimensional torus surface. Outside the synchronization region, the phase trajectory covers the torus surface, closing nowhere, this corresponding to quasiperiodic oscillations. A projection of the phase portrait of a two-dimensional torus is shown in Fig. 13.14. When we cross the borders l_1 and l'_1, quasiperiodic oscillations arise in a rigid manner and the amplitude of modulation (the 'thickness' of the torus in its phase portrait) takes a large finite value immediately after the bifurcation point. Here, besides the attracting torus T, there is an unstable limit cycle C_R in the system phase space. This restructuring of the system phase portrait on the boundaries of regions A and C illustrates the bifurcation mechanism of synchronization via locking. Region A of the synchronization tongue is called the locking region.

The phase-parameter diagram is plotted in Fig. 13.15 for limit cycles C_N, C_R, and C_S of the system (13.18), showing the dependence on the amplitude of the external force b for fixed $\omega = 1.01$ and $\lambda = 0.1$. This case corresponds to moving along the line ef in the external amplitude–external frequency parameter plane in Fig. 13.11.

When the values of b are small and fall below the line l_1, then, in addition to the attracting torus, an unstable limit cycle C_R also exists. It has been born from the equilibrium at the origin. As b increases, the amplitude of C_R grows. When we cross the line l_1, nothing happens to it, while a saddle-node bifurcation occurs on the surface of the attracting torus and a pair of two limit cycles is born, viz., the stable cycle C_N and the saddle cycle C_S. As b increases further, C_S approaches C_R.

When we cross the line l_2, a saddle-repeller bifurcation occurs. The two cycles C_S and C_R merge and disappear. The disappearance of the saddle cycle C_S leads to the destruction of a resonance torus. Only one stable limit cycle C_N remains in the phase space and it corresponds to the synchronization mode. The synchronization region B located above the line l_2 in Fig. 13.11 is called the suppression region. In

Fig. 13.15 Phase-parameter diagram of the limit cycles of the system (13.18) when the amplitude of the external force b is varied for $\omega = 1.01$ and $\varepsilon = 0.1$

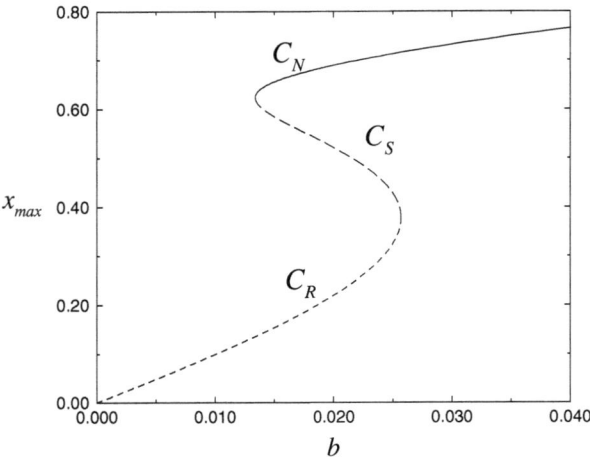

Fig. 13.16 Phase-parameter diagram of the limit cycle of the system (13.18) when the external frequency ω is varied for fixed values $b = 0.04$ and $\varepsilon = 0.1$. The *solid line* corresponds to a stable state and the *dashed line* to an unstable state of the cycle C_N

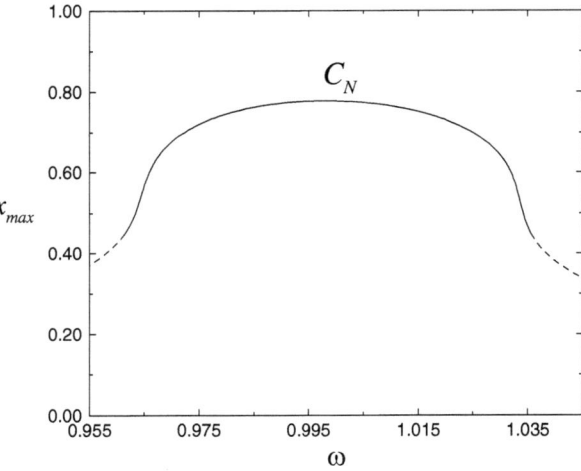

contrast to the locking region A, when we exit the suppression region and when we vary the external frequency, quasiperiodic oscillations arise softly. The amplitude of modulation grows gradually from zero and is conditioned by the character of bifurcations of the limit cycle C_N.

Figure 13.16 shows the phase-parameter diagram of the limit cycle C_N when the external force frequency ω is varied for fixed values $b = 0.04$ and $\varepsilon = 0.1$. This situation corresponds to moving along the line ab in the external amplitude–external frequency parameter plane in Fig. 13.11. The limit cycle is stable (solid line in the bifurcation diagram in Fig. 13.16) within the interval of values bounded by the lines l_3 and l'_3. One of the cycle multipliers is always equal to $+1$. The other two multipliers have modulus less than unity in the indicated interval. When we cross the bifurcation lines l_3 and l'_3, the two multipliers are complex conjugate and their modulus now exceeds unity. The limit cycle loses its stability

13.2 Forced Synchronization of the van der Pol Oscillator: Truncated...

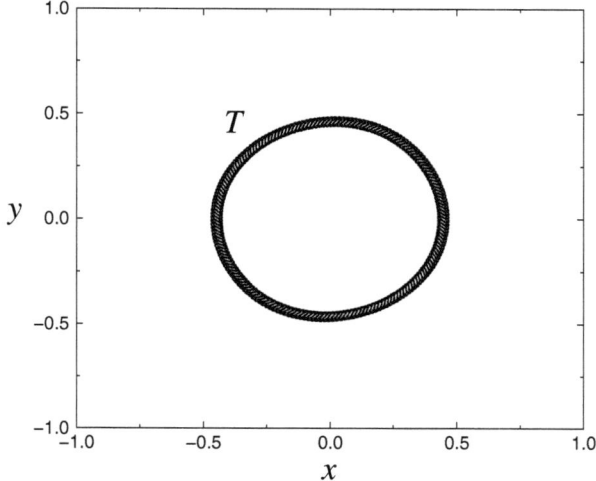

Fig. 13.17 Projection of a two-dimensional torus for $\varepsilon = 0.1, b = 0.04$, and $\omega = 1.0357$ (near the line l_3 in Fig. 13.11)

and gives birth to an attracting two-dimensional torus. Its projection on the plane (x, y) is shown in Fig. 13.17. The 'thickness' of the torus grows gradually as the supercriticality increases. This rebuilding of the phase portrait is called the *Neimark–Saker bifurcation*. It determines the mechanism of synchronization via suppression.[2]

In the synchronization region B, the oscillator oscillates at the frequency of the external force. However, in contrast to the case of weak external action, the self-sustained oscillatory mode is not adjusted to the rhythm of the external force, but is in fact completely suppressed by it.

Only when the frequency detuning increases and when we cross the boundaries l_3 and l'_3 of the synchronization region (Fig. 13.11) is the self-sustained oscillatory mode excited, whereupon it gradually begins to increase. For a moderate supercriticality, the spectral components corresponding to the external force continue to prevail in the spectrum of quasiperiodic oscillations with two independent frequencies. For this reason, this mode of slightly modulated oscillations is often also referred to as the synchronization regime.

Figure 13.18 illustrates the dependence of the mean beat frequency Ω, i.e., the difference between the mean frequency of the self-sustained oscillations and the frequency of the external force, of the system (13.18) on the external frequency ω for $\varepsilon = 0.1$ and $b = 0.02$. The horizontal part corresponds to the synchronization region when the oscillator oscillates at the frequency of the external force. The width

---

[2]It should be noted here that the boundaries l_3 and l'_3 in the plane of the parameters β, ω represent the lines of torus birth on which the synchronization tongues with different winding numbers rest. The winding number is defined by the ratio of the beat frequency to the frequency of the external force. The beat frequency is the difference between the frequencies of the self-sustained oscillations and the external force. However, in the framework of this chapter, we do not consider the structure of the control parameter space over a wide range of values.

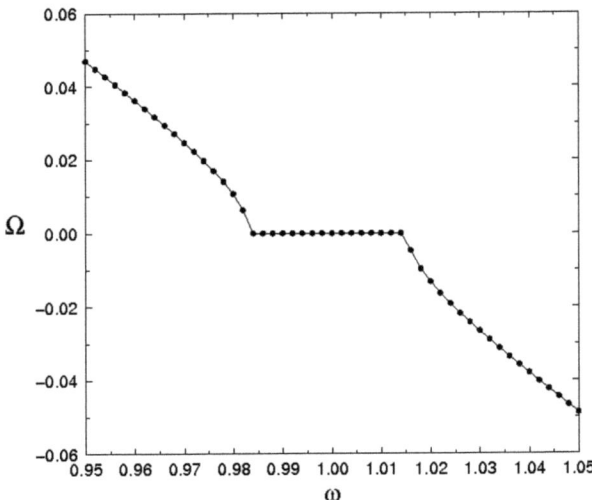

Fig. 13.18 Dependence of the mean beat frequency Ω on the external frequency ω of the system (13.18) for $\varepsilon = 0.1$ and $b = 0.02$

of this plateau depends on the external amplitude b. The value $b = 0.02$ for $\varepsilon = 0.1$ corresponds to the locking region.

13.3 Mutual Synchronization: Effect of Oscillation Death in Dissipatively Coupled van der Pol Oscillators

Let us now analyze the dynamics of two dissipatively coupled van der Pol oscillators described by the system of ordinary differential equations

$$\ddot{x}_1 - \left(\varepsilon_1 - x_1^2\right)\dot{x}_1 + \omega_1^2 x_1 = \gamma\left(\dot{x}_2 - \dot{x}_1\right),$$
$$\ddot{x}_2 - \left(\varepsilon_2 - x_2^2\right)\dot{x}_2 + \omega_2^2 x_2 = \gamma\left(\dot{x}_1 - \dot{x}_2\right), \tag{13.19}$$

where $x_{1,2}$ are the dynamical variables of the first and second subsystems, $\varepsilon_{1,2}$ are the parameters responsible for the generation of self-sustained oscillations in the partial oscillators, $\omega_{1,2}$ are the natural (basic) frequencies of the uncoupled oscillators, and γ is the coupling strength.

The system (13.19) demonstrates typical effects and bifurcation mechanisms encountered in various coupled self-sustained oscillatory systems with limit cycles. The range of observed phenomena and the peculiarities of bifurcational transitions between oscillatory modes are mainly defined by the character of coupling between interacting oscillators. Besides synchronization, an interesting phenomenon can be observed in the case of mutual diffusive coupling in the system of coupled van der Pol oscillators.

13.3 Mutual Synchronization: Effect of Oscillation Death in Dissipatively... 237

By introducing the change of variables $\dot{x}_{1,2} = y_{1,2}$, we rewrite (13.19) as a system of first-order differential equations:

$$\begin{aligned}
\dot{x}_1 &= y_1 , \\
\dot{y}_1 &= \left(\varepsilon_1 - x_1^2\right) y_1 - \omega_1^2 x_1 + \gamma \left(y_2 - y_1\right) , \\
\dot{x}_2 &= y_2 , \\
\dot{y}_2 &= \left(\varepsilon_2 - x_2^2\right) y_2 - \omega_2^2 x_2 + \gamma \left(y_1 - y_2\right) .
\end{aligned} \quad (13.20)$$

The system (13.20) has a four-dimensional phase space. Without interaction ($\gamma = 0$), the partial oscillators do not generate self-sustained oscillations for negative values of the control parameters $\varepsilon_{1,2}$. The oscillations are excited when $\varepsilon_{1,2}$ pass through zero, which corresponds to the supercritical Andronov–Hopf bifurcation.

When the coupling is introduced ($\gamma > 0$) and the parameters ε_1 and ε_2 are positive, the interacting oscillators can exhibit quasiperiodic oscillations and the synchronization effect, which depends on the coupling strength and the detuning between the basic frequencies ω_1 and ω_2. If the coupling is of a dissipative type, the system can also exhibit an interesting phenomenon known as *oscillation death* (quenching). For sufficiently large coupling and frequency detuning, the self-sustained oscillations in the two subsystems are suppressed (or die out) due to coupling. For positive $\varepsilon_{1,2}$, there are regions of coupling and detuning values where the equilibrium state of the system becomes stable again.

Consider the dynamics of the interacting oscillators when the values of the control parameters $\varepsilon_1 = \varepsilon_2 = \varepsilon = 0.1$ are fixed and identical and the coupling γ and the detuning $p = \omega_2/\omega_1$ are varied. Bifurcation lines are plotted in Fig. 13.19 for the basic region of synchronization in the plane of the parameters γ, p. Synchronous self-sustained oscillations are observed in regions A and B which surround regions C of quasiperiodic oscillations and regions D of self-sustained oscillation suppression (or so-called amplitude death).

When parameter values are taken from the region A bounded by the bifurcation lines l_{SN}, l'_{SN}, and l_{SR}, an unstable fixed point P_R, three saddle limit cycles C_S, C_P, C_R, and a stable limit cycle C_N all exist in the phase space of the system. Phase portraits typical for region A are shown in Fig. 13.20a, b. The limit cycles correspond to synchronous motions with different phase shifts between time series in the partial oscillators. Obviously, only stable limit cycles correspond to synchronization regimes observed experimentally. The stable cycle C_N is related to the regime of in-phase synchronization of periodic self-sustained oscillations (strictly in-phase oscillations are observed for $p = 1$). The saddle cycle C_S corresponds to unstable anti-phase synchronous motions. When the image point evolves on the cycles C_R and C_P, the partial systems execute synchronous oscillations and the phase shift between them lies in the range from 0 to π for C_R and from π to 2π for C_P. It should be noted that the saddle cycles C_P and C_R are symmetrically located in the phase space relative to the cycle C_S. This is well illustrated in Fig. 13.20b. Calculating the eigenvalues of the fixed point P_R, we find that it is

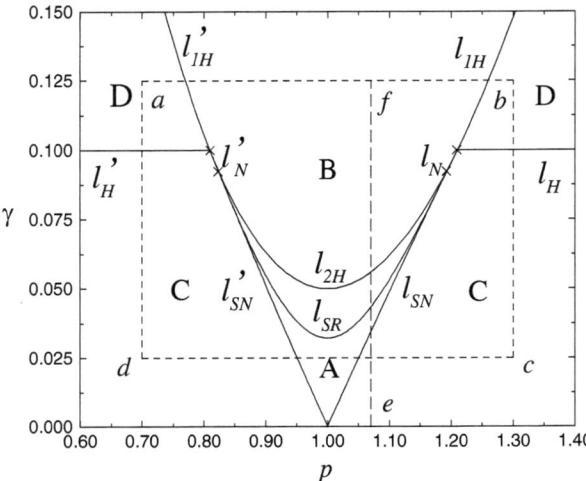

Fig. 13.19 Bifurcation lines for the basic region of synchronization in the plane of the control parameters γ, p for $\varepsilon_1 = \varepsilon_2 = 0.1$

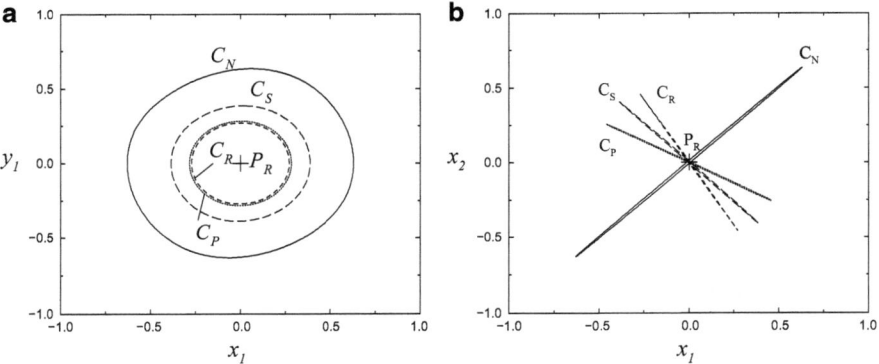

Fig. 13.20 Projections of the phase portrait on the planes (y_1, x_1) (**a**) and (x_2, x_1) (**b**) for $\varepsilon_1 = \varepsilon_2 = 0.1$, $\gamma = 0.03$, and $p = 1.001$. The unstable fixed point P_R, saddle cycles C_S, C_P, C_R, and the stable cycle C_N all exist in the phase space

unstable in all directions because it has two pairs of complex-conjugate eigenvalues with positive real parts. Analysis of the cycle multipliers indicates that the saddle cycles C_P and C_R have one-dimensional stable manifolds and three-dimensional unstable manifolds, while the saddle cycle C_S has two-dimensional stable and two-dimensional unstable manifolds. The unstable manifolds of the cycle C_S close on the stable cycle C_N, forming an attracting torus. In other words, the resonance cycles C_N and C_S lie on the two-dimensional torus.

Let us now follow bifurcations of the limit cycles and the fixed point when we move along the plane of the parameters γ, p. Figure 13.21 shows the phase-

13.3 Mutual Synchronization: Effect of Oscillation Death in Dissipatively...

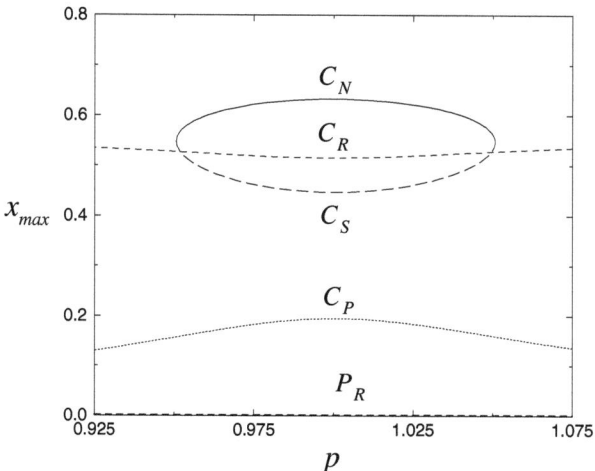

Fig. 13.21 Phase-parameter diagram for the limit cycles and the fixed point showing the dependence on the detuning p for $\varepsilon_1 = \varepsilon_2 = 0.1$ and $\gamma = 0.025$

parameter diagram for the limit cycles and the fixed point depending on the detuning parameter p and for the fixed coupling strength $\gamma = 0.025$. This case corresponds to moving along the line cd in the γ, p parameter plane in Fig. 13.19. Maximal values $x = x_{\max}$ of the dynamical variable of the corresponding limit cycles and the fixed point (the dynamical variables for point P_R are equal to zero) are plotted as ordinates in the diagram. When the synchronization region A is exited by varying p, saddle-node bifurcations occur on the lines l_{SN} and l'_{SN}. The stable cycle C_N and saddle cycle C_S, which lie on the torus, come close together as the parameter p approaches the bifurcation lines l_{SN} and l'_{SN}. On these lines, they merge, becoming a structurally unstable saddle-node cycle, and then disappear beyond the point of bifurcation (in region C in Fig. 13.19). There is now an attracting ergodic two-dimensional torus in the phase space. Phase trajectories cover it, closing nowhere. When the image point moves on this torus, the system exhibits quasiperiodic oscillations with two incommensurable frequencies. When we leave the synchronization region A by varying p, no bifurcations occur with saddle cycles C_R and C_P and fixed point P_R. Thus, in the region C of quasiperiodic oscillations, the phase portrait of the system comprises the attracting ergodic torus, saddle limit cycles C_R and C_P, and an unstable fixed point P_R located at the origin. As the coupling γ increases, bifurcations of these limit cycles depend on the value of the detuning p.

The phase-parameter diagram for the limit cycles and the fixed point is shown in Fig. 13.22 for increasing coupling γ and fixed parameters $\varepsilon_1 = \varepsilon_2 = 0.1$ and $p = 1.07$. This picture corresponds to moving along the line ef in Fig. 13.19. For weak coupling, the attracting torus, saddle cycles C_P and C_R, and unstable fixed point P_R exist in the phase space. As γ increases and when we cross the line l_{SN} (see Fig. 13.19), a saddle-node bifurcation occurs on the torus and a stable limit cycle C_N and a saddle cycle C_S are born. In the diagram (Fig. 13.22), this corresponds to the point b_{SN}, from which two branches issue out. The solid line is formed by

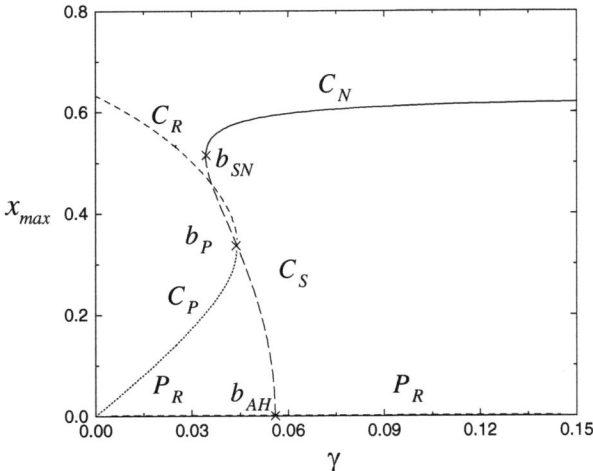

Fig. 13.22 Phase-parameter diagram for the limit cycles and the fixed point as a function of the coupling γ for fixed $\varepsilon_1 = \varepsilon_2 = 0.1$ and $p = 1.07$

the values of a dynamical variable on the stable cycle C_N, while the dotted line is formed by the values on the saddle cycle C_S. When the coupling grows further, a subcritical pitchfork bifurcation takes place on the line l_{SR} in the plane of parameters γ, p (Fig. 13.19). The saddle cycles C_P and C_R approach C_S and are stuck in it on the line l_{SR}. Above the bifurcation point, there remains the saddle limit cycle C_S which has a different kind of stability. Analysis of the cycle multipliers shows that, before the bifurcation point, the saddle cycles C_P and C_R have one-dimensional stable manifolds and three-dimensional unstable manifolds. The saddle cycle C_S has two-dimensional stable and two-dimensional unstable manifolds which close on the stable limit cycle C_N and form a two-dimensional torus. Beyond the subcritical pitchfork bifurcation point, the saddle cycle C_S is transformed into a saddle limit cycle with one-dimensional stable and three-dimensional unstable manifolds. This leads to the destruction of the two-dimensional torus on which the resonance cycles lie. The pitchfork bifurcation point is denoted by symbols b_p in the diagram of Fig. 13.22. When the coupling values are taken from the region bounded by the lines l_{SR} and l_{2H} (Fig. 13.19), the phase portrait of the system includes the unstable fixed point P_R, the saddle cycle C_S, and the stable cycle C_N. Above the line l_{SR}, the two-dimensional torus no longer exists. It has been destroyed as a result of the subcritical pitchfork bifurcation of the saddle cycle C_S. When γ increases further, the radius of the saddle cycle C_S decreases and contracts into the unstable fixed point P_R on the line l_{2H}, whereupon the Andronov–Hopf bifurcation occurs. Before the bifurcation, the fixed point has two pairs of complex-conjugate eigenvalues with positive real parts. After the bifurcation it has a pair of eigenvalues with positive real parts and a pair of eigenvalues with negative real parts. For strong coupling in the synchronization region B, there is a stable limit cycle C_N and an unstable focus P_R in the phase space of the system. In the case of strong coupling, growth of the detuning p leads to the effect of self-sustained oscillation quenching

13.3 Mutual Synchronization: Effect of Oscillation Death in Dissipatively...

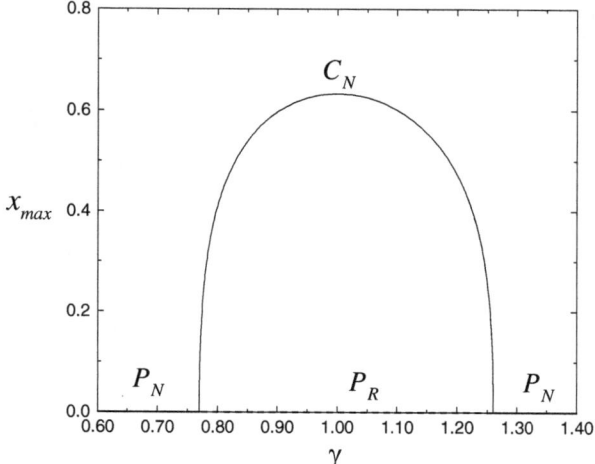

Fig. 13.23 Phase-parameter diagram for the limit cycle and the fixed point as a function of the detuning p for $\varepsilon_1 = \varepsilon_2 = 0.1$ and $\gamma = 0.125$

(or so-called amplitude death). As the detuning between the natural frequencies ω_1 and ω_2 increases, the amplitude of self-sustained oscillations in each oscillator vanishes smoothly.

The phase-parameter diagram is shown in Fig. 13.23 for the limit cycle and the fixed point as a function of the detuning p for the fixed parameter values $\varepsilon_1 = \varepsilon_2 = 0.1$ and $\gamma = 0.125$. This case corresponds to moving along the line ab in the γ, p parameter plane in Fig. 13.19. When the natural frequencies of the partial oscillators coincide ($p = 1$), the radius of the stable cycle C_N corresponding to the synchronization mode has its maximal value. If p increases or decreases, the cycle radius is smoothly reduced, and on the bifurcation lines l_{1H} and l'_{1H} (Fig. 13.19), the cycle contracts into the equilibrium at the origin. When we pass from region B to region D, the fixed point P_S becomes a stable focus. The supercritical Andronov–Hopf bifurcation takes place here. In region D, only the stable fixed point P_S exists in the phase space of the system. The system does not generate self-sustained oscillations as the interaction causes oscillation quenching. However, each of the partial oscillators generates oscillations without coupling.

When we pass from region D to region B (Fig. 13.19), for example, varying p with fixed coupling γ, periodic synchronous self-sustained oscillations are smoothly excited. However, when we pass from region D to region C, for example, by decreasing the coupling at a fixed detuning p, the regime of mutual synchronization of the oscillators does not appear. Instead, quasiperiodic self-sustained oscillations are smoothly generated in the system. The bifurcation line l_H is the boundary of the transition from the regime of oscillation quenching to the regime of quasiperiodic self-sustained oscillations. In region D, before the line l_H, the fixed point P_R has two pairs of complex-conjugate eigenvalues with negative real parts, which vanish on the line l_H (two pairs of purely imaginary eigenvalues) and become positive below it. As a result, an attracting torus T and the saddle cycles C_P and C_R are softly born

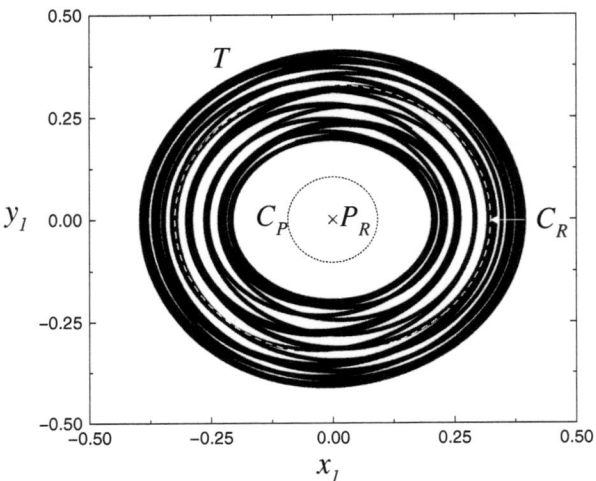

Fig. 13.24 Projection of the phase portrait on the plane (y_1, x_1) for $\varepsilon_1 = \varepsilon_2 = 0.1$, $\gamma = 0.075$, and $p = 1.3$. T is the attracting torus, C_R and C_P are saddle limit cycles, and P_R is the unstable equilibrium

in the neighborhood of the fixed point P_R. As the coupling decreases, these limit sets (the fixed point, the two limit cycles, and the attracting torus) diverge from each other. A projection of a typical phase portrait is shown in Fig. 13.24 for parameter values $\varepsilon_1 = \varepsilon_2 = 0.1$, $\gamma = 0.075$, and $p = 1.3$, which corresponds to a point on the line bc in the γ, p parameter plane shown in Fig. 13.19.

The above bifurcation transition with birth of the attracting torus and two saddle cycles from the fixed point refers to the degenerate case. This is due to the identity of the partial oscillators: the excitation parameters are assumed to be equal, i.e., $\varepsilon_1 = \varepsilon_2$. The identity of the oscillators through the parameters controlling the cycle birth bifurcation in the uncoupled oscillators leads not only to this degenerate situation, but also to the pitchfork bifurcation which occurs on the line l_{SR} of the basic synchronization tongue (Fig. 13.19) and in which the saddle cycles C_S, C_P, and C_R are involved (see the phase-parameter diagram in Fig. 13.22). Introducing the nonidentity of the oscillators by means of the parameter ε eliminates these degeneracies in a well-defined way and leads to the typical bifurcation transitions.

13.4 Summary

In this chapter we have described the main ideas of the classical theory of synchronization of periodic self-sustained oscillations. The theoretical description is applied to the simplest self-sustained oscillatory system with one degree of freedom, namely, the van der Pol oscillator driven by an external harmonic force. The results are presented for the effects of external (forced) and mutual synchronization.

The effect of external synchronization is analyzed in the (one-dimensional) phase approximation using the truncated equations and for the full nonautonomous van der Pol oscillator. The fundamental bifurcation mechanisms of synchronization are presented. They include frequency (phase) locking and suppression of basic oscillations of the oscillator. The results are obtained for the basic region of synchronization when the frequencies of the basic self-sustained oscillations and the external force are close (synchronization at the basic tone). The detailed analysis of external synchronization enables one to compare the observed effect for the quasiharmonic oscillator with synchronization effects of quasiperiodic and chaotic oscillations, something that will be discussed in the following chapters.

The results for the classical problem of synchronization of periodic self-sustained oscillators can be found in [1–3].

References

1. Anishchenko, V.S., Astakhov, V.V., Neiman, A.B., Vadivasova, T.E., Schimansky-Geier, L.: Nonlinear Dynamics of Chaotic and Stochastic Systems. Springer, Berlin (2002)
2. Balanov, A.G., Janson, N.B., Postnov, D.E., Sosnovtseva, O.: Synchronization: From Simple to Complex. Springer, Berlin (2009)
3. Pikovsky, A., Rosenblum, M., Kurths, J.: Synchronization: A Universal Concept in Nonlinear Science. Cambridge University Press, Cambridge (2003)

Chapter 14
Synchronization of Two-Frequency Self-Sustained Oscillations

14.1 Introduction

Two-frequency oscillations are the simplest case of quasiperiodic oscillations. In an autonomous regime, these oscillations are accompanied by the effect of mutual synchronization that corresponds to rational values of the Poincaré winding number. In this case synchronization regions are characterized by the so-called Arnold tongues, where the winding number Θ satisfies the condition $\Theta = m : n$, with m and n positive integers.

To investigate the synchronization of quasiperiodic self-sustained oscillations, one must analyze the response of a two-frequency oscillator to an external periodic force (external synchronization) or study the dynamics of two coupled two-frequency oscillators (mutual synchronization). Due to the presence of Arnold resonances, one must analyze the influence of forcing, both on ergodic oscillations and on oscillations corresponding to the resonance conditions. In any case it is clear that the synchronization of quasiperiodic two-frequency oscillations is a more complex problem than the synchronization of a limit cycle. In the present chapter we consider the fundamental effects by studying the simple models of two-frequency oscillators discussed in Chap. 12.

14.2 Influence of an External Periodic Force on a Resonant Limit Cycle in a System of Coupled Oscillators

We start by studying the synchronization of the stable resonant limit cycle on the torus (see Fig. 12.2b) in the presence of an external harmonic force. For this purpose we introduce the harmonic signal $b \sin\left[(2\pi f_e)t\right]$ in the second equation of the system (12.6) to obtain

$$\dot{x}_1 = y_1,$$
$$\dot{y}_1 = (\varepsilon - x_1^2)y_1 - \omega_1^2 x_1 + \gamma(x_2 - x_1) + b\sin[(2\pi f_e)t],$$
$$\dot{x}_2 = y_2,$$
$$\dot{y}_2 = (\varepsilon - x_2^2)y_2 - \omega_2^2 x_2 + \gamma(x_1 - x_2),$$
(14.1)

and then analyze the system's response to this external force.

We consider the oscillatory mode of the autonomous system (14.1) ($b = 0$), which corresponds to resonance region I (see Fig. 12.1), and set the parameters as follows: $\varepsilon = 0.1$, $\omega_1 = 1$, $\omega_2 = 1.0015$, $\gamma = 0.02$. This regime is indicated by the point on the parameter plane in Fig. 12.1. The autonomous system generates stable periodic self-sustained oscillations. Their image is a stable limit cycle L_0. We note that, from an experimental point of view, this is a typical periodic motion with frequency f_1 and a spectrum that includes odd harmonics $f_n = 2n + 1$ ($n = 0, 1, 2, \ldots$), since the nonlinearity is quadratic. The fact that the limit cycle L_0 lies on the surface of a two-dimensional torus is in no way manifested in full-scale experiment. However, more detailed studies have shown that this fact is fundamental and leads to very significant differences when analyzing external synchronization of a resonant limit cycle on a two-dimensional torus.

Figure 14.1a, b illustrates the dependence of the oscillator frequencies f_1 and f_2 on the external signal frequency f_e. The external amplitude is set to be $b = 0.025$, and the frequencies f_1 and f_2 are normalized on f_e. Figure 14.1c shows the winding number $\Theta = f_1 : f_2$ as a function of f_e. As can be seen from the plots, four regions A, B, C, and D can be distinguished in Fig. 14.1, in which the system dynamics is qualitatively different. To analyze this, we use the calculation results for the spectrum of Lyapunov characteristic exponents, as presented in Fig. 14.2.

In region A the external frequency is sufficiently far from the frequency of the limit cycle $f_1 = f_2 \approx 0.158$. The system exhibits quasiperiodic oscillations with frequencies f_e and $f_1 = f_2$. Their image in the phase space is a two-dimensional torus. This is confirmed by the presence of two zero exponents in the LCE spectrum (see Fig. 14.2). The resonance condition $f_1 = f_2$ still holds in region A.

The regime of mutual synchronization (see Figs. 12.1 and 12.2) is destroyed in region B. Frequencies f_1 and f_2 become different, as can be seen from the plot of the winding number Θ (Fig. 14.1c). A regime of quasiperiodic oscillations with three independent frequencies f_1, f_2, and f_e is born. It is characterized by three zero exponents in the LCE spectrum (Fig. 14.2) and corresponds to an attractor in the form of a three-dimensional torus. The system dynamics in region B is quite complicated. If f_e is varied, high-order partial resonances in the form of two-dimensional tori T^2 and even chaotic regimes may occur on the three-dimensional torus.

In region C, we observe the first phenomenon among those of special interest here: the basic frequency of the first oscillator is locked by the external signal, $f_e = f_1$, but $f_1 \neq f_2$. A resonance structure in the form of a two-dimensional torus emerges on the three-dimensional torus, and this is proved by the presence of two

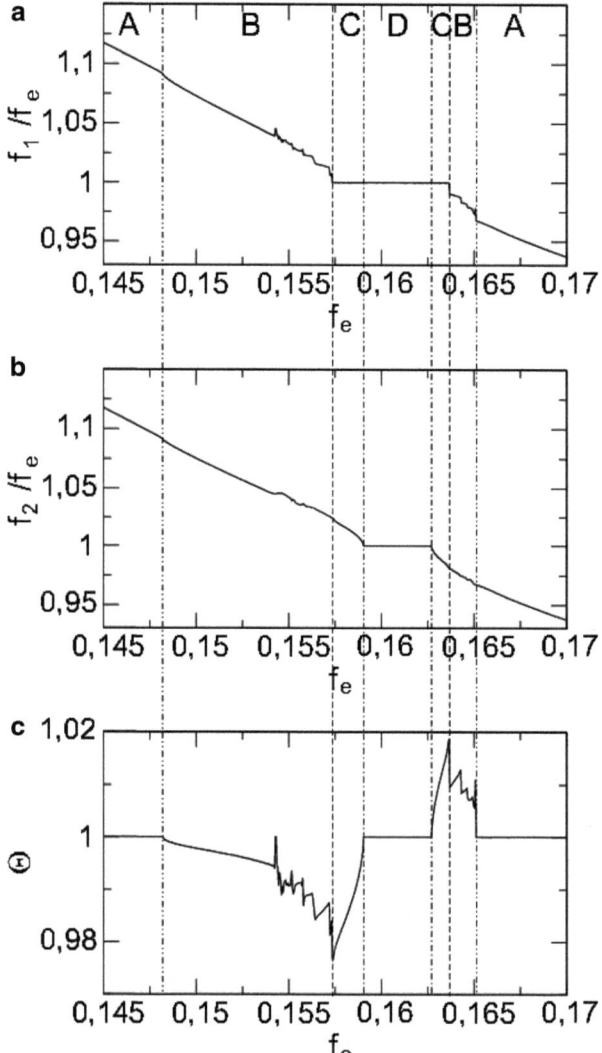

Fig. 14.1 Frequency ratios (**a**) f_1/f_e and (**b**) f_2/f_e, and (**c**) the winding number $\Theta = f_1 : f_2$ as functions of the frequency of the external force f_e for $b = 0.025$

zero exponents in the LCE spectrum (Fig. 14.2). Calculations have shown that, in this regime, the Poincaré section looks like a closed invariant curve.

Finally, in region D the regime of complete synchronization is realized: both the frequencies of the interacting oscillators are locked by the external signal and the condition $f_e = f_1 = f_2$ is satisfied. Only one Lyapunov exponent is zero in region D, and the phase portrait represents the attractor in the form of a limit cycle.

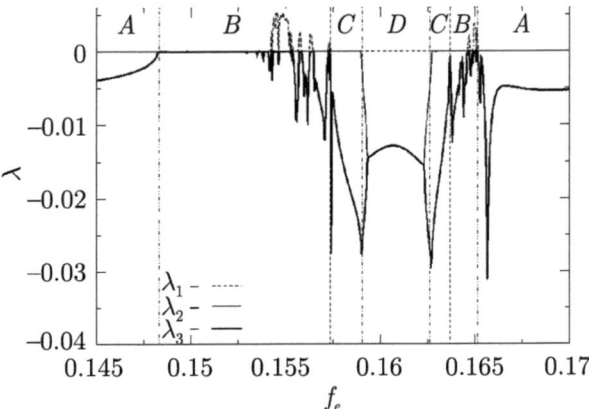

Fig. 14.2 Dependence of the three maximal Lyapunov exponents on the frequency of the external force f_e for $b = 0.025$

The results described above show a very important difference between synchronization of a resonant limit cycle on a two-dimensional torus and the classic case. This difference lies in the fact that, when the frequency detuning $f_e - f_1$ decreases, the external force first destroys the original mutual synchronization and then consistently locks one basic frequency followed by the other. As a result, the effect of complete synchronization is realized when the winding number is locked (Fig. 14.1c, region D).

14.3 Basic Bifurcations of Quasiperiodic Regimes When Synchronizing a Resonant Limit Cycle

To understand in more detail the mechanisms that underlie the rebuilding of oscillatory modes in the system (14.1) when the frequency of the external force is varied, a bifurcation diagram is constructed on the amplitude–frequency parameter plane of the external force. This is shown in Fig. 14.3.

The plots presented in Figs. 14.1 and 14.2 correspond to motion along the straight line $b = 0.025$ of the diagram in Fig. 14.3. Bifurcation lines l_{T^3} relate to transitions from regions A to regions B, lines l_p from B to C and lines l_f from C to D. As can be seen from the diagram, transitions to region D (complete synchronization of a resonant cycle) can be achieved from regions A or B through lines l_f. Let us consider in detail the bifurcation phenomena that correspond to the above indicated bifurcation lines l_{T^3}, l_p, and l_f.

We start with region B, where a stable three-dimensional torus T^3 exists. Bifurcations of a three-dimensional torus can be better illustrated by using a double Poincaré section. To do that, the usual Poincaré section is first computed and then

14.3 Basic Bifurcations of Quasiperiodic Regimes When Synchronizing a...

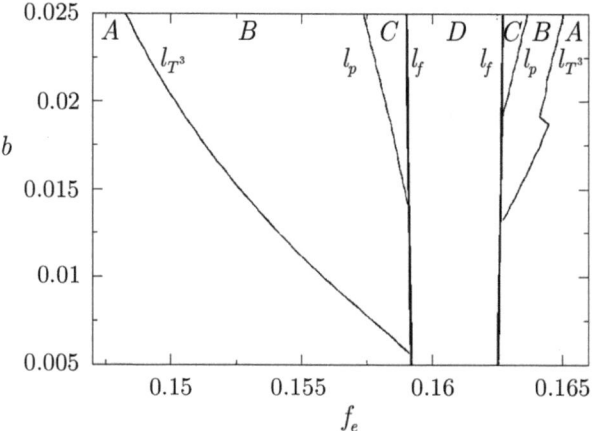

Fig. 14.3 Bifurcation diagram for the system (14.1) on the parameter plane (f_e, b) for the fixed values $\varepsilon = 0.1$, $\omega_1 = 1$, $\omega_2 = 1.0015$, and $\gamma = 0.02$

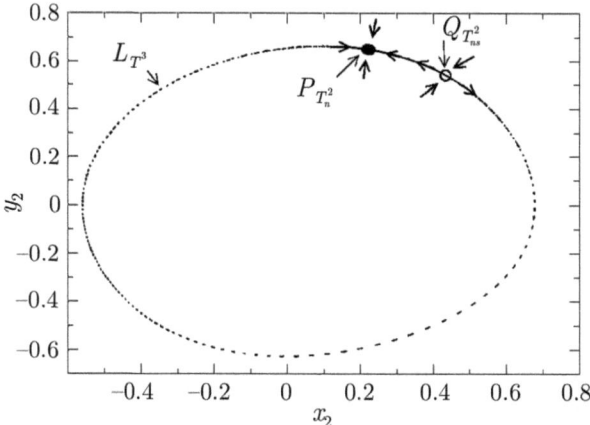

Fig. 14.4 The saddle-node bifurcation in the double Poincaré section when crossing the line l_{T^3} from region B to region A. L_{T^3} is an invariant curve, $P_{T_n^2}$ is a stable node, and $Q_{T_{ns}^2}$ is a saddle. The parameters are $f_e = 1.482$, $b = 0.025$, $\varepsilon = 0.1$, $\omega_1 = 1$, $\omega_2 = 1.0015$, and $\gamma = 0.02$

an additional secant plane is introduced to the resulting point sequence. Further, since the resulting points are unlikely to be close to the additional secant plane, solutions are linearly approximated in its vicinity. The usual Poincaré section of T^3 gives a two-dimensional torus T_{T^3}. The double Poincaré section represents an invariant closed curve L_{T^3}. A fixed point on this curve is the image of a resonant two-dimensional torus lying on T^3.

Let us explore the transition from region B to region A of the diagram in Fig. 14.3 when we cross the line l_{T^3}. Corresponding calculations are carried out by applying the double Poincaré section and the results are shown in Fig. 14.4. The curve L_{T^3}

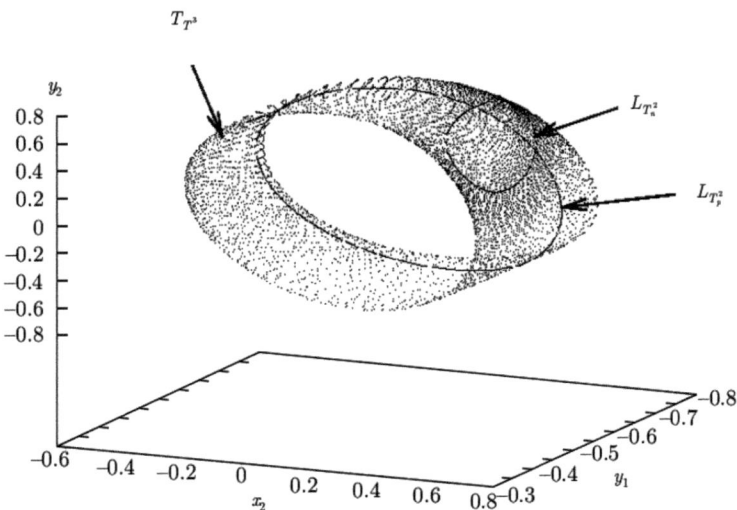

Fig. 14.5 Projections of Poincaré sections of the three-dimensional torus T_{T^3} ($f_e = 0.15$) and of the resonant two-dimensional tori $L_{T_n^2}$ ($f_e = 0.1482$) and $L_{T_p^2}$ ($f_e = 0.158$)

here is the image of T^3 and indicates the regime of T^3 in region B. When the bifurcation point (the intersection point of line l_{T^3} when we go from region B to region A) is reached, a fixed point of saddle-node type is born on the curve L_{T^3}. When the parameters are shifted to region A, the saddle node splits into a stable node and a saddle. The classical saddle-node bifurcation is realized in the double Poincaré section representation.

In the full phase space of the system (14.1), the picture presented in Fig. 14.4 corresponds to the birth (disappearance) of a pair of two-dimensional tori on T^3. One of them is stable (T_s^2) and the other is a saddle (T_{ns}^2).

We now study the bifurcation transition from region B to region C when we cross bifurcation lines l_p in the diagram shown in Fig. 14.3. Calculations have shown that the saddle-node bifurcation also occurs on the line l_p, and stable and saddle resonant[1] two-dimensional tori are also born in region C and lie on the three-dimensional torus T^3. Calculation results are illustrated in Fig. 14.5. The oscillatory mode in region B is represented by the three-dimensional torus T^3 whose Poincaré section is denoted by T_{T^3} in Fig. 14.5. A stable resonant two-dimensional torus in region C corresponds to the invariant closed curve indicated by $L_{T_p^2}$ in Fig. 14.5. The image of the stable resonant two-dimensional torus $L_{T_n^2}$ in region A is also shown in Fig. 14.5 for comparison. Saddle tori are not represented in Fig. 14.5.

---

[1] A resonant two-dimensional torus is understood here as a partial resonance on a three-dimensional torus, when two of three independent frequencies become equal.

14.3 Basic Bifurcations of Quasiperiodic Regimes When Synchronizing a...

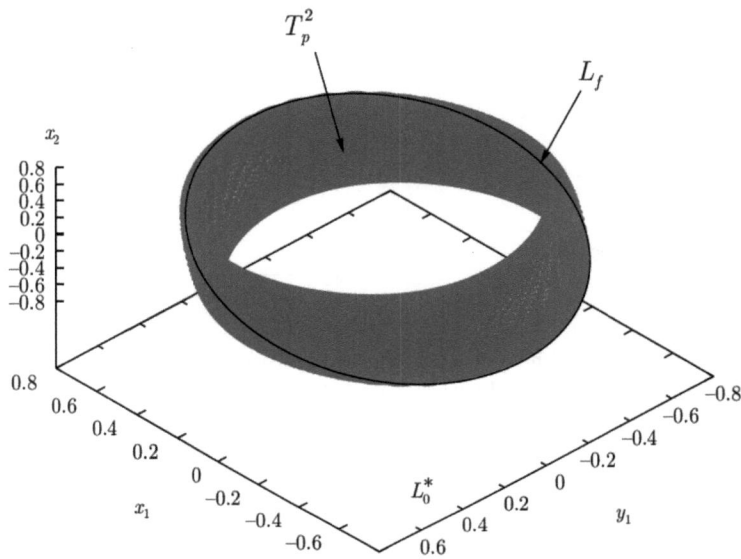

Fig. 14.6 Projections of phase portraits of the two-dimensional torus T_p^2 (*gray*) and of the resonant limit cycle L_f (*black*) on it for the frequency values $f_e = 0.1587$ (T_p^2) and $f_e = 0.1592$ (L_f)

The tori T_n^2 and T_p^2 are different since, from a physical standpoint, they correspond to different conditions of partial synchronization. We have $f_1 = f_2$ and $f_e \neq f_1$ in region A and $f_e = f_1$ and $f_1 \neq f_2$ in region C.

Finally, we consider the bifurcation transition from region C to region D when we cross the line l_f. The second frequency is then locked, i.e., $f_2 = f_e$, and the regime of complete synchronization $f_1 = f_2 = f_e$ is realized. Our studies have shown that l_f is a bifurcation line of the classical saddle-node bifurcation of resonant cycles which lie on the two-dimensional torus T_p^2. This torus exists in region C and represents a resonant structure on T^3. At the bifurcation point (on the line l_f), stable and saddle cycles are born on the torus T_p^2. When the line l_f is crossed in the direction of region D, there appears a stable periodic motion which corresponds to the regime of complete synchronization. This situation is illustrated in Fig. 14.6, which shows phase projections of the two-dimensional torus T_p^2 (in region C) and of the stable resonant limit cycle L_f on it (region D).

14.3.1 Peculiarities in the Synchronization of Resonant Limit Cycles

The study of synchronization effects described above has shown that the considered resonance with winding number $\Theta = 1 : 1$ is complicated enough to constitute

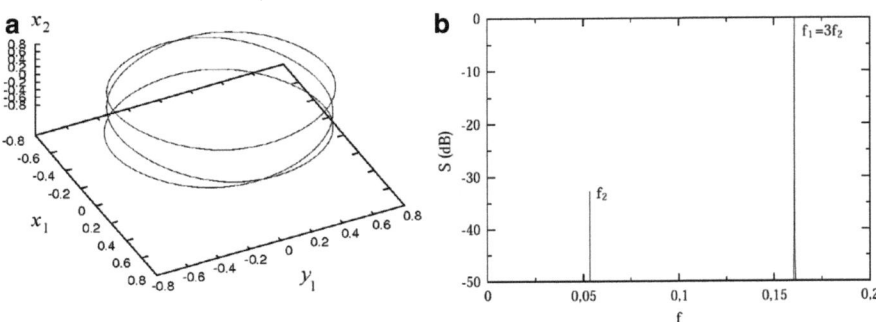

Fig. 14.7 The limit cycle in the system (12.6) for the resonance condition $\Theta = 1:3$ (**a**) and the corresponding power spectrum (**b**) for $\omega_1 = 1$, $\omega_2 = 0.328$, $\varepsilon = 0.1$, and $\gamma = 0.005$

the most general case as far as the theory of bifurcations is concerned. It is important to explore the effects of synchronization for different values of the winding number, which correspond to resonances $\Theta = m:n$, where $m, n = 1, 2, \ldots$. Besides, synchronization phenomena in the system (14.1) must depend on the coupling coefficient γ of the oscillators. This dependence is important for understanding the synchronization mechanisms of quasiperiodic oscillators in which the coupling parameter may be implicitly included in the dynamical model or is not an independent parameter due to constructional features of the system. In this connection, it is interesting to explore the peculiarities of the bifurcation properties of the system (14.1) for different values of the coupling γ.

Let us consider the regime of the resonant limit cycle in the system (12.6) with winding number $\Theta = 1:3$ and try to synchronize it by the external periodic signal (14.1). The effect of mutual synchronization is realized in the autonomous system (14.1) ($b = 0$) for $\omega_1 = 1$, and a stable resonant cycle with winding number $\Theta = 1:3$ exists on a two-dimensional torus. The phase portrait of this cycle and the power spectrum are shown in Fig. 14.7. Now we introduce the external force in the system (14.1) and vary the external frequency f_e near the frequency f_2.

Results of the external action are presented in Fig. 14.8. The main difference between the results obtained here and the case of $\Theta = 1:1$ (see Fig. 14.1) is that the effect of complete synchronization is no longer realized. The second frequency is locked ($f_e = f_2$) but f_1 is not changed by the external force. As in the case of the 1 : 1 resonance, provided that f_e is far from f_2, there exists a region A in which there is a resonant two-dimensional torus $T^2_{n1:3}$ on the surface of the three-dimensional torus T^3. There is also a region B where T^3 exists as the 1 : 3 resonance is destroyed. Further, the transition from regions B to C occurs where a resonant two-dimensional torus $T^2_{p1:3}$ emerges, this corresponding to the regime of partial synchronization $f_e = f_2$, $f_1 \neq 3f_2$.

There is no region D in this case. A resonant cycle can be synchronized completely by applying an optional external signal with frequency close to f_1.

14.3 Basic Bifurcations of Quasiperiodic Regimes When Synchronizing a... 253

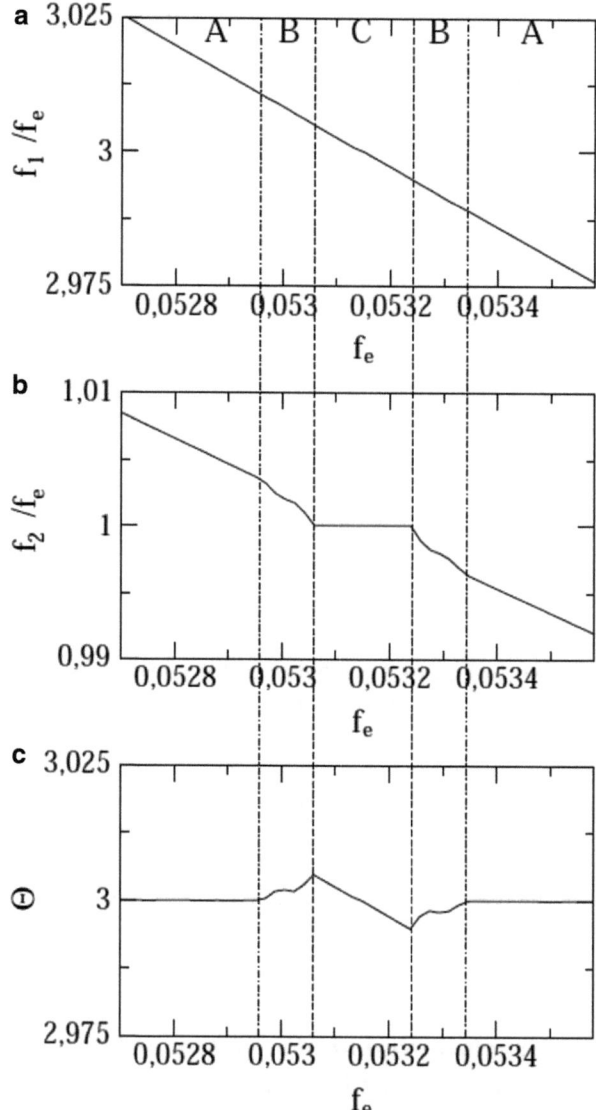

Fig. 14.8 Dependences of the frequency ratios (**a**) f_1/f_e and (**b**) f_2/f_e and (**c**) of the winding number Θ on the frequency of the external force f_e for $b = 0.005$ in the case of the 1 : 3 resonance

Figure 14.9 shows projections of Poincaré sections of two-dimensional tori corresponding to regions A ($T^2_{n1:3}$) and C ($T^2_{p1:3}$) (Fig. 14.8) in the form of cycles L. Both cycles lie on the three-dimensional torus T^3 represented in the Poincaré section in the form of a two-dimensional torus T_{T^3}.

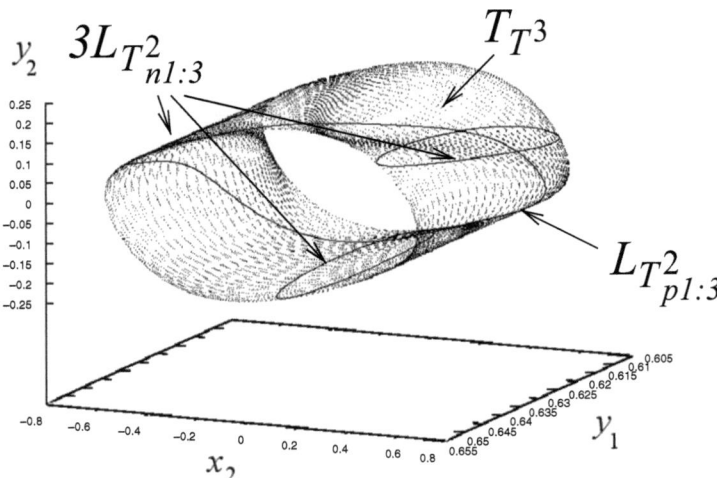

Fig. 14.9 Projections of Poincaré sections of two-dimensional tori in the form of cycles: $3L_{T^2_{n1:3}}$ in region A (Fig. 14.8), $f_e = 0.0527$, and $L_{T^2_{p1:3}}$ in region C (Fig. 14.8), $f_e = 0.0531$, both lying on the surface of T_{T^3}

The results shown in Fig. 14.8 are obtained for a relatively small value of the coupling coefficient, viz., $\gamma = 0.005$. It is interesting to find out how an increase in the degree of interrelation between the oscillators can influence the effects of synchronization. For this purpose, calculations have been carried out for $\gamma = 0.02$, and numerical data are presented in Fig. 14.10. As can be seen from Fig. 14.10, when the coupling increases, region D appears where complete synchronization of the resonance cycle on the torus is realized. Studies have shown that, for $\gamma > 0.02$, the width of region D increases and the synchronization picture is qualitatively similar to that for the 1 : 1 resonance (see Fig. 14.1).

14.3.2 Phase Synchronization of a System of Coupled van der Pol Oscillators by an External Harmonic Signal

As has been shown above, under certain conditions the synchronization effect can be analyzed in the framework of the phase approach. Now we explore external synchronization of self-sustained oscillations in a system of coupled van der Pol oscillators by using the phase dynamics approach. For this purpose we consider the differential equations of the system:

$$\ddot{x}_1 + \omega_1^2 x_1 = \left(\varepsilon - x_1^2\right)\dot{x}_1 + \gamma\left(\dot{x}_2 - \dot{x}_1\right) + b\cos(\omega_e t) ,$$
$$\ddot{x}_2 + \omega_2^2 x_2 = \left(\varepsilon - x_2^2\right)\dot{x}_2 + \gamma\left(\dot{x}_1 - \dot{x}_2\right) .$$
(14.2)

14.3 Basic Bifurcations of Quasiperiodic Regimes When Synchronizing a...

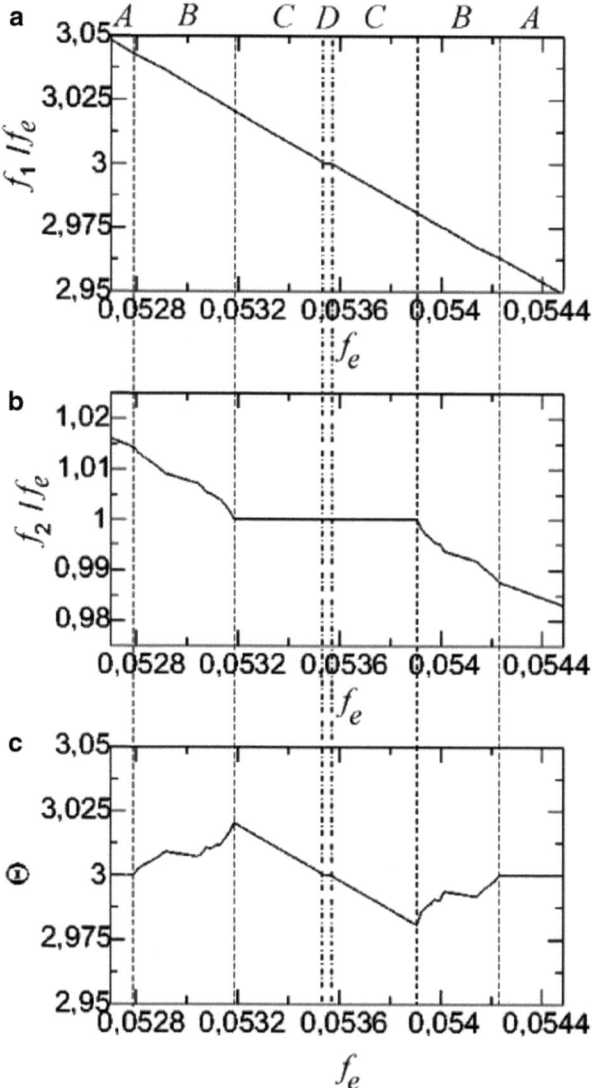

Fig. 14.10 Dependences of the frequency ratios (**a**) f_1/f_e and (**b**) f_2/f_e and (**c**) of the winding number Θ on the frequency of the external force f_e for $b = 0.005$ and $\gamma = 0.02$ in the case of the 1 : 3 resonance

Here ε is the excitation parameter, identical for the two oscillators, while the coupling function is $F(\dot{x}_1, \dot{x}_2) = \gamma(\dot{x}_{21} - \dot{x}_{12})$, with γ the coupling coefficient, and b and ω_e are the amplitude and frequency of the external signal. As can be seen from (14.2), the oscillators interact through the derivatives $\dot{x}_{1,2}$, and this corresponds to the resistive coupling in the experiment.

The solution of (14.2) is sought in the form $x_{1,2} = \rho_{1,2}(t)\cos\Psi_{1,2}(t)$, where $\Psi_{1,2}(t) = \omega_e t + \varphi_{1,2}(t)$. We assume $\rho_{1,2}(t)$ and $\varphi_{1,2}(t)$ to be slowly varying functions of time (in full accordance with the classical description of the van der Pol oscillator dynamics in the quasiperiodic approximation). This means that $\dot\rho_{1,2}(t) \ll \rho_{1,2}(t)$ and $\dot\varphi_{1,2}(t) \ll \varphi_{1,2}(t)$.

We rewrite Eqs. (14.2) in the form of a dynamical system

$$\begin{cases} \dot x_1 = y_1, \\ \dot y_1 = (\varepsilon - x_1^2) y_1 - \omega_1^2 x_1 + \gamma(y_2 - y_1) + b\cos(\omega_e t), \\ \dot x_2 = y_2, \\ \dot y_2 = (\varepsilon - x_2^2) y_2 - \omega_2^2 x_2 + \gamma(y_1 - y_2), \end{cases} \quad (14.3)$$

and, in accordance with what was said above, introduce the change of variables:

$$x_1 = \rho_1(t)\cos\left[\omega_e t + \varphi_1(t)\right], \quad (14.4)$$

$$x_2 = \rho_2(t)\cos\left[\omega_e t + \varphi_2(t)\right]. \quad (14.5)$$

Using the standard condition

$$\dot\rho_{1,2}\cos\left[\omega_e t + \varphi_{1,2}(t)\right] - \rho_{1,2}(t)\dot\varphi_{1,2}(t)\sin\left[\omega_e t + \varphi_{1,2}(t)\right] = 0, \quad (14.6)$$

and averaging over the period of the external force, we derive the first approximation equations for the amplitudes $\rho_{1,2}(t)$, viz.,

$$\begin{aligned} \dot\rho_1 &= \rho_1\left(\frac{\varepsilon}{2} - \frac{\rho_1^2}{8}\right) + \frac{\gamma}{2}\left[\rho_2\cos(\varphi_2 - \varphi_1) - \rho_1\right] - \frac{b}{2\omega_e}\sin\varphi_1, \\ \dot\rho_2 &= \rho_2\left(\frac{\varepsilon}{2} - \frac{\rho_2^2}{8}\right) + \frac{\gamma}{2}\left[\rho_1\cos(\varphi_2 - \varphi_1) - \rho_2\right], \end{aligned} \quad (14.7)$$

and for the phases $\varphi_{1,2}(t)$ of the oscillations in the first and second oscillators, viz.,

$$\begin{aligned} \dot\varphi_1 &= \Delta_1 + \frac{\gamma}{2}\frac{\rho_2}{\rho_1}\sin(\varphi_2 - \varphi_1) - \frac{b}{2\rho_1\omega_e}\cos\varphi_1, \\ \dot\varphi_2 &= \Delta_2 - \frac{\gamma}{2}\frac{\rho_1}{\rho_2}\sin(\varphi_2 - \varphi_1), \end{aligned} \quad (14.8)$$

where $\Delta_{1,2} = (\omega_{1,2}^2 - \omega_e^2)/2\omega_e$.

We assume that the basic frequency detuning of the oscillators is very small, i.e., $\omega_2 - \omega_1 = \delta \ll 1$, and the external signal frequency is close to the values of the basic frequencies, i.e., $\Delta_{1,2} \approx \omega_{1,2} - \omega_e$. In this approach, the following equality is valid:

14.3 Basic Bifurcations of Quasiperiodic Regimes When Synchronizing a...

$$\Delta_2 = \Delta_1 + \delta. \tag{14.9}$$

From the amplitude equations (14.7), it follows that, for a weak coupling $\gamma \ll 1$ and a small external amplitude $b \ll 1$, we may take it that

$$\rho_1 = \rho_2 = 2\sqrt{\varepsilon}. \tag{14.10}$$

The amplitudes of the limit cycles in the oscillators will be constant and equal to each other. In this approach, the system dynamics can be analyzed using the phase equations (14.8), whence

$$\dot{\varphi}_1 = \Delta_1 + g \sin(\varphi_2 - \varphi_1) - \frac{C}{1 - \Delta_1} \cos \varphi_1,$$
$$\dot{\varphi}_2 = \Delta_1 + \delta - g \sin(\varphi_2 - \varphi_1), \tag{14.11}$$

where $g = \gamma/2$, $C = b/4\sqrt{\varepsilon}$, $\Delta_1 = (\omega_1^2 - \omega_e^2)/2\omega_e \approx \omega_1 - \omega_e$, $\delta = \omega_2 - \omega_1$, and $\omega_1 = 1$.

Equations (14.11) describe the phase dynamics of the original system (14.2) and enable us to conduct the bifurcation analysis of phase synchronization effects when the amplitude C and frequency ω_e of the external force are varied for chosen values of the coupling g and the frequency detuning δ.

14.3.3 Bifurcations of Equilibrium States

The coordinates of equilibrium states of the system (14.11) can be found from the conditions $\dot{\varphi}_1 = \dot{\varphi}_2 = 0$, i.e., from the equations

$$\Delta_1 + g \sin(\varphi_2 - \varphi_1) - \frac{C}{1 - \Delta_1} \cos \varphi_1 = 0,$$
$$\Delta_1 + \delta - g \sin(\varphi_2 - \varphi_1) = 0. \tag{14.12}$$

Conditions for the existence of real solutions have the form

$$\left|\frac{\Delta_1 + \delta}{g}\right| \leq 1, \quad \left|\frac{(1 - \Delta_1)(2\Delta_1 + \delta)}{C}\right| \leq 1. \tag{14.13}$$

Under these conditions, Eqs. (14.12) have four solutions that are defined by the following four pairs of coordinates of fixed points:

$$\begin{cases} \varphi_1^{(1)} = \arccos \dfrac{(1-\Delta_1)(2\Delta_1+\delta)}{C}, \\ \varphi_2^{(1)} = \varphi_1^{(1)} + \arcsin \dfrac{\Delta_1+\delta}{g}, \end{cases}$$

$$\begin{cases} \varphi_1^{(2)} = -\arccos \dfrac{(1-\Delta_1)(2\Delta_1+\delta)}{C}, \\ \varphi_2^{(2)} = \varphi_1^{(2)} + \arcsin \dfrac{\Delta_1+\delta}{g}, \end{cases}$$

(14.14)

$$\begin{cases} \varphi_1^{(3)} = \arccos \dfrac{(1-\Delta_1)(2\Delta_1+\delta)}{C}, \\ \varphi_2^{(3)} = \varphi_1^{(3)} - \arcsin \dfrac{\Delta_1+\delta}{g} + \pi, \end{cases}$$

$$\begin{cases} \varphi_1^{(4)} = -\arccos \dfrac{(1-\Delta_1)(2\Delta_1+\delta)}{C}, \\ \varphi_2^{(4)} = \varphi_1^{(4)} - \arcsin \dfrac{\Delta_1+\delta}{g} + \pi. \end{cases}$$

The system (14.11) is characterized by four equilibrium states whose stability and bifurcations must be explored. Solving the characteristic equation of the linearization matrix of the system (14.11), we obtain the following eigenvalues for an equilibrium:

$$\varepsilon_{1,2} = \frac{1}{2}\left(-\left[2g\cos(\varphi_2-\varphi_1) - \frac{C}{1-\Delta_1}\sin\varphi_1\right]\right. \tag{14.15}$$

$$\left. \pm \left\{\left[2g\cos(\varphi_2-\varphi_1) - \frac{C}{1-\Delta_1}\sin\varphi_1\right]^2 + 4g\frac{C}{1-\Delta_1}\cos(\varphi_2-\varphi_1)\sin\varphi_1\right\}^{1/2}\right).$$

The bifurcation diagram for the system (14.11) is drawn in Fig. 14.11. Four equilibrium states exist in the region D bounded by bifurcation lines L_{T_1} and L_{T_2}. One of them is a stable node, two of them are saddles, and one of them is an unstable node (repeller). When exiting region D by crossing lines L_{T_1} and L_{T_2}, saddle-node bifurcations of the equilibria occur. Let us discuss this in more detail.

Figure 14.12 illustrates phase portraits of the equilibria on the 2π-periodic phase plane (φ_1, φ_2) before crossing the lines L_{T_1} (Fig. 14.12a) and on the bifurcation line L_{T_1} (Fig. 14.12b). As can be seen from Fig. 14.12b, on the line L_{T_1}, the equilibria merge in pairs in structurally unstable equilibria of saddle-node type and then disappear. At the moment of bifurcation (on the line L_{T_1}), the eigenvalues of the corresponding pairs of fixed points $\lambda_1^{(1)}$, $\lambda_1^{(2)}$ and $\lambda_2^{(3)}$, $\lambda_2^{(4)}$ vanish, indicating a saddle-node bifurcation. A qualitatively equivalent picture is realized when crossing bifurcation lines L_{T_2}. This is illustrated in Fig. 14.13.

14.3 Basic Bifurcations of Quasiperiodic Regimes When Synchronizing a...

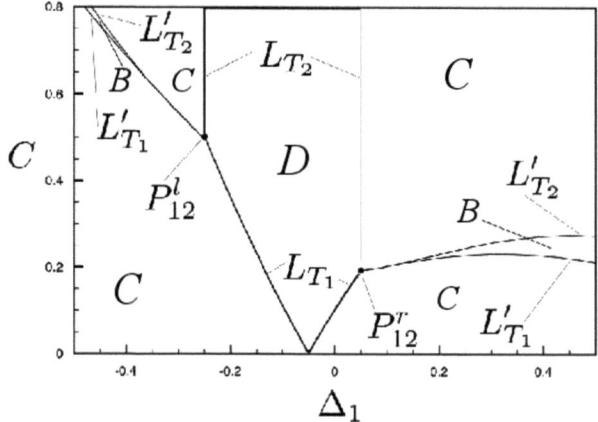

Fig. 14.11 Codimension 1 bifurcation lines of the system (14.11) on the parameter plane (Δ_1, C) for $g = 0.15$, $\delta = 0.1$. L_{T_1} are lines of saddle-node bifurcations of equilibria $\varphi^{(1)} \leftrightarrow \varphi^{(2)}$ and $\varphi^{(3)} \leftrightarrow \varphi^{(4)}$, while L_{T_2} are lines of saddle-node bifurcations of equilibria $\varphi^{(1)} \leftrightarrow \varphi^{(3)}$ and $\varphi^{(2)} \leftrightarrow \varphi^{(4)}$, and L'_{T_1} and L'_{T_2} are lines of tangent bifurcations of stable and saddle invariant closed curves

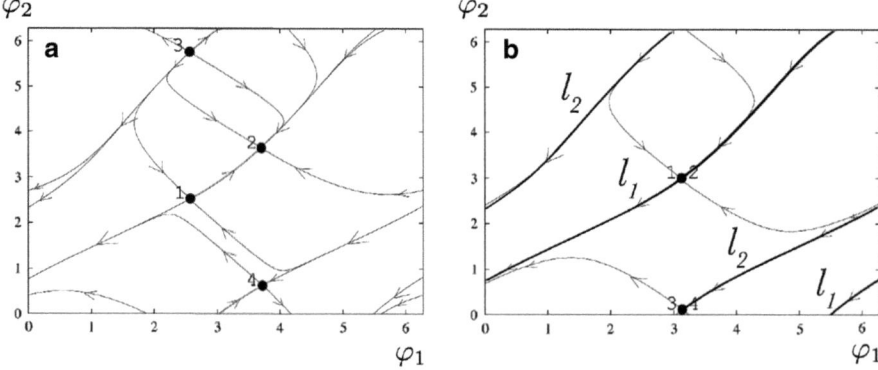

Fig. 14.12 Saddle-node bifurcations of equilibria when crossing the curve L_{T_1} on the parameter plane (Fig. 14.11). (**a**) Equilibria and their invariant manifolds before the bifurcation. (**b**) A nonrobust 'saddle-node' state at the moment of bifurcation. $\lambda_i^{(j)}$ is the ith eigenvalue of the jth equilibrium. The parameters are $g = 0.15$, $\delta = 0.1$, and $C = 0.15$. For the state in (**a**), $\lambda_1^{(1)} = 0.041$, $\lambda_2^{(1)} = -0.268$, $\lambda_1^{(2)} = -0.032$, $\lambda_2^{(2)} = -0.34$, $\lambda_1^{(3)} = 0.34$, $\lambda_2^{(3)} = 0.032$, $\lambda_1^{(4)} = 0.268$, $\lambda_2^{(4)} = -0.041$, $\Delta_1 = -0.10713$. For the state in (**b**), $\lambda_1^{(1)} = 0$, $\lambda_2^{(1)} = -0.297$, $\lambda_1^{(2)} = 0$, $\lambda_2^{(2)} = -0.299$, $\lambda_1^{(3)} = 0.299$, $\lambda_2^{(3)} = 0$, $\lambda_1^{(4)} = 0.297$, $\lambda_2^{(4)} = -0$, $\Delta_1 = -0.11713$

Let us consider the differences between Figs. 14.12 and 14.13. One is that in the first case the saddle-node bifurcation is realized for points 1, 2 and 3, 4, while in the second case it is realized for points 3, 1 and 4, 2. As can be seen from Figs. 14.12 and 14.13, the equilibria disappear through saddle-node bifurcations,

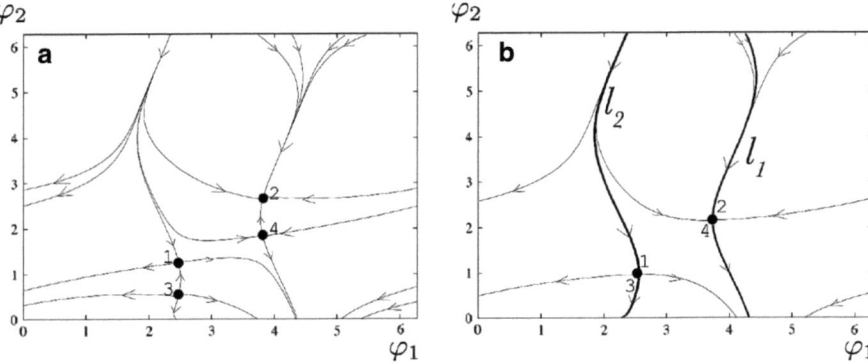

Fig. 14.13 Saddle-node bifurcations of equilibria when crossing the curve L_{T_2} on the parameter plane (Fig. 14.11). (**a**) Equilibria and their invariant manifolds before the bifurcation. (**b**) A nonrobust 'saddle-node' state at the moment of bifurcation. $\lambda_i^{(j)}$ is the ith eigenvalue of the jth equilibrium. The parameters are $g = 0.15$, $\delta = 0.1$, and $C = 0.6$. For the state in (**a**), $\lambda_1^{(1)} = 0.255$, $\lambda_2^{(1)} = -0.064$, $\lambda_1^{(2)} = -0.045$, $\lambda_2^{(2)} = -0.363$, $\lambda_1^{(3)} = 0.363$, $\lambda_2^{(3)} = 0.045$, $\lambda_1^{(4)} = 0.064$, $\lambda_2^{(4)} = -0.255$, $\Delta_1 = -0.2399$. For the state in (**b**), $\lambda_1^{(1)} = 0.265$, $\lambda_2^{(1)} = 0$, $\lambda_1^{(2)} = 0$, $\lambda_2^{(2)} = 0.266$, $\lambda_1^{(3)} = 0.266$, $\lambda_2^{(3)} = 0$, $\lambda_1^{(4)} = 0$, $\lambda_2^{(4)} = -0.265$, $\Delta_1 = -0.249999$

and appropriate pairs of stable (l_1) and unstable (l_2) limit cycles[2] are born outside region D (in regions C).

In the full system of equations (14.2), region D shown in Fig. 14.11, where there is a stable fixed point, corresponds to the existence of a stable limit cycle of the complete synchronization regime. At the same time, both frequencies of the interacting oscillators[3] ω_1' and ω_2' are locked by the external signal and the equality $\omega_1' = \omega_2' = \omega_e$ holds. Outside region D, there are beats, i.e., two-frequency oscillations whose image is the stable invariant curve l_1.

14.3.4 Bifurcations of Invariant Curves

Bifurcations of equilibrium states when going from region D into regions C are studied analytically using (14.14) and (14.15). Bifurcations of the invariant curves l_1, l_2 are analyzed numerically. Calculation results are presented in Fig. 14.14 for the invariant curves shown in Fig. 14.12b.

---

[2]From a qualitative point of view, these limit cycles are the images of invariant curves in the Poincaré section of the original system (14.2). In this connection, in what follows we will refer to these cycles as stable and unstable invariant curves.

[3]One must distinguish between the natural frequencies ω_1 and ω_2 of the autonomous oscillators, which are defined as the parameters of the system, and the frequencies ω_1' and ω_2' of the interacting oscillators.

14.3 Basic Bifurcations of Quasiperiodic Regimes When Synchronizing a... 261

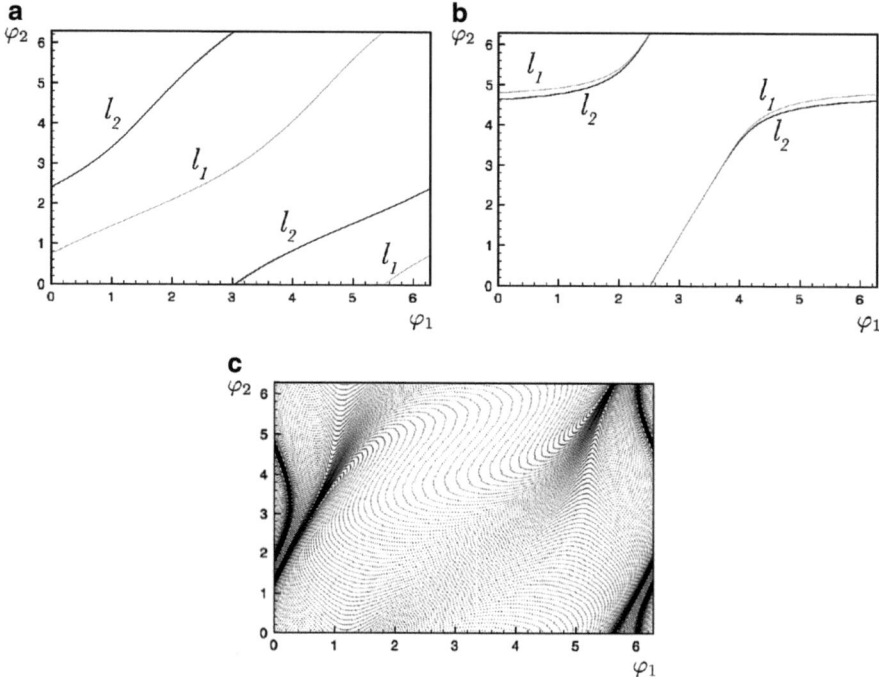

Fig. 14.14 Transformations of the phase portrait as a result of the saddle-node bifurcation of invariant curves (transition from region C to region B in the diagram of Fig. 14.11) for (**a**) $\Delta_1 = -0.12$, $C = 0.15$, (**b**) $\Delta_1 = -0.3418$, $C = 0.610235$, and (**c**) $\Delta_1 = -0.4064$, $C = 0.7$. The other parameters are $g = 0.15$, $\delta = 0.1$

The following picture is observed when we move from region C to region B in the parameter space of the diagram in Fig. 14.11. The invariant curves l_1 and l_2 approach each other, merge in a single saddle-node invariant curve on the bifurcation lines L'_{T_1}, and disappear when they cross them. As a result, trajectories cover the entire phase plane everywhere densely in regions B. This is illustrated in Fig. 14.14c. Thus, a bifurcation occurs which may be called a saddle-node bifurcation of invariant curves, by analogy with the saddle-node bifurcation of equilibrium states. A similar bifurcation is realized for the invariant curves, as shown in Fig. 14.13b, when we cross the lines L_{T_2} from regions C to region B. Regarding the dynamics of the original differential system (14.2), the above bifurcation corresponds to the transition from the two-frequency oscillation mode to the regime of three-frequency oscillations when the three frequencies of the system are not multiples of one another and are not equal to one another ($\omega'_1 \neq \omega'_2 \neq \omega_e$).

The saddle-node bifurcation of invariant curves can be conveniently illustrated as follows. Introduce mean frequencies of phase oscillators (14.11)[4] as follows:

---

[4] They represent the beat frequencies of the self-sustained oscillators in (14.2).

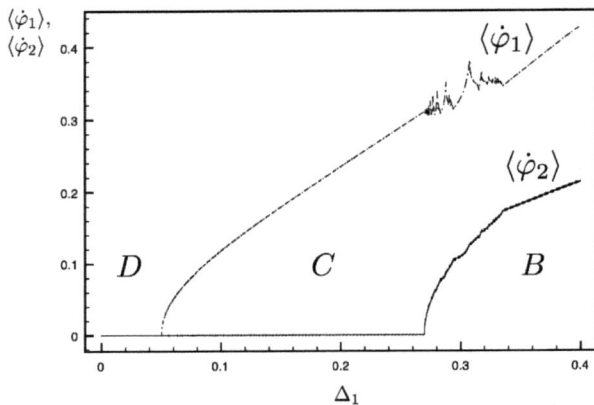

Fig. 14.15 Frequencies $\Omega_{1,2} = \langle \dot{\varphi}_{1,2} \rangle$ calculated from (14.11) as functions of the parameter Δ_1 for $C = 0.25$, $\delta = 0.1$, and $g = 0.15$

$$\Omega_{1,2} = \langle \dot{\varphi}_{1,2} \rangle = \lim_{T \to \infty} \frac{1}{T} \int_0^T \dot{\varphi}_{1,2}(t) dt . \tag{14.16}$$

Using the system (14.11), we calculate frequencies $\Omega_{1,2}$ depending on the parameters. We fix the parameter values $C = 0.25$, $\delta = 0.1$, $g = 0.15$ and move on the plane of the bifurcation diagram in Fig. 14.11 by changing the parameter Δ_1. As Δ_1 increases, we cross regions D, C, then B of the bifurcation diagram. Calculation results are shown in Fig. 14.15.

In region D ($\Delta_1 \leq 0.05$), we have $\Omega_1 = \Omega_2 = 0$, which corresponds to a stable fixed point in the synchronization region D. When we cross the line L_{T_2} (the transition from region D to regions C), there is a bifurcation giving rise to the birth of invariant curves l_1 and l_2. This bifurcation is indicated in Fig. 14.15 by the appearance of the nonzero frequency $\Omega_2 \neq 0$. At the same time, the other frequency is $\Omega_1 = 0$. Further, when we intersect the bifurcation line L'_{T_2} (the transition to region B), a saddle-node bifurcation of invariant curves l_1 and l_2 occurs, and this results in the birth of a two-dimensional torus. In Fig. 14.15, this bifurcation corresponds to the appearance of a second nonzero frequency Ω_1 for $\Delta_1 = 0.27$. Thus, in region B of the diagram in Fig. 14.11, there are oscillations with two independent frequencies $\Omega_1 \neq 0$ and $\Omega_2 \neq 0$, which arise as a result of the saddle-node bifurcation of invariant curves l_1 and l_2.

We note the following important result. The existence of four equilibria in the system (14.11) follows from the fact that this system describes the phase dynamics of two interacting oscillators (14.2). The pairwise merging of fixed points 1, 2 and 2, 4 (Fig. 14.13) due to the saddle-node bifurcation gives rise to two invariant curves (stable l_1 and unstable l_2). As a result, a saddle-node bifurcation of invariant curves becomes possible. This represents the next level of complexity in bifurcation type after the classical saddle-node bifurcation of equilibrium states.

The above bifurcation analysis of the dynamics of the phase approximation system (14.11) has shown the following. If there is a frequency detuning $\delta \neq 0$ and a mismatch between the external frequency and the first oscillator frequency $\Delta_1 \neq 0$, the system can generate ergodic quasiperiodic oscillations with two independent frequencies (in region B of the diagram in Fig. 14.11). Their image on the phase portrait is a two-dimensional torus. Going from region B to regions C leads to the appearance of a pair of invariant curves, stable l_1 and saddle l_2. Further, the transition from regions C to D is accompanied by the classical saddle-node bifurcation that results in the emergence of four fixed points in region D, one of which is stable.

The bifurcation transitions described here lead to the following rebuilding of oscillatory modes in the original differential system (14.2). In region B, the system (14.2) exhibits quasiperiodic oscillations with three independent frequencies ω_e, ω_1', and ω_2'. The transition from regions B to C is characterized by the emergence of a two-dimensional torus as a partial resonance on a three-dimensional torus. The partial resonance corresponds to the locking of one of the partial frequencies of the system (14.2). The following cases are possible here: $\omega_e = \omega_1'$ and $\omega_2' \neq \omega_1'$, or $\omega_e = \omega_2'$ and $\omega_1' \neq \omega_2'$. Partial synchronization takes place. Stable periodic motions are observed when we move from regions C to region D. A stable limit cycle is born and lies on the surface of the two-dimensional torus. This cycle indicates the regime of complete synchronization when both frequencies of the partial oscillators are locked by the external signal, i.e., $\omega_e = \omega_1' = \omega_2'$.

We note that the results of the above analysis of the phase approximation equations (14.11) are in full qualitative agreement with the data described in Sect. 13.3.

14.3.5 *Synchronization of Two-Frequency Oscillations in a Self-Sustained Quasiperiodic Oscillator*

The results presented above show that a resonant cycle on a two-dimensional torus cannot generally be synchronized by an external periodic signal. The main reason is that the external signal causes the resonance regime to destroy itself. Oscillations become quasiperiodic with three independent frequencies. If the parameters are varied, partial synchronization is observed when one of the basic frequencies of the system is locked by the external force. The effect of partial synchronization is realized through the saddle-node bifurcation giving rise to birth of resonant two-dimensional tori on a three-dimensional torus. After that, the second basic frequency can be locked in the classical way through the saddle-node bifurcation of limit cycles on a two-dimensional torus.

The effects observed here depend on the basic frequency detuning (on the winding number) and on the degree of interaction of the two oscillators. When the original basic frequencies differ significantly, a large detuning leads to a situation where only one basic frequency of the system is synchronized. For a very strong coupling between the two oscillators, the external force does not destroy their mutual synchronization and in fact synchronizes both oscillators simultaneously. In this case the forced synchronization of a resonant cycle on a torus is realized according to the classical scenario. These are the main conclusions. The question is whether they are likely to be representative of a general situation. To answer this question, we study synchronization of quasiperiodic oscillations using a two-frequency oscillator.

The oscillator is described by the following equations:

$$\begin{aligned} \dot{x} &= mx + y - x\phi - px^3 \,, \\ \dot{y} &= -x \,, \\ \dot{z} &= \phi \,, \\ \dot{\phi} &= -\gamma\phi + \gamma\Phi(x) - gz \,. \end{aligned} \qquad (14.17)$$

Using the bifurcation diagram of the system (14.17), we choose the control parameter values in such a way as to realize resonant oscillations with winding number $\Theta = 1 : 4$ and ergodic oscillations with a close but irrational value of the winding number. The parameters are set at $m = 0.096$, $\gamma = 0.2$, and $p = 0.001$, while the function $\Phi(x)$ is $\Phi(x) = I(x)x^2$. An ergodic two-dimensional torus is obtained for $g = 0.257$ and a resonant[5] torus with $\Theta = 1 : 4$ is realized for $g = 0.263$. The calculation results are shown in Fig. 14.16.

We now investigate the dynamics of the system (14.17) in the regimes indicated in Fig. 14.16 in the presence of external harmonic forcing:

$$\begin{aligned} \dot{x} &= mx + y - x\phi - px^3 + b\sin(2\pi f_e) \,, \\ \dot{y} &= -x \,, \\ \dot{z} &= \phi \,, \\ \dot{\phi} &= -\gamma\phi + \gamma\Phi(x) - gz \,. \end{aligned} \qquad (14.18)$$

We compute the power spectrum of $x(t)$ oscillations when the frequency of the external signal f_e is varied. Numerical results are presented in Fig. 14.17. As can be seen from Fig. 14.17a, frequency locking, i.e., synchronization, takes place in the range $f_e \approx 0.0381 \div 0.0385$. In the synchronization region (Fig. 14.17a), the modulation frequency f_1 is locked by the external force and the condition $f_1/f_e = 1$ holds. The results shown in Fig. 14.17b indicate that the frequency f_0 is not synchronized by the external signal. As can be seen from the plot, f_0 is almost unchanged, both inside and outside the synchronization region, i.e., it does not respond to the change in external frequency f_e.

---

[5] The term 'resonant torus' is used here in the classical sense and characterizes the regime of a stable limit cycle on a torus.

14.3 Basic Bifurcations of Quasiperiodic Regimes When Synchronizing a...

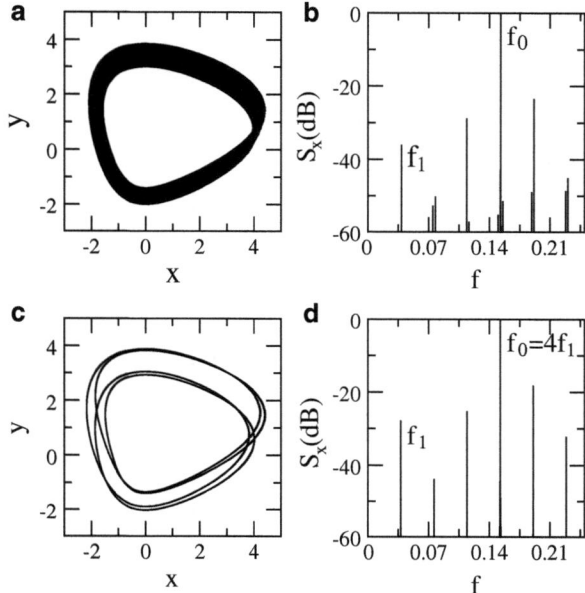

Fig. 14.16 Ergodic (**a, b**) and resonant (**c, d**) quasiperiodic oscillations with two frequencies (f_1 is the modulation frequency, f_0 is the carrier frequency). (**a**) Projections of an ergodic torus on the (x, y) plane. (**b**) Corresponding power spectrum of $x(t)$ oscillations. (**c**) Limit cycle on a torus for the $f_1 : f_0 = 1 : 4$ resonance. (**d**) Power spectrum of the resonant cycle on the torus

Completely analogous results are obtained when the oscillator in the regime of ergodic beats is subjected to the external signal (Fig. 14.17). This is quite understandable. As established when we studied the system of two coupled van der Pol oscillators (14.1), the external action first destroys the resonance regime. Indeed, in our case we observe that the 1 : 4 resonance is first destroyed and then one of the basic frequencies is synchronized (Fig. 14.17). If we destroy the resonance when we leave the synchronization region $\Theta = 1 : 4$ (Fig. 14.16a), then only one of the basic frequencies is synchronized in the presence of the external force. This is the frequency f_1 in the case shown in Fig. 14.17a, since the external frequency is close to it. If the frequency of the external signal approaches f_0, the effect of frequency locking $f_0 = f_e$ is observed, but f_1 remains unlocked.

The above results confirm the general nature of the conclusions we drew when we studied the system of two coupled van der Pol oscillators. Indeed, near the $\Theta = 1 : 4$ resonance, when f_1 is four times less than f_0 (the basic frequency detuning is large), only one of the two basic frequencies is locked by the external force. The two frequencies can never both be synchronized. The reason is not only the big difference between them, but also the fact that the oscillator (14.17) does not provide any way to increase the degree of interconnection between two

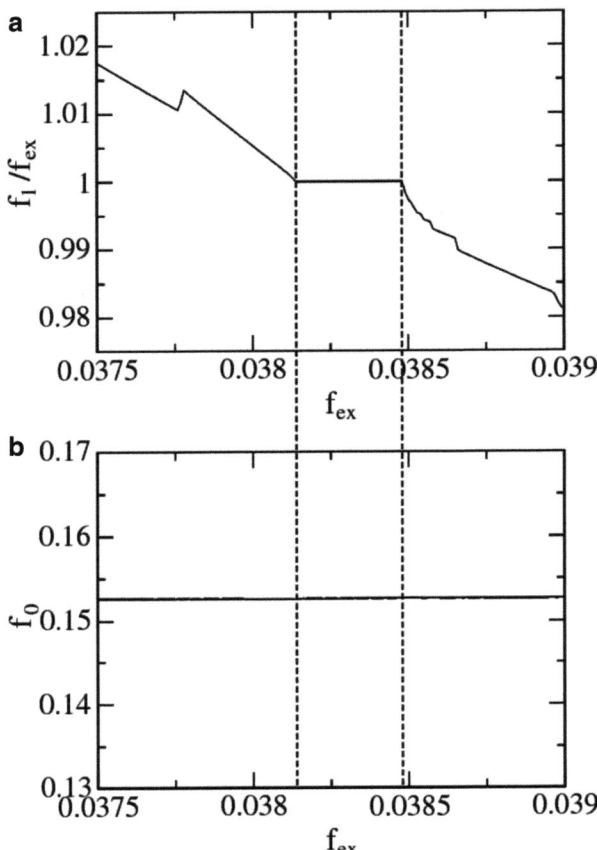

Fig. 14.17 Numerical simulation. (**a**) Frequency ratio f_1/f_e and (**b**) frequency f_0 as functions of the external frequency in the nonautonomous system (14.18) for $m = 0.096$, $g = 0.263$, $\gamma = 0.2$, $p = 0.001$, and $b = 0.01$

self-sustained oscillatory modes of the system. This interconnection is given by the intrinsic properties of the system (14.17). The above numerical results for the synchronization effect are confirmed by experimental data obtained for an electronic model of the oscillator (14.17).

The limit cycle synchronization corresponding to the 1 : 3 resonance has also been studied experimentally. The measurement results are shown in Fig. 14.18a, b and are similar to the calculations presented in Fig. 14.17. The effect of external synchronization of the frequency f_1 is illustrated in Fig. 14.18a. The fact that f_0 is independent of the external frequency f_e is confirmed in Fig. 14.18b.

The results described above confirm that, regardless of the resonance conditions, synchronization of two-frequency oscillations by an external harmonic signal is

14.3 Basic Bifurcations of Quasiperiodic Regimes When Synchronizing a...

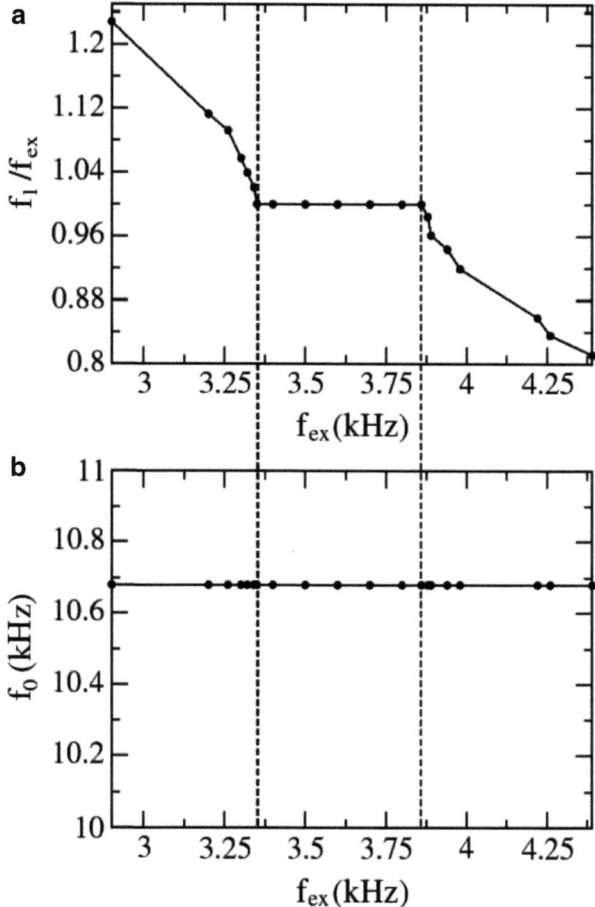

Fig. 14.18 Physical experiment. (a) Frequency ratio f_1/f_e and (b) frequency f_0 as functions of the external frequency

manifested by first locking one and then (possibly) the other basic frequency of an oscillator with quasiperiodic oscillations. The two basic frequencies can both be locked under certain special conditions, e.g., large coupling coefficient for partial subsystems, or basic frequencies almost equal. However, this cannot be achieved in the general case. It is natural to assume that the two basic frequencies can be locked by an external two-frequency signal if the winding number of the external oscillator is close to that of the synchronized oscillator.

Consider the case when one quasiperiodic oscillator unidirectionally influences a second quasiperiodic oscillator. The system of two unidirectionally interacting oscillators (14.17) is described by the following equations:

$$\begin{aligned}
\dot{x}_1 &= mx_1 + y_1 - x_1\phi_1 - px_1^3 + bx_2 , \\
\dot{y}_1 &= -x_1 , \\
\dot{z}_1 &= \phi_1 , \\
\dot{\phi}_1 &= -\gamma\phi_1 + \gamma\Phi(x_1) - g_1 z_1 , \\
\dot{x}_2 &= mx_2 + y_2 - x_2\phi_2 - px_2^3 , \\
\dot{y}_2 &= -x_2 , \\
\dot{z}_2 &= \phi_2 , \\
\dot{\phi}_2 &= -\gamma\phi_2 + \gamma\Phi(x_2) - g_2 z_2 .
\end{aligned} \qquad (14.19)$$

This system describes the case when the second oscillator affects the first one (the term bx_2 in the first equation). The intensity of this action is defined by the parameter b. We fix the other parameters as follows: $m = 0.06$, $d = 0.001$, $\gamma = 0.2$, $g_1 = 0.55$. For $g_1 = 0.55$, the first oscillator in its autonomous regime generates quasiperiodic oscillations with winding number $\Theta_1 = f_{11}/f_{01}$ (the first digit 0 corresponds to the carrier frequency, the first digit 1 indicates the modulation frequency, the second digit specifies the oscillator number). The winding number of the second oscillator Θ_2 is controlled by the parameter g_2. Without coupling for $g_1 \neq g_2$, the winding numbers will be different: $\Theta_1 \neq \Theta_2$. By analogy with synchronization of a limit cycle when the frequency detuning $\Delta f = f_1 - f_2$ is introduced, we consider the winding number detuning (mismatch) $\Delta\Theta \sim g_1 - g_2$. In this case the problem for the study of synchronization is to analyze oscillation regimes on the plane of the two control parameters b and g_2. The numerical solution of this task is illustrated in Fig. 14.19, which shows the structure of the synchronization regions.

The locking of the basic quasiperiodic oscillation frequencies, $f_{01} = f_{02}$, occurs inside the large synchronization 'beak' bounded by the bifurcation lines l_c. In other words, synchronization of the carrier frequencies takes place. However, the modulation frequencies are still different, i.e., $f_{11} \neq f_{12}$. Partial synchronization of the quasiperiodic oscillations is observed. The modulation frequencies and thus the winding numbers $\Theta_1 = \Theta_2$ are locked inside the region bounded by the bifurcation lines l_m. This effect is illustrated in Fig. 14.20. As can be seen from the figure, there is a finite region with winding number detuning Δg_2 in which $\Theta_2/\Theta_1 = 1$. The winding number of the second oscillator locks the winding number of the first oscillator. As in the case of the limit cycle, the width of winding number locking grows as the intensity of action b increases.

The described numerical experiments on the external synchronization of two oscillators with quasiperiodic oscillations have shown that the region of modulation frequency locking lies within the synchronization beak of the carrier frequencies. When values of the control parameters of the oscillators approach each other, the oscillations in the system contain first four independent frequencies and then three independent frequencies, after the carrier frequencies are locked. Finally, the modulation frequencies and hence also the winding numbers are locked.

14.3 Basic Bifurcations of Quasiperiodic Regimes When Synchronizing a...   269

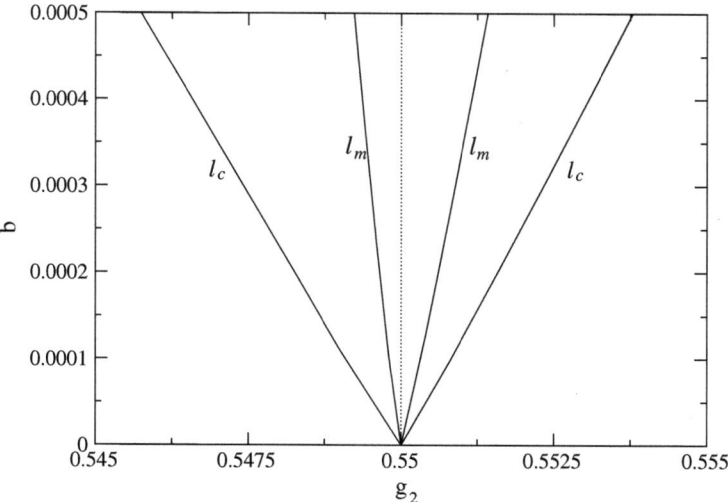

Fig. 14.19 Synchronization regions of the carrier (l_c) and of the envelope (l_m) for $m = 0.06$, $\gamma = 0.2$, $g_1 = 0.55$, and $p = 0.001$

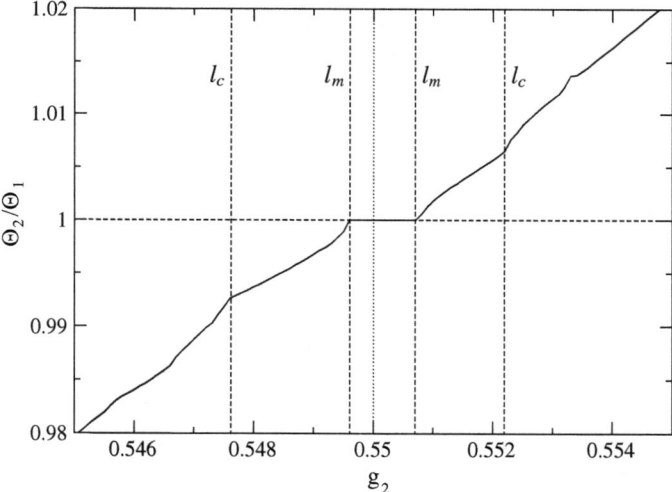

Fig. 14.20 Effect of winding number locking for $m = 0.06$, $\gamma = 0.2$, $g_1 = 0.55$, $p = 0.001$, and $b = 0.0003$

Let us study the influence of two-frequency oscillations in a physical experiment. Take the oscillator with quasiperiodic oscillations described by (14.17) to be a synchronized one. The external signal $F(t)$ represents amplitude-modulated oscillations which are generated by two oscillators of standard signals and a modulator:

$$F(t) = b\big[1 + K_{\mathrm{mod}} \sin(2\pi f_{1e} t)\big] \sin(2\pi f_{0e} t) ,$$

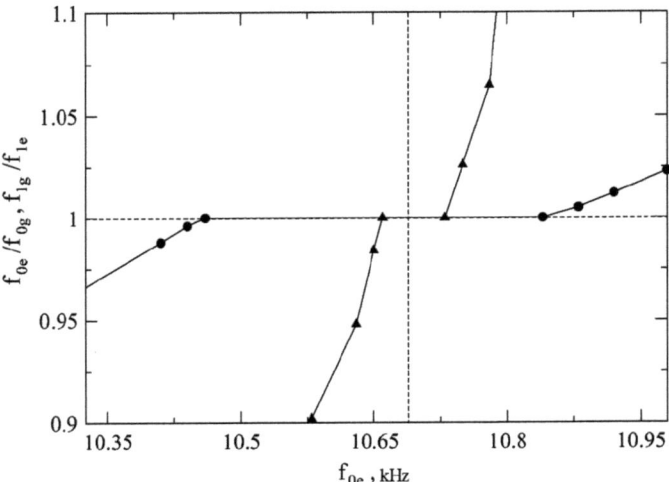

Fig. 14.21 Normalized modulation and carrier frequencies of the oscillator as functions of the carrier frequency of the external force

where b is the amplitude multiplier, K_{mod} is the modulation coefficient, f_{1e} is the modulation frequency, and f_{0e} is the carrier frequency. The parameters of the quasiperiodic oscillator are chosen as follows: $f_{1g} = 2.82\,\text{kHz}$, $f_{0g} = 10.69\,\text{kHz}$, and $K_{\text{mod}} \simeq 0.3$. The parameters of the external signal $F(t)$ are set at $f_{1e} = 3.06\,\text{kHz}$, $f_{0e} = 10.69\,\text{kHz}$, and $K_{\text{mod}} = 0.5$. In the experiments, the carrier frequency f_{0e} varies over the range $10.0 \leq f_{0e} \leq 11.0\,\text{kHz}$. The winding number of the external quasiperiodic signal $\Theta = f_{1e}/f_{0e}$ varies accordingly.

Normalized frequencies f_{0e}/f_{0g} (black circles) and f_{1g}/f_{1e} (triangles) are shown in Fig. 14.21 as functions of the carrier frequency f_{0e}. As shown by the plots, the carrier frequency of the oscillator (f_{0g}) is first locked by the carrier frequency of the external signal (f_{0e}). There is a range in the frequency interval $10.5 < f_{0e} < 10.85$ within which $f_{0g} = f_{0e}$. Then the modulation frequency is locked within a significantly smaller frequency range ($10.65 < f_{0e} < 10.77$). It is in this region that both frequencies of the oscillator f_{0g} and f_{1g} are locked by the external signal $F(t)$. Winding number locking is thus realized.

14.4 Summary

Analysis of the synchronization of two-frequency quasiperiodic oscillations enables us to draw the following conclusions. Quasiperiodic self-sustained oscillations with two independent frequencies can result from the interaction of two nonlinear active oscillators, each of which is characterized by its own independent basic frequency of self-sustained oscillations. Resonances relating to the so-called Arnold

14.4 Summary

tongues correspond to effects of mutual synchronization when basic frequencies are mutually locked (captured). Despite the fact that resonances are formally associated with regimes of stable periodic oscillations with a certain basic frequency, from the physical point of view, two basic modes of oscillation continue to exist in the system. If we attempt to synchronize a resonant limit cycle with an external harmonic signal, resonance regimes are destroyed and three-frequency oscillations arise in the system. In this case the effect of synchronization is observed for each of the existing modes independently. First, one of the two independent frequencies is locked, followed by the second one. Specific conditions of synchronization depend on the winding number (on the initial basic frequency detuning of the interacting oscillators) and on their degree of interrelation.

The results described in this chapter are relatively new and have not yet been described in the monographic literature. They can only be found in periodic publications.

Chapter 15
Synchronization of Chaotic Oscillations

15.1 Introduction

Periodic and quasiperiodic oscillations only represent some of the possible oscillatory regimes of dynamical systems with phase space dimension $N \geq 3$. In connection with the development of nonlinear dynamics and the theory of dynamical chaos, the question of synchronization of chaotic oscillations inevitably arises. Being the fundamental property of self-sustained oscillatory systems, synchronization must also be observed in one form or another in the regime of dynamical chaos. Chaotic self-sustained oscillations differ from periodic and quasiperiodic ones primarily in that they have a continuous spectrum resembling the spectrum of color noise. For this reason, it is impossible to introduce a strict period for chaotic oscillations and to unambiguously define their phase. In addition, if they are allocated in the power spectrum of chaotic oscillations on the background of a continuous component, spectral lines have a finite width.

Several definitions of chaos synchronization have been put forward. One of the first understands synchronization of chaos to involve establishing a periodic regime under the influence of an external harmonic force on the system in the mode of chaotic self-sustained oscillations. In this picture, a transition from chaotic to regular oscillations is observed for a sufficiently high intensity of the external influence, i.e., there is a certain threshold for synchronization.

Another type of chaos synchronization takes place when two similar chaotic systems interact. As the coupling coefficient increases, oscillations in the two interacting systems become completely identical. This means that the time series of the relevant dynamical variables of the two subsystems exactly repeat each other. This kind of synchronization is usually called *complete synchronization*.

Studies have shown that the classical ideas of synchronization as effects of frequency locking or frequency suppression enable us to generalize this phenomenon to the case of chaotic oscillations. In our opinion, this approach to the problem of chaos synchronization is fully consistent. The appropriate type of synchronization is called *phase–frequency synchronization* of chaos. It can be observed only for self-sustained oscillators in the regime of a so-called spiral attractor.

Besides the aforementioned types of chaos synchronization, some other effects of partial synchronization are also known. They occur when nonidentical chaotic systems interact and include *lag synchronization* and *generalized synchronization*. Lag synchronization is observed when there is strong interaction of one-type chaotic systems with a minor parameter mismatch. It is quite close to complete synchronization in its characteristics: the time series of the relevant dynamical variables of the two systems repeat each other with some time delay. Generalized synchronization of chaos implies that a functional relationship exists between the instantaneous states of interacting chaotic systems. In this chapter we shall focus mainly on the effects of phase–frequency and complete synchronization.

15.2 Phase–Frequency Synchronization of Chaotic Self-Sustained Oscillations

The regime of a spiral attractor can be called phase-coherent chaos. Indeed, in this case chaotic self-sustained oscillations are characterized by the presence of a pronounced spectral line at a certain basic frequency ω_0 which corresponds to the mean rotation frequency of phase trajectories around an equilibrium state in a suitably chosen projection. The basic frequency of chaotic self-sustained oscillations, like the frequency of quasiperiodic self-sustained oscillations, can be locked or suppressed when oscillators interact. Besides, it is possible to introduce the instantaneous phase of chaotic self-sustained oscillations in a certain way and to observe the effect of phase locking. In these terms, synchronization of chaos can be encountered in a wide class of dynamical systems which can exhibit a spiral attractor mode. Both mutual and forced synchronization of phase-coherence chaos can be realized, including the case of periodic influence. Although a spiral attractor is a particular case of a chaotic attractor, it can be observed in many dynamical systems which have a saddle-focus separatrix loop. Therefore, phase–frequency synchronization of chaos is not something exceptional, but represents a widespread phenomenon.

We consider the regime of a spiral attractor in the well-known dynamical Rössler system:

$$\dot{x} = -y - z, \qquad \dot{y} = x + \alpha y, \qquad \dot{z} = \beta + z(x - \mu). \qquad (15.1)$$

A phase trajectory projection on the plane of variables (x, y) and the spectral power density $S_x(\omega)$ of $x(t)$ oscillations are shown in Fig. 15.1. As can be seen from the figure, the trajectory rotates in its projection around the origin of coordinates (Fig. 15.1a), while the spectrum $S_x(\omega)$ has a fairly narrow peak at a certain frequency $\omega = \omega_0$ and its second harmonic on the background of the continuous pedestal (Fig. 15.1b).

15.2 Phase–Frequency Synchronization of Chaotic Self-Sustained Oscillations

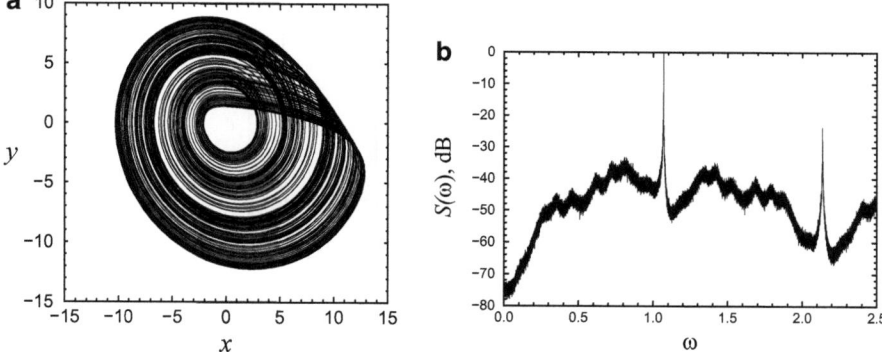

Fig. 15.1 (a) Projection of the spiral attractor on the plane of variables (x, y) and (b) the normalized power spectrum $S_x(\omega)$ of $x(t)$ oscillations in the system (15.1) for $\alpha = \beta = 0.2$ and $\mu = 6.5$

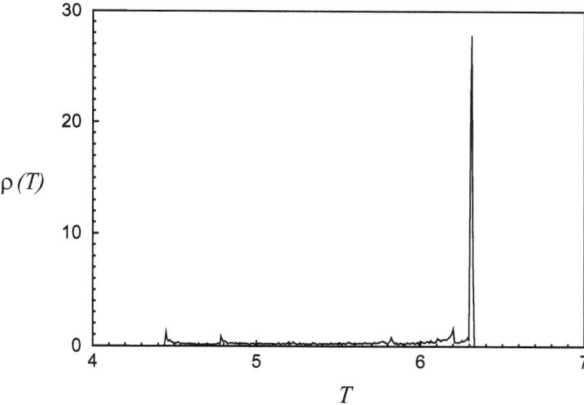

Fig. 15.2 Distribution of time intervals between successive intersections of the secant plane $y = 0$ by the phase trajectory in one direction, calculated for the Rössler attractor presented in Fig. 15.1a

Spiral attractors of the type shown in Fig. 15.1a arise through a sequence of period-doubling bifurcations of a limit cycle. The original stable limit cycle becomes unstable (saddle), but trajectories on the spiral attractor rotate near this saddle cycle with a high degree of regularity.

If we introduce a secant surface into the phase space, one which a trajectory intersects everywhere transversally (for example, the plane $y = 0$), and if we calculate a sequence of time intervals T_i between intersections of this surface by the trajectory in the same direction, then we observe the following result. The distribution of time intervals $p(T)$ shown in Fig. 15.2 has a pronounced maximum at $T = T_0$, which indicates a nearly regular rotation with an almost constant period. For this reason, this type of chaos is called phase-coherent. A consequence of the regular rotation of phase trajectories is the presence of a narrow spectral line with a

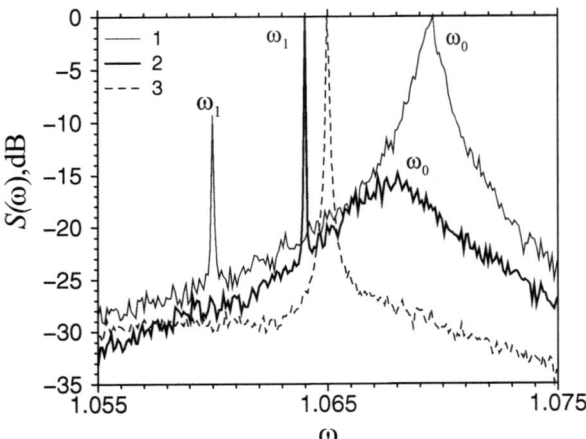

Fig. 15.3 Locking of the basic frequency ω_0 in the system (15.2). Normalized power spectrum of $x(t)$ oscillations for $\omega_1 = 1.06$ (*curve 1*), $\omega_1 == 1.064$ (*curve 2*), and $\omega_1 = 1.065$ (*curve 3*)

maximum at the frequency $\omega_0 = 2\pi/T_0$ in the power spectrum. The width of this line is finite but small, i.e., $\Delta\omega/\omega_0 \ll 1$. Thus, without taking into account a wide but low pedestal (at the level of -50 dB), an oscillator with spiral chaos is similar to a noisy oscillator with periodic oscillations.

Basic frequency locking of chaotic self-sustained oscillations was first established in work by V.S. Anishchenko and coauthors. The existence of a locking region (region of synchronous chaos) in the parameter space was also shown. These results initiated ideas on chaos synchronization based on the classical theory of oscillations. In the aforementioned work, the phenomenon of chaos synchronization was studied for the Anishchenko–Astakhov oscillator (the modified oscillator with inertial nonlinearity). In this section we illustrate those findings for the Rössler attractor.

We thus introduce an external harmonic force in the system (15.1):

$$\dot{x} = -y - z + b\sin(\omega_1 t), \qquad \dot{y} = x + \alpha y, \qquad \dot{z} = \beta + z(x - \mu). \qquad (15.2)$$

Here, b and ω_1 are the amplitude and the cyclic frequency of the external force, respectively. The system parameters are fixed as follows: $\alpha = \beta = 0.2$, $\mu = 6.5$. In this case the basic frequency value of the spiral attractor without external force is $\omega_0 = 1.0683 \pm 10^{-4}$ (see Fig. 15.1b).

We choose the amplitude to be $b = 0.05$ and study the evolution of the power spectrum $S_x(\omega)$ of the $x(t)$ variable when the frequency ω_1 of the external force is varied over the interval $1.060 \leq \omega_1 \leq 1.065$. Calculation results for the power spectra are shown in Fig. 15.3 for three different values of ω_1. As can be seen from the data presented there, when the detuning between the external frequency and the basic frequency of the autonomous system is relatively large, the power spectrum of oscillations (curve 1) contains the lines of the external frequency ω_1 and the basic frequency ω_0. When the frequency detuning is decreased, these lines approach each other (curve 2) and finally the basic frequency of chaotic self-sustained oscillations

15.2 Phase–Frequency Synchronization of Chaotic Self-Sustained Oscillations

is locked by the external signal: $\omega_0 = \omega_1$ (curve 3). The effect of frequency locking occurs for the frequency value $\omega_1 = 1.065$, which is not equal to the initial value $\omega_0 = 1.0683$. Thus, the locking effect shifts the spectral line of the basic frequency until it becomes equal to the frequency of the external force. This is completely analogous to the effect of frequency locking in the classical theory of limit cycle synchronization. Further studies have shown that this analogy is fundamental. Effects of frequency and phase locking can also be considered for spiral chaos, as well as for the case of a limit cycle. However, the phase approach can be applied if the instantaneous phase of chaotic self-sustained oscillations is correctly defined.

The high degree of regularity of the phase trajectory rotation on the Rössler attractor enables one to represent $x(t)$ and $y(t)$ oscillations in the same form as previously used in analyzing the synchronization of the van der Pol oscillator:

$$x(t) = \rho(t)\cos\Phi(t), \qquad y(t) = \rho(t)\sin\Phi(t). \tag{15.3}$$

The change of variables (15.3) implies the following. On the plane of variables (x, y) (Fig. 15.1a), a representative point on the trajectory is associated with a vector originating from an equilibrium state. The equilibrium point for the Rössler system is located nearly at the coordinate origin of the parameter values corresponding to the spiral attractor. Thus, the origin of the vector under consideration can be put at point $(0, 0)$. The vector length is the instantaneous amplitude $\rho(t)$ of oscillations and the vector rotation angle with respect to the x-axis is the instantaneous phase $\Phi(t)$. A positive angle corresponds to counterclockwise rotation. The representative point rotates exactly that way on the plane (x, y) in the Rössler system. In the case when the image point rotates clockwise in the considered projection (as, for example, in the Anishchenko–Astakhov oscillator), a minus sign should be inserted before the sine function in the second equation of (15.3). The change of variables (15.3) corresponds to changing to a polar system of coordinates on the plane of dynamical variables (x, y). Thus, knowing the current values of x and y, the instantaneous values of the amplitude and phase can be easily found as follows:

$$\rho(t) = \sqrt{x^2(t) + y^2(t)},$$
$$\Phi(t) = \arctan\frac{y(t)}{x(t)} \pm \pi k, \quad k = 0, 1, 2, \ldots. \tag{15.4}$$

An integer number of rotations through the angle π must be added or subtracted to complement the phase with respect to the full angle of rotation. The value of the integer variable k is determined by the condition of continuity of the function $\Phi(t)$. Introduced in this way, the instantaneous phase $\Phi(t)$ is a complete phase, taking values over the whole real line: $\Phi(t) \in [-\infty, +\infty]$. Besides, we may need to consider the phase with values in a bounded interval: $\tilde{\Phi}(t) \in [-\pi, +\pi]$ or $\tilde{\Phi}(t) \in [0, 2\pi]$. It is obvious that $\Phi(t) = \tilde{\Phi}(t) \pm 2m\pi$, where m is an integer.

The above definition of the instantaneous amplitude and phase is not unique. Other pairs of variables may be used in (15.3) instead of the dynamical variables $x(t)$ and $y(t)$. The main thing in selecting such variables is to ensure a regular rotation of a phase trajectory projection about a certain point which is taken as the origin for the radius vector. Derivatives dx/dt and dy/dt or variables $x(t)$ and $x_H(t)$ may be suitable, where $x_H(t)$ is a Hilbert conjugate process

$$x_H(t) = \frac{1}{\pi} \int_{-\infty}^{\infty} \frac{x(\theta)}{t - \theta} d\theta . \tag{15.5}$$

The phase can also be defined using a sequence of times t_i corresponding to crossings of a secant plane by a trajectory in a selected direction. In a piecewise linear approximation, the phase at an arbitrary time can be defined as follows:

$$\Phi(t) = 2\pi \frac{t - t_i}{t_{i+1} - t_i} \pm 2\pi i , \quad i = 0, 1, 2, \ldots . \tag{15.6}$$

Here i is the number of crossings in a given direction during the observation time $t - t_0$.

Any definition of the instantaneous phase $\Phi(t)$ can be used to define the instantaneous and mean frequency of oscillations:

$$\omega(t) = \frac{d\Phi(t)}{dt} , \quad \omega_m = \left\langle \frac{d\Phi(t)}{dt} \right\rangle = \langle \omega(t) \rangle . \tag{15.7}$$

Here the brackets $\langle \ldots \rangle$ can denote both ensemble and time averaging since the process $\omega(t)$ typically has the property of ergodicity. By time averaging, it is easy to obtain

$$\omega_m = \lim_{T \to \infty} \frac{\Phi(t_0 + T) - \Phi(t_0)}{T} . \tag{15.8}$$

Obviously, the instantaneous values of the amplitude, phase, and frequency are different when different methods are used to introduce the amplitude and phase. However, the amplitude–phase presentation of the dynamics of a chaotic system is meaningful as long as statistical characteristics (at least those related to the observed quantities and phenomena) do not depend on the accepted basic phase definition. When studying phase synchronization of chaotic self-sustained oscillations, it is very important to define the instantaneous phase correctly. The coincidence of the mean frequency ω_m (15.8) and the basic frequency of oscillations ω_0 serves as a criterion for correct determination of the instantaneous phase in the regime of a spiral attractor.

Making the change of variables (15.3) in (15.2), we obtain the equations of the Rössler system in the following form:

15.2 Phase–Frequency Synchronization of Chaotic Self-Sustained Oscillations

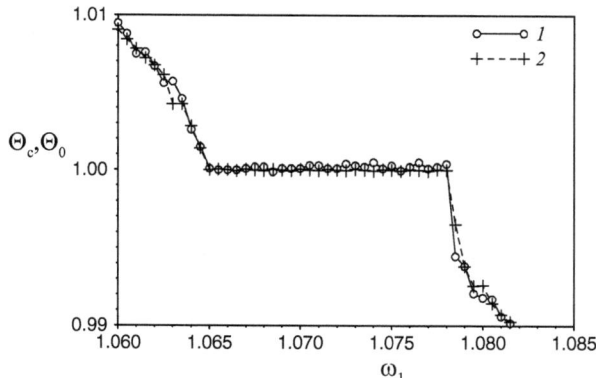

Fig. 15.4 Frequency ratios $\Theta_m = \omega_m/\omega_1$ (curve 1) and $\Theta_0 = \omega_0/\omega_1$ (curve 2) as functions of the external frequency ω_1 for the system (15.9)

$$\dot{\rho} = \frac{1}{2}\alpha\rho - \frac{1}{2}\alpha\rho\cos 2\Phi - z\cos\Phi + b\cos\Phi\sin(\omega_1 t),$$

$$\dot{\Phi} = 1 + \frac{1}{2}\alpha\sin 2\Phi + \frac{1}{\rho}z\sin\Phi - \frac{1}{\rho}b\sin\Phi\sin(\omega_1 t), \quad (15.9)$$

$$\dot{z} = \beta + z(\rho\cos\Phi - \mu).$$

We fix the parameters $\alpha = \beta = 0.2$, $\mu = 6.5$, and $b = 0.05$ and integrate the system (15.9) numerically, for different values of the frequency ω_1 of the external force.

Using the data obtained for $\Phi(t)$, we compute the corresponding values of the mean frequency ω_m (15.8). Dependences of the frequency ratios $\Theta_m = \omega_m/\omega_1$ (curve 1) and $\Theta_0 = \omega_0/\omega_1$ (curve 2) on the external frequency ω_1 are plotted in Fig. 15.4. As can be seen from the plots, the mean frequency ω_m is locked by the external signal in the finite frequency interval $1.065 \leq \omega_1 \leq 1.078$. In addition, the calculated values of ω_m and ω_0 coincide within the calculation error.

Using the calculation data for the instantaneous phase $\Phi(t)$, we can plot temporal dependences of the instantaneous phase difference of self-sustained oscillations and the external forcing $\Delta\Phi(t) = \Phi(t) - \omega_1 t$. They are presented in Fig. 15.5 for three different values of the external signal frequency. As can be seen from the plot, the effect of phase locking by the external signal takes place in the synchronization region (curve 3): the absolute value of the phase difference $|\Delta\Phi|$ does not grow in time, but remains bounded near some constant value.

According to the concept of phase synchronization of chaotic self-sustained oscillations, the classical definition of the phase-locking effect should be transformed as follows:

$$\left|m\Phi_1(t) - n\Phi_2(t)\right| \leq K, \quad (15.10)$$

where K is a constant which depends on the initial phase difference, and m and n are integers. Unlike synchronization of periodic oscillations, the phase difference is

Fig. 15.5 Temporal dependences of the instantaneous phase difference $\Delta\Phi(t) = \Phi(t) - \omega_1 t$ in the system (15.9) for three different values of the external frequency: $\omega_1 = 1.06$ (*curve 1*), $\omega_1 = 1.064$ (*curve 2*), and $\omega_1 = 1.065$ (*curve 3*)

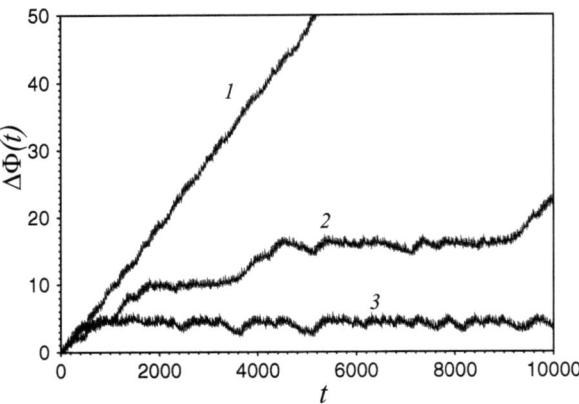

not a constant in the phase locking of chaotic self-sustained oscillators. It fluctuates in time, but its deviations from a certain constant value are strictly bounded. The absolute value of these deviations should not exceed π. It is easy to see that the phase-locking condition (15.10) provides the rational relationship between the mean frequencies, i.e.,

$$m\omega_{m1} = n\omega_{m2}, \quad \text{where } \omega_{m1} = \left\langle \frac{d\Phi_1(t)}{dt} \right\rangle, \quad \omega_{m2} = \left\langle \frac{d\Phi_2(t)}{dt} \right\rangle. \quad (15.11)$$

Since in the spiral attractor mode the basic frequencies of partial oscillators coincide with the corresponding mean frequencies, the equality (15.11) must also be satisfied for the basic frequencies. Thus, phase and frequency synchronizations of chaos are not two different effects. Similarly to the classical case, they represent two sides of the same phenomenon, namely, phase–frequency synchronization.

Synchronization of a chaotic self-sustained oscillator is observed in a certain region of control parameter values called the *synchronization region*. Consider the systems (15.2) and (15.9) for $\alpha = \beta = 0.2$ and $\mu = 6.5$. We vary the parameters ω_1 and b of the external force and construct a domain of synchronous modes on the plane (ω_1, b) using the aforementioned criteria to diagnose chaos synchronization. Calculation results are shown in Fig. 15.6.

The area labeled by 1 in the diagram corresponds to the regime of synchronous chaos. Inside this region, there is an effect of basic frequency ω_0 locking by the external signal at the fundamental tone, $\omega_m = \omega_1$. Simultaneously, in this region, phase locking occurs with the condition $|\Delta\Phi| = |\Phi(t) - \omega_1 t| \leq K$.

The transition from regions 1 to 2 is the transition from synchronous chaos to nonsynchronous chaos. Here, the regime of frequency and phase locking is violated. Since the spectrum of nonsynchronous chaos (region 2) contains spectral lines of the two basic frequencies ω_0 and ω_1 besides a broadband pedestal, nonsynchronous chaos is called *toroidal chaos* or *torus chaos*. In region 3, there are stable periodic oscillations (a limit cycle) of period $T = 8\pi/\omega_0$ with the locked basic frequency

15.2 Phase–Frequency Synchronization of Chaotic Self-Sustained Oscillations

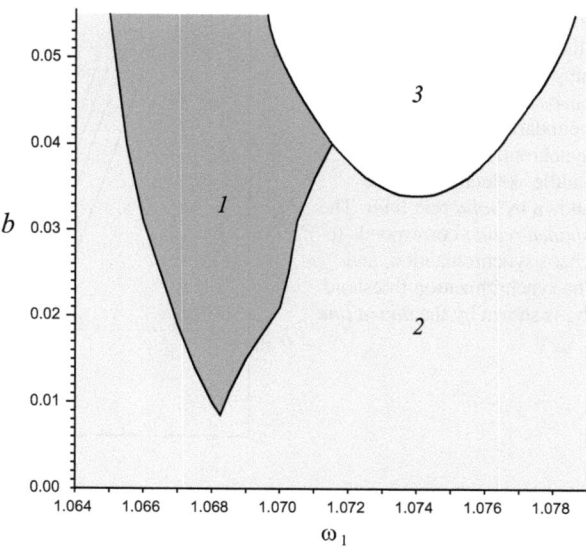

Fig. 15.6 Diagram showing the regimes of the system (15.2) on the parameter plane (ω_1, b). The region of synchronous chaos is denoted by *1*, the region of nonsynchronous chaos by *2*, and the region of synchronous periodic oscillations by *3*

$\omega_0 = \omega_1$ and subharmonics $\omega_1/2$ and $\omega_1/4$. Thus, the synchronization region shown in Fig. 15.6 consists of two domains with different kinds of synchronous oscillations: the region of synchronous chaos (1) and the region of synchronous periodic oscillations (3). In the general case, the structure of the synchronization region of chaotic self-sustained oscillations can be very complicated and can include many domains corresponding to different chaotic and periodic attractors.

If only small values of the amplitude of the external force are taken into account ($b \leq 0.34$ in the case considered), the picture is comparatively simple. It looks like a typical synchronization region in the form of a 'beak', as in the case of periodic self-sustained oscillations. The only difference is that the beak of chaos synchronization does not originate from the x-axis corresponding to zero amplitude of the external force. Phase locking of chaotic self-sustained oscillations can be realized if the amplitude of the external force (or coupling for mutual synchronization) exceeds a certain threshold value. The reason for the existence of the threshold is closely related to the bifurcation mechanism of phase–frequency chaos synchronization. This mechanism is still insufficiently studied. However, it is clear that saddle limit cycles embedded in a chaotic attractor will play a major role. It has been shown that, on the control parameter plane, the lines of tangent bifurcations of saddle cycles with different periods accumulate at the boundary of chaos synchronization. The boundary itself is a critical line to which the points of tangent bifurcations for cycles with increasing periods converge.

This problem can be analyzed by introducing an artificially constructed two-dimensional irreversible map which models synchronization of a chaotic oscillator by an external periodic force. As though they were the 'skeleton' of a synchronous chaotic attractor, saddle cycles undergo tangent bifurcations together with corresponding periodic repellers. The latter form the skeleton of a chaotic repeller which

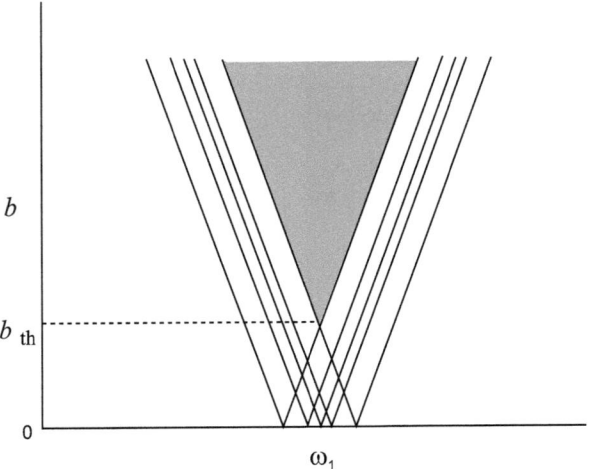

Fig. 15.7 Qualitative illustration of how the amplitude threshold arises in chaos synchronization. The boundaries of synchronization beaks of saddle skeletal cycles are shown by *solid thin lines*. The *shaded region* corresponds to chaos synchronization, and the synchronization threshold b_{th} is shown by the *dotted line*

touches the chaotic attractor at separate points, namely, at points of saddle and repeller cycles at the moment when they merge.[1] Each pair of skeletal cycles belongs to an unstable invariant curve (corresponding to a saddle torus of the flow system). As a result of tangent bifurcation, the motion on the invariant curve becomes ergodic, i.e., a direction arises along which the phase point leaves the synchronous attractor, makes a circuit around the invariant curve, and returns again. At the same time, there is a jump of 2π in the phase difference.

The existence of a chaos synchronization threshold can be explained as follows. Skeletal cycles of a chaotic attractor have close, but still slightly different basic frequencies. Each of those cycles is characterized by its own beak of synchronization, whose boundaries are represented by lines of saddle-repeller bifurcations. Synchronization beaks of different cycles originate from different points of the x-axis corresponding to zero amplitude of the external force. Thus, the chaos synchronization region that represents the intersection of synchronization beaks of all kinds of skeletal cycles cannot already have a common point with the abscissa axis. Its low boundary is located at some distance from the x-axis, a distance equal to the threshold value of the external amplitude (Fig. 15.7).

Numerical experiments with self-sustained oscillators of coherent chaos described by ordinary differential equations tend to confirm the presumed bifurcation mechanism for phase locking of chaotic oscillations.

Two synchronization mechanisms are distinguished in the classical theory of synchronization of periodic self-sustained oscillations. They are phase locking and self-sustained oscillation suppression in one of the interacting systems. A

---

[1] In a modeled flow system, repeller cycles obviously belong to saddle cycles whose unstable manifolds have a higher dimension than those of the skeletal cycles of an attractor. A chaotic saddle exists instead of a chaotic repeller.

transition to the synchronization region via suppression of self-sustained oscillations is observed for large values of the frequency detuning and for strong interaction between the systems (see Chap. 13). Suppression may also take place in the case of chaotic self-sustained oscillations. Studies show that, for mutual synchronization of chaotic self-sustained oscillators or in the case of forced synchronization of chaos by a harmonic signal, the effect of self-sustained oscillation suppression is preceded by a transition from nonsynchronous chaos to a quasiperiodic regime. In this case periodic oscillations are observed in the synchronization region and its boundary corresponds to the torus birth bifurcation (Neimark–Saker bifurcation).[2] The transition nonsynchronous chaos → synchronous chaos is observed when one chaotic oscillator is unidirectionally driven by another one. When the frequency detuning and the coefficient of unidirectional coupling are appropriately selected, both classical mechanisms of synchronization can be observed in the region of chaotic dynamics of that system. They are related to locking and suppression of self-sustained oscillations.

Generalized synchronization of chaos, which can be implemented in unidirectionally coupled systems (the coupling is of master–slave type), is simply suppression of self-sustained oscillations of the driven system. In this case the state of this system is completely defined by the state of the driving system. Such a phenomenon is also observed when interacting chaotic systems have, not only a similar type of behavior, but also a completely different structure of chaos. Chaotic self-sustained oscillations can also be suppressed by a periodic signal (we have discussed this above) and even by a random force.

15.3 Experimental Investigation of Forced Synchronization of an Oscillator with Spiral Chaos

We now describe experimental results on forced (external) synchronization in the Anishchenko–Astakhov oscillator. Experiments employ a radio-technical oscillator with inertial nonlinearity (the Anishchenko–Astakhov oscillator), an oscillator with a harmonic signal which is fed to the oscillator as an external force, and a computer with a high-speed analog–digital converter (ADC). For certain values of the control parameters, the oscillator exhibits a spiral chaotic attractor which can be observed experimentally. The corresponding projection of the attractor on the plane (x, y) and the spectrum of the $x(t)$ process are shown in Fig. 15.8.[3]

---

[2] The region of periodic oscillations depicted in Fig. 15.6 is not related to the suppression effect. This is evidenced by the character of the boundary between regions 2 and 3, which is connected with a saddle-node bifurcation of cycles.

[3] In practice, voltages are measured at certain points of the setup, and they are proportional to the dimensionless variables x and y.

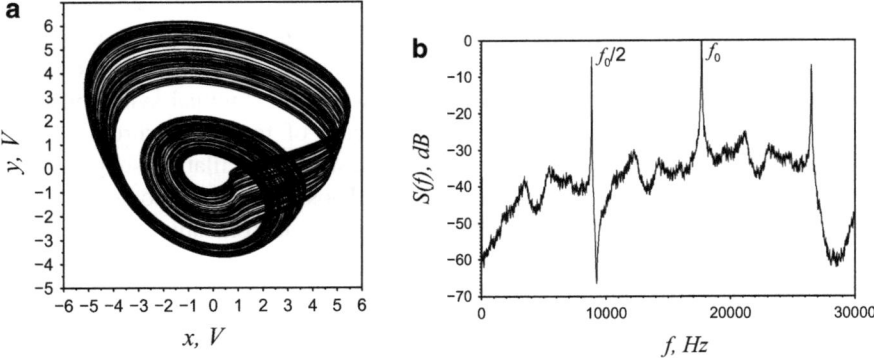

Fig. 15.8 Experimentally obtained projection of a spiral attractor (**a**) and normalized power spectrum $S_x(f)$ of $x(t)$ oscillations (**b**) in the regime of spiral chaos in the Anishchenko–Astakhov radio-technical oscillator

The spiral attractor under study is two-connected, i.e., it has the form of a double-round Möbius strip. Accordingly, the power spectrum contains a subharmonic at the frequency $f_0/2$ and lines at multiple frequencies. This fact is not of key importance in studying the effect of synchronization. If the basic frequency f_0 is locked by an external force, then all its harmonics and subharmonics will also appear to be locked. This choice of experimental regime is substantiated by the circumstance that a developed (one-coupled or one-band) spiral attractor in the Anishchenko–Astakhov oscillator is highly sensitive to the kind of weak noise that is inevitably present in the experimental setup.

Forced synchronization is explored experimentally when the oscillator is driven by a harmonic force with frequency f_1 close to the basic frequency of chaotic self-sustained oscillations. The experimental data obtained for the Anishchenko–Astakhov oscillator are in full correspondence with the results of numerical simulation of forced chaos synchronization in the Rössler oscillator, as presented in the last section. Figure 15.9 shows fragments of power spectra of the oscillations which illustrate the effect of basic frequency f_0 locking.

The frequency ratio $\Theta_0 = f_0/f_1$ is measured experimentally and the dependence of Θ_0 on the external frequency f_1 is plotted in Fig. 15.10. This indicates that the effect of frequency locking is realized in a finite region.

Figure 15.11 depicts a fragment of the forced synchronization region for the Anishchenko–Astakhov chaotic oscillator on the control parameter plane spanned by the frequency f_1 and the amplitude b of the external force. The shape of the synchronization region is close to that of the van der Pol oscillator. However, the threshold of synchronization is clearly visible in this case.

15.4 Complete Synchronization of Interacting Chaotic Systems

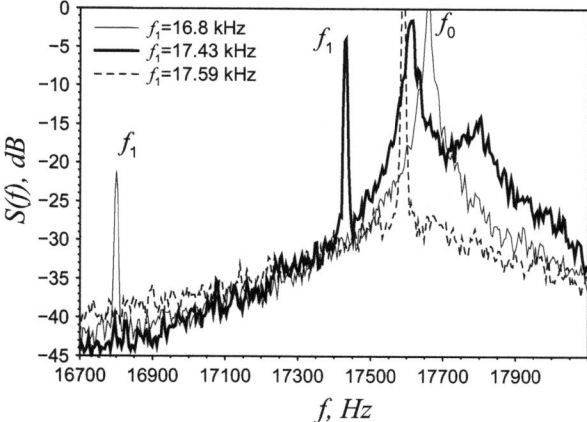

Fig. 15.9 Effect of basic frequency locking in the Anishchenko–Astakhov chaotic oscillator when the frequency of the harmonic force is varied (physical experiment)

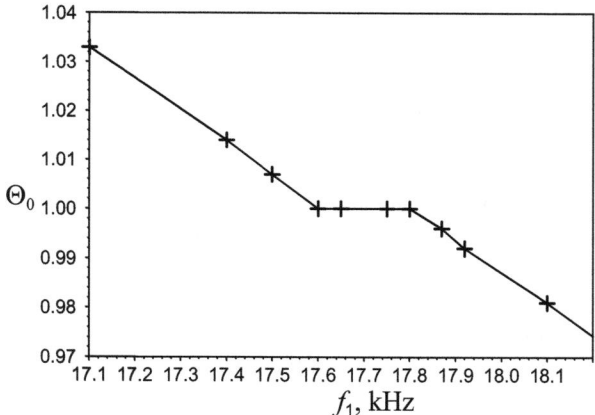

Fig. 15.10 Frequency ratio $\Theta_0 = f_0/f_1$ as a function of the external frequency f_1 in the Anishchenko–Astakhov chaotic oscillator (physical experiment)

15.4 Complete Synchronization of Interacting Chaotic Systems

The effect of complete synchronization of chaos can be observed when two absolutely identical chaotic self-sustained oscillators interact. Starting from a certain value of the coupling coefficient, the oscillations of partial systems become completely identical.

Consider a system of two interacting and completely identical chaotic subsystems in the general form

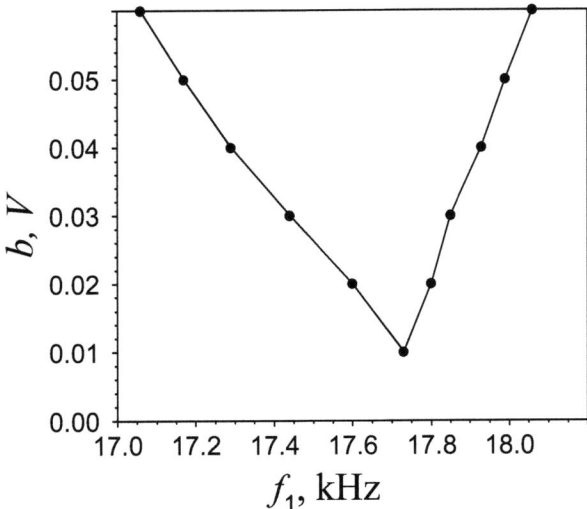

Fig. 15.11 Synchronization region of the Anishchenko–Astakhov oscillator on the parameter plane (f_1, b) (physical experiment)

$$\dot{\mathbf{x}}_1 = \mathbf{F}(\mathbf{x}_1, \boldsymbol{\alpha}_1) + \gamma \mathbf{g}(\mathbf{x}_1, \mathbf{x}_2) ,$$
$$\dot{\mathbf{x}}_2 = \mathbf{F}(\mathbf{x}_2, \boldsymbol{\alpha}_2) + \gamma \mathbf{g}(\mathbf{x}_2, \mathbf{x}_1) .$$
(15.12)

Here $\mathbf{x}_{1,2}$ are the state vectors and $\boldsymbol{\alpha}_{1,2}$ are the vector parameters of the subsystems. If $\boldsymbol{\alpha}_1 = \boldsymbol{\alpha}_2$, the partial oscillators are completely identical. The vector function $\mathbf{g}(\ldots)$ determines the coupling through $\mathbf{g}(\mathbf{x}_1, \mathbf{x}_1) = \mathbf{g}(\mathbf{x}_2, \mathbf{x}_2) = 0$. In the case of complete identity of the partial subsystems, there is an invariant manifold $\mathbf{U}$ ($\mathbf{x}_1 = \mathbf{x}_2$) in the phase space of the full system. It is referred to as a *symmetric subspace*. Phase trajectories lying in $\mathbf{U}$ correspond to completely synchronized oscillations. If a limit set in $\mathbf{U}$ is attracting in the phase space of the system (15.12), i.e., if it is an attractor, then *the effect of complete synchronization is realized*.

We exemplify the effect of complete synchronization using two symmetrically coupled identical Rössler oscillators. The system equations have the form

$$\dot{x}_1 = -y_1 - z_1 + \gamma(x_2 - x_1) ,$$
$$\dot{y}_1 = x_1 + \alpha y_1 ,$$
$$\dot{z}_1 = \beta + z_1(x_1 - \mu) ,$$
$$\dot{x}_2 = -y_2 - z_2 + \gamma(x_1 - x_2) ,$$
$$\dot{y}_2 = x_2 + \alpha y_2 ,$$
$$\dot{z}_2 = \beta + z_2(x_2 - \mu) .$$
(15.13)

15.4 Complete Synchronization of Interacting Chaotic Systems

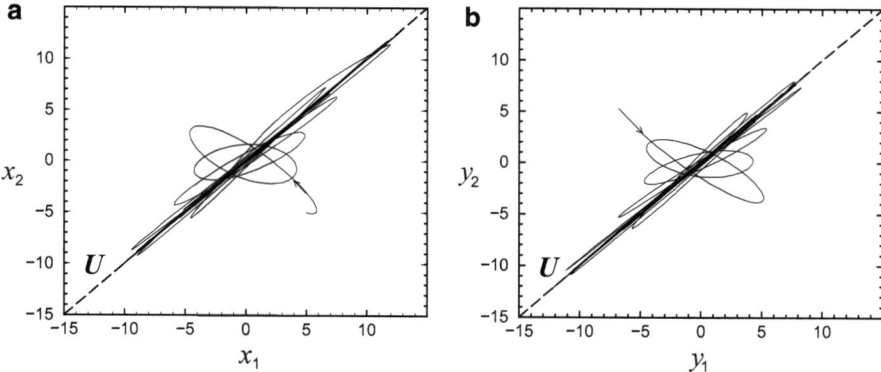

Fig. 15.12 Phase trajectory on the attractor of the system (15.13) when projected on the plane of variables (**a**) (x_1, x_2) and (**b**) (y_1, y_2)

We choose the regime of spiral chaos in each of the interacting subsystems by setting $\alpha = \beta = 0.2$ and $\mu = 6.5$. We follow the evolution of the limit set of phase trajectories of the system (15.13) as the coupling parameter increases. For this purpose, a projection of the attractor is drawn on the plane of one-type variables (x_1, x_2) or (y_1, y_2). In this case, the regime of complete synchronization is associated with the attractor located in the symmetric subspace **U**. Correspondingly, projections of trajectories on the attractor lie on the diagonal $x_1 = x_2$ (or $y_1 = y_2$).

Calculation results are shown in Fig. 15.12 for the coupling coefficient $\gamma = 0.2$. It can be seen that, after some transient process, the phase trajectory starting from an initial point that does not belong to **U** reaches the invariant subspace **U** and remains there. Thus, for the selected values of the coupling coefficient, the regime of complete chaotic synchronization is realized in the system: for any time $t > t_t$ (t_t is the transient process time), $x_1(t) \equiv x_2(t)$, $y_1(t) \equiv y_2(t)$, and $z_1(t) \equiv z_2(t)$.

Numerous studies have shown that complete synchronization (unlike phase–frequency synchronization) can be observed, not only for spiral chaos, but also in the case of more complex chaotic self-sustained oscillations, such as quasiperiodic chaos in the Lorenz system or the double scroll regime in Chua's circuit. Complete chaotic synchronization can be implemented, not only for self-sustained oscillators, but also for interacting nonlinear chaotic oscillators driven by the same external force, and for model return maps.

As in the case of phase–frequency synchronization, the mechanism for the appearance and destruction of complete synchronization is related to bifurcations of saddle and repeller cycles in a chaotic limit set. Unlike the mechanism of phase–frequency synchronization, this one has been fully explored. It has been clearly established how the invariant manifold **U** of the system (15.12), in which the 'in-phase' chaotic attractor is located, ceases to be attracting as the coupling parameter decreases.

To simplify our investigation, we consider model return maps rather than the system (15.12). These represent coupled one-dimensional irreversible maps:

$$x_{n+1} = f(x_n) + \gamma g(y_n - x_n) ,$$
$$y_{n+1} = f(y_n) + \gamma g(x_n - y_n) .$$
(15.14)

The stability of trajectories in the system (15.14) which belong to the invariant manifold **U** (the bisectrix $x = y$) is defined by two Lyapunov exponents λ_{tn} and λ_{tr}. The exponent λ_{tn} characterizes the evolution of perturbations lying in **U**, while λ_{tr} is responsible for perturbations transverse to **U**. Thus, the destruction of the complete chaotic synchronization regime in (15.14) is diagnosed by the sign of the *transverse exponent*, which is defined as follows:

$$\lambda_{tr} = \lim_{n \to \infty} \frac{1}{n} \ln \left| \frac{d f(x)}{dx} \right|_{x=x_n} .$$
(15.15)

If the exponent λ_{tr} becomes positive, this means that the limit set lying in **U** loses its stability in the transverse direction. In this case, when the symmetry in initial conditions is even slightly violated, the trajectory leaves the invariant manifold and goes to an attractor not lying in **U**. In a multi-dimensional case, the evolution of a perturbation that is transverse to **U** is described by N Lyapunov exponents, where N is the phase space dimension of a partial system. If at least one of them becomes positive, the invariant manifold is no longer stable. As a result, the complete chaos synchronization regime is destroyed via a so-called *blowout bifurcation*. This process is usually accompanied by the phenomenon of transient (on finite time intervals) or 'true' intermittency (*Yamada–Fujisaka intermittency* or *on–off intermittency*).

However, Lyapunov exponents are averaged over attractor characteristics which cannot reflect all the local changes in the structure of the limit set. It is known that, before λ_{tr} becomes positive, a countable set of points of the invariant manifold may appear at which the transverse instability occurs. These points belong to unstable cycles lying in the invariant subspace. For the two-dimensional map (15.14), these cycles are repellers and generally saddles. Having fallen in the neighborhood of such a cycle, the phase point, if it does not strictly lie in **U**, moves away from the invariant manifold. If for these parameter values the system does not possess another attractor, except the 'in-phase' one, i.e., lying in **U**, then after a certain time, the trajectory will return to its neighborhood again and then reach the invariant manifold. However, if the number of unstable cycles embedded in the 'in-phase' chaotic attractor is large, a long transient process of on–off intermittency may be observed. The effect of weak noise on 'in-phase' chaos causes the intermittency process to be constantly renewed. Due to the influence of noise, an experimentally observed chaotic attractor no longer lies in the invariant manifold, and one can observe dynamics with temporal bursting. This phenomenon is referred to as *attractor bubbling*.

15.4 Complete Synchronization of Interacting Chaotic Systems

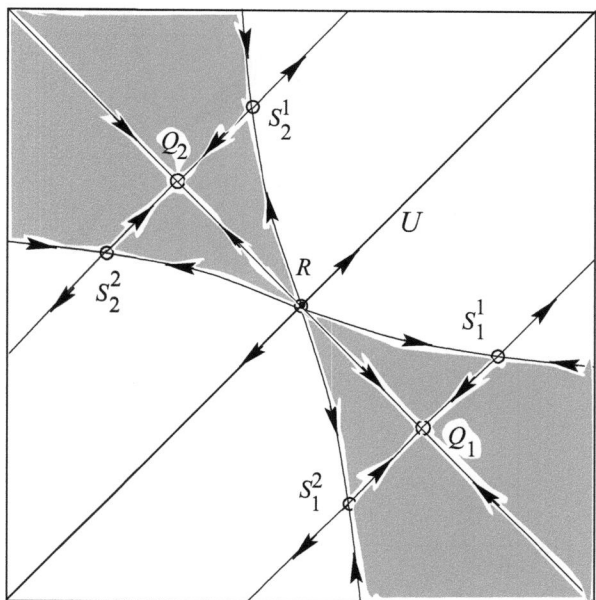

Fig. 15.13 Emergence of a tongue in the basin of an 'out-of-phase' period-2 cycle (points $Q_{1,2}$) which originates from the repeller R in the invariant manifold **U** (the bisectrix). $S^1_{1,2}$ and $S^2_{1,2}$ are points of saddle period-2 cycles

If the system has a regular or chaotic attractor not lying in **U**, the appearance of unstable cycles in the 'in-phase' attractor is responsible for the formation of 'tongues' in the basin of attraction of the 'out-of-phase' attractor. These tongues originate from the points of these cycles. A typical tongue on the phase plane is shown in Fig. 15.13. The tongue is bounded by the stable manifolds of the 'out-of-phase' saddle cycles $S^{1,2}$.

The emergence of a countable set of tongues belonging to the attractor not lying in **U** leads to a phenomenon referred to as the *riddling* of the local neighborhood of the invariant manifold. This process implies that, arbitrarily close to any point of the 'in-phase' attractor, there is a point belonging to the basin of another attractor. As a result of riddling, the chaotic attractor in **U** is no longer attracting in the usual sense. It becomes a so-called *weak attractor* or *Milnor attractor*.

Figure 15.14 shows the riddled neighborhood of the chaotic attractor lying in **U** (on the bisecting line) for the system of coupled logistic maps. The picture corresponds to the case when the transverse Lyapunov exponent is still negative, i.e., a blowout bifurcation has not yet occurred. When the transverse direction becomes unstable, on average, over the attractor, the attractor lying in the invariant manifold is no longer attractive even in the Milnor sense. This situation corresponds to the blowout bifurcation.

We note once again that complete chaos synchronization can be realized only for completely identical interacting systems. If there is a parameter mismatch of

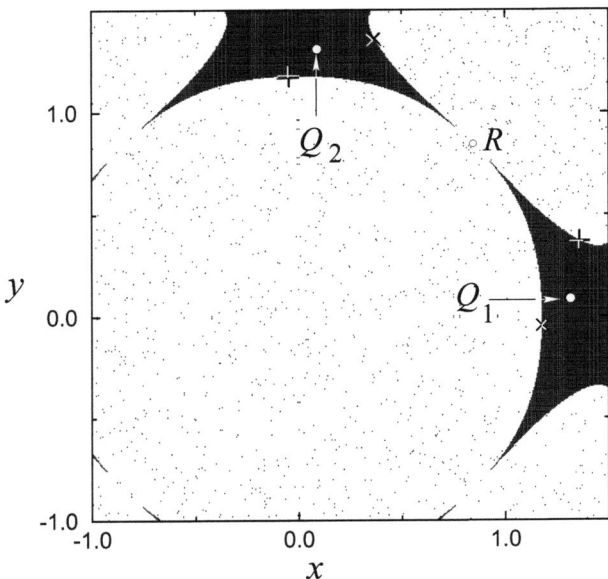

Fig. 15.14 Bistability and riddling in a system of coupled logistic maps. Besides the 'in-phase' chaotic attractor located on the bisectrix $x = y$, there exists the 'out-of-phase' stable period-2 cycle (points $Q_{1,2}$). The basin of attraction of the 'out-of-phase' attractor is marked by *black dots*

partial systems, the symmetric subspace **U** no longer exists and, as a consequence, complete synchronization cannot be observed at any finite value of the coupling parameter. However, if the difference between interacting systems is negligible, an effect close to complete synchronization can be observed for sufficiently strong coupling. This consists in the oscillations of the partial systems fully repeating each other with a certain time delay (lag) τ_d: $\mathbf{x}_1(t) = \mathbf{x}_2(t + \tau_d)$. This effect is called *lag synchronization*. Although there is no invariant subspace **U** in the case of lag synchronization, the chaotic attractor is topologically equivalent to the attractor in the regime of complete synchronization. Hence, lag synchronization can be treated as a generalization of the complete synchronization notion to systems with a slight parameter mismatch.

15.5 Quantitative Characteristics of the Degree of Synchronization of Chaotic Self-Sustained Oscillations

Three degrees of chaos synchronization can be distinguished for interacting chaotic self-sustained oscillators of the same type, but with a frequency detuning. The boundary of a synchronization region on the plane of the parameters controlling

15.5 Quantitative Characteristics of the Degree...

the coupling and the detuning corresponds to phase–frequency synchronization of the partial systems. When the detuning is decreased and the coupling parameter increased, a stronger synchronization effect may appear, namely, lag synchronization. The transition from phase–frequency synchronization to lag synchronization is a complicated process whose bifurcation mechanism is still insufficiently understood, but it is likely to be similar to the mechanism of appearance–destruction of complete synchronization. Without detuning and starting from a certain value of the coupling parameter, the complete synchronization effect can be observed when the chaotic limit set in the symmetric subspace becomes attractive in the full phase space of the system. In the scientific literature, one can find a lot of different quantitative characteristics of the degree of synchronization of interacting chaotic self-sustained oscillators. Some of these can be used to determine the boundary of phase–frequency locking, while others can diagnose lag or complete synchronization.

Mutual phase–frequency synchronization of chaotic self-sustained oscillators, as well as the forced synchronization considered above, are easily diagnosed by the ratio of characteristic frequencies (basic or mean frequencies of interacting self-sustained oscillators) Θ. The equality $\Theta = 1$ holds in the main region of synchronization. Furthermore, phase locking can be determined by using the effective diffusion coefficient $B_{\text{eff}\Delta\Phi}$ of the phase difference. This coefficient evaluates the rate of linear growth of the variance of the phase difference in time. In the case of strong phase locking, the effective diffusion coefficient must be zero according to the condition (15.10). In numerical experiments, due to the finite accuracy of calculations, it is impossible to achieve a zero value for this coefficient. However, $B_{\text{eff}\Delta\Phi}$ dramatically decreases (by several orders of magnitude) at the boundary of the locking region.

The degree of synchronization is sometimes evaluated by means of the following characteristic:

$$v = \frac{\sigma^2_{x_1+x_2}}{2(\sigma^2_{x_1} + \sigma^2_{x_2})}, \qquad (15.16)$$

where $x_1(t)$ and $x_2(t)$ are the one-type dynamical variables of the self-sustained oscillators, $\sigma^2_{x_1}$ and $\sigma^2_{x_2}$ are the variances of the variables $x_1(t)$ and $x_2(t)$, and $\sigma^2_{x_1+x_2}$ is the variance of their sum. Obviously, $v \in [0.5, 1]$, and $v = 1$ corresponds to the case of complete synchronization. When the processes $x_1(t)$ and $x_2(t)$ are completely independent, we have $v = 0.5$. However, neither phase locking nor lag synchronization can be clearly diagnosed by means of this characteristic.

To reveal complete and lag synchronization of interacting chaotic systems, one can use the following quantity:

$$\kappa = \min_\tau G(\tau), \qquad (15.17)$$

where $G(\tau)$ is the similarity function derived from the relation

$$G^2(\tau) = \frac{\langle [x_2(t+\tau) - x_1(t)]^2 \rangle}{\sqrt{\langle x_1^2(t) \rangle \langle x_2^2(t) \rangle}}, \quad (15.18)$$

and $x_1(t)$ and $x_2(t)$ are the one-type dynamical variables of the two partial systems. In the case of complete and lag synchronization, $\kappa = 0$. As the detuning grows and the coupling decreases, κ increases. The degree of synchrony can be evaluated using the mutual normalized correlation function (the mutual correlation coefficient):

$$R_{x_1 x_2}(\tau) = \frac{\langle x_1(t) x_2(t+\tau) \rangle - \langle x_1(t) \rangle \langle x_2(t+\tau) \rangle}{\sqrt{\left[\langle x_1^2(t) \rangle - \langle x_1(t) \rangle^2 \right] \left[\langle x_2^2(t+\tau) \rangle - \langle x_2(t+\tau) \rangle^2 \right]}}. \quad (15.19)$$

The quantity $\eta = \max_\tau R_{x_1 x_2}(\tau)$ is equal to unity in the case of lag synchronization (and thus, complete synchronization) and vanishes when the $x_1(t)$ and $x_2(t)$ oscillations become statistically independent.

The degree of synchrony can also be estimated in the context of the spectral approach. For this purpose, the coherence function is used:

$$r_{x_1 x_2}(\omega) = \frac{|S_{x_1 x_2}(\omega)|}{\sqrt{S_{x_1}(\omega) S_{x_2}(\omega)}}, \quad (15.20)$$

where S_{x_1}, S_{x_2} are the power spectra of fluctuations $x_1(t) - \langle x_1 \rangle$ and $x_2(t) - \langle x_2 \rangle$, while $S_{x_1 x_2}$ is the mutual fluctuation spectrum. If processes are statistically independent, then $r_{x_1 x_2} \equiv 0$. If $x_1(t)$ and $x_2(t)$ are linearly interrelated, we have $r_{x_1 x_2} \equiv 1$. A quantitative frequency-independent characteristic can be introduced by considering the mean value $\bar{r}$ of the coherence coefficient in the relevant frequency interval.

As an example consider a system of two coupled Rössler oscillators with a minor frequency detuning. The system equations read

$$\begin{aligned}
\dot{x}_1 &= -y_1 - z_1 + \gamma(x_2 - x_1), \\
\dot{y}_1 &= x_1 + \alpha y_1, \\
\dot{z}_1 &= \beta + z_1(x_1 - \mu), \\
\dot{x}_2 &= -(1-\Delta) y_2 - z_2 + \gamma(x_1 - x_2), \\
\dot{y}_2 &= (1-\Delta) x_2 + \alpha y_2, \\
\dot{z}_2 &= \beta + z_2(x_2 - \mu),
\end{aligned} \quad (15.21)$$

where Δ is the parameter that determines the mismatch of the partial self-sustained oscillators, and γ is the coupling parameter. For this system, the above

15.5 Quantitative Characteristics of the Degree...

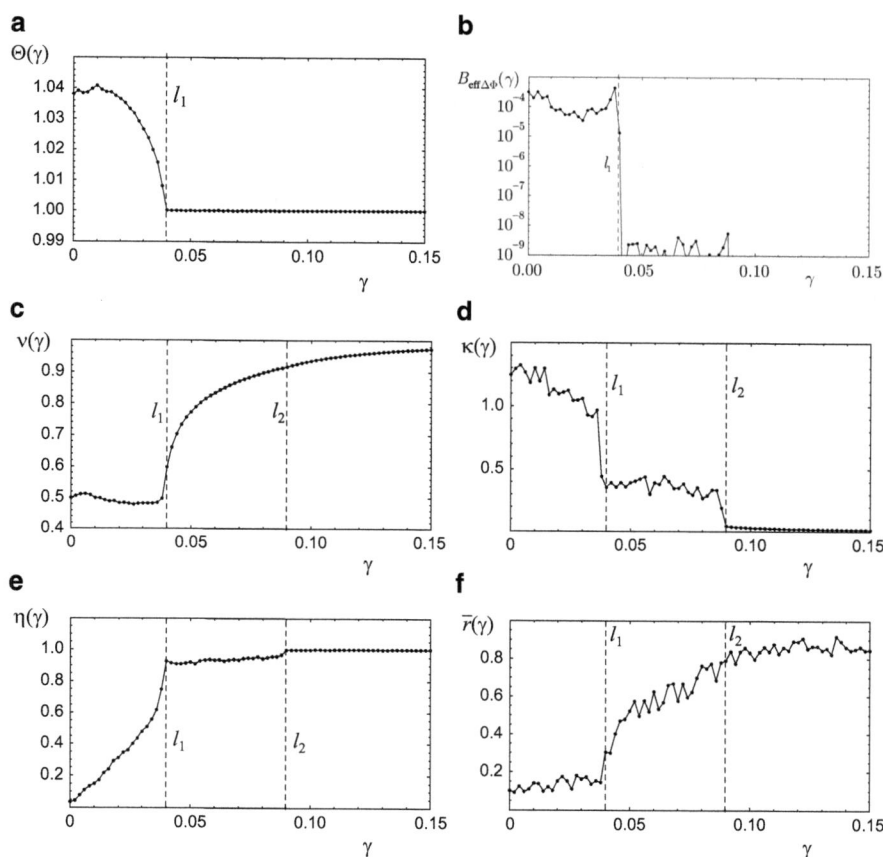

Fig. 15.15 Dependences on the coupling parameter γ of different quantitative characteristics of the degree of synchronization of interacting chaotic oscillators (15.21) for $\alpha = \beta = 0.2$, $\mu = 6.5$, and $\Delta = 0.02$. (**a**) Mean frequency ratio $\Theta = \Omega_2/\Omega_1$. (**b**) Effective diffusion coefficient $B_{\text{eff}\Delta\Phi}$. (**c**) The quantity ν given by (15.16). (**d**) Minimum of the similarity function κ. (**e**) Maximum of the mutual correlation coefficient η. (**f**) Mean value $\bar{r}$ of the coherence coefficient. The *dashed lines* l_1 and l_2 indicate the boundaries of phase and lag synchronization, respectively

synchronization characteristics are calculated and plotted in Fig. 15.15 as functions of the coupling parameter.

As can be seen from the plots, the different characteristics respond differently depending on which boundary l_1 or l_2 has been crossed, and accordingly one must take into account the kind of synchronization effect that is being studied when applying them. Besides the above quantities, other characteristics of the degree of synchronization are also used, e.g., those based on the calculation of mutual or conditional information.

15.6 Summary

The effects considered in this chapter show that chaotic self-sustained oscillatory systems can also exhibit synchronization. However, the way the fundamental property of synchronization manifests itself and the kind of synchronization effects that can be observed in a particular system depend on specific features of the chaotic attractor. The classical theory of synchronization can be most consistently applied to self-sustained oscillators in the regime of a spiral attractor. These systems can exhibit phase–frequency synchronization. Such chaotic self-sustained oscillators are similar to noisy periodic oscillators in their spectral and correlation characteristics. At the same time, unlike noisy oscillators, phase–frequency synchronization is strict in a deterministic regime of spiral chaos, i.e., the instantaneous phase remains locked for an arbitrarily long time, and the effective diffusion of the phase difference is strictly zero in the synchronization region.

Strict phase–frequency synchronization cannot be observed for other types of chaotic attractors (funnel attractor, Lorenz attractor, Chua's double scroll, and other attractors of switching type). However, these systems can exhibit partial phase synchronization, synchronization of switchings, and complete and generalized synchronization. Each synchronization effect must be explored by applying special numerical and physical methods and using the most appropriate diagnostic tools and characteristics for the degree of synchronization. The problem of synchronization of chaotic self-sustained oscillation is described most completely in the monographs [1–3].

References

1. Anishchenko, V.S., Astakhov, V.V., Neiman, A.B., Vadivasova, T.E., Schimansky-Geier, L.: Nonlinear Dynamics of Chaotic and Stochastic Systems. Springer, Berlin (2002)
2. Balanov, A.G., Janson, N.B., Postnov, D.E., Sosnovtseva, O.: Synchronization: From Simple to Complex. Springer, Berlin (2009)
3. Pikovsky, A., Rosenblum, M., Kurths, J.: Synchronization: A Universal Concept in Nonlinear Science. Cambridge University Press, Cambridge (2003)

MIX
Papier aus verantwortungsvollen Quellen
Paper from responsible sources
FSC® C105338

If you have any concerns about our products,
you can contact us on
ProductSafety@springernature.com

In case Publisher is established outside the EU,
the EU authorized representative is:
**Springer Nature Customer Service Center GmbH
Europaplatz 3, 69115 Heidelberg, Germany**

Printed by Libri Plureos GmbH
in Hamburg, Germany